RENEWABLE ENERGY SYSTEMS

Simulation with Simulink® and SimPowerSystems™

RENEWABLE ENERGY SYSTEMS

Simulation with Simulink® and SimPowerSystems™

Viktor Perelmuter

CRC Press
Taylor & Francis Group
Boca Raton London New York

CRC Press is an imprint of the
Taylor & Francis Group, an **informa** business

CRC Press
Taylor & Francis Group
6000 Broken Sound Parkway NW, Suite 300
Boca Raton, FL 33487-2742

Printed by CPI on sustainably sourced paper
Version Date: 20160718

International Standard Book Number-13: 978-1-4987-6598-5 (Hardback)

Library of Congress Cataloging-in-Publication Data

Names: Perelmuter, V. M. (Viktor Moiseevich), author.
Title: Renewable energy systems : simulation with Simulink® and SimPowerSystems™ / Viktor Perelmuter.
Description: Boca Raton : Taylor & Francis, CRC Press, 2017. | Includes bibliographical references and index.
Identifiers: LCCN 2016027253| ISBN 9781498765985 (hardback : alk. paper) | ISBN 9781315316246 (ebook)
Subjects: LCSH: Electric power systems. | Renewable energy sources.
Classification: LCC TK1005 .P45535 2017 | DDC 621.0420285/53--dc23
LC record available at https://lccn.loc.gov/2016027253

Visit the Taylor & Francis Web site at
http://www.taylorandfrancis.com

and the CRC Press Web site at
http://www.crcpress.com

Dedicated to my family and friends.

Contents

Preface

The development of renewable sources of electrical energy, which do not pollute the environment and do not use the limited natural resources, is the mainstream in electrical engineering today. Progress in the creation of such systems is achieved, mainly, by their electrical parts and, in the first place, by power electronics. The electrical devices that are used in the renewable sources have a very complicated inner structure, and the methods of computer simulation make the development of such systems easier and quicker.

MATLAB® is a high-performance programming language developed for the solution of technical problems. It found extensive application in the educational process and in scientific and engineering activities. MATLAB contains the graphical programming language Simulink® that is intended for dynamic system simulation. It is supposed that the system under investigation is developed as a functional diagram that consists of the blocks, which are equivalent, by their functions, to the program blocks in the Simulink library. When the model is working out, the developer transfers the needed blocks from the Simulink library to his or her block diagram and connects them according to the system functional diagram.

These features of Simulink made it very popular among scientists and engineers. There are also many books in which various aspects of Simulink application are described. The list of these books can be found on the site http://www.mathworks.com/support/books/. This list contains more than 1500 books and shows an extensive employment of MATLAB and Simulink.

In order to simplify and to precipitate elaborations and investigations in diverse electrical fields, the toolbox SimPowerSystems™ is developed that uses the Simulink environment, which gives the possibility to make the system model with the help of the simple procedure of click and drag.

This book considers the methods of simulation of electrical energy renewable sources with SimPowerSystems employment. There are not many books devoted to this issue, and this book intends to fill that gap. About 140 models of renewable energy sources are appended to the book, which are made with the use of SimPowerSystems. These models are in https://www.crcpress.com/Renewable-Energy-Systems-Simulation-with-Simulink-and-SimPowerSystems/Perelmuter/p/book/9781498765985. The models are called out in text where applicable and the models are listed in alphabetical order in the appendix. The aims of these models are to help understand SimPowerSystems; to help understand the various electrical engineering fields: industrial electronics, electrical machines, electrical drives, production and distribution of the electrical energy, etc.; to facilitate the understanding of various renewable energy system functions; and to be a basis for the development for the reader of his or her own systems in this field. In this connection, not all technical decisions, which were made under the creation of these models, are optimal ones, because the author tried to show various approaches to solve the problems that are admissible by SimPowerSystems. Moreover, the author tried to avoid the intricate models that are of interest, mainly, for the specialists in a certain technical field, in order to do the examples intelligibly for a wider circle of readers. During the work with the aforementioned models, the reader can investigate them with the given or the chosen parameters, use them as a basis for the analogous models developed by him or her and so on. Interactive work with a computer is supposed to be done during reading the book.

The content of this book can be summarized as follows:

Chapter 1 covers some common characteristics of SimPowerSystems and describes the graphical user interface, **Powergui**, which makes simulation and analysis much easier. Models of the main circuit elements that are used in the full model of the system, electrical sources, resistors and loads, transformers, and transmission line models, are described in Chapter 2. A number of models of measuring and control blocks that are often used together with the power blocks included in SimPowerSystems are described in Chapter 2, too, together with the blocks that are intended for the connection of SimPowerSystems to the blocks of Simulink.

Chapter 3 describes the models of the semiconductor devices that are used in power electronics: diodes, thyristors and gate turn-off thyristors, power transistors, and the control blocks intended for functioning with these devices. The models of the high-voltage direct current transmission line, of two- and three-level voltage source inverters (VSIs), of cascaded inverters, of four-leg VSI, of Z-source inverters, of modular inverters, and of direct and indirect matrix converters are elaborated with the employment of the semiconductor devices, power and control blocks mentioned earlier. Some models of flexible alternating current transmission systems devices are considered in this chapter as well.

The models of various types of generators that are used in the renewable sources of electrical energy, both contained in SimPowerSystems and developed by the author, including induction generators (IG) with pole number switching and IG with self-excitation, 6-phase IG, brushless doubly fed IG, 6- and 12-phase synchronous generators, and switched reluctance generators are described in Chapter 4.

The simulation of a number of renewable energy sources, together with the sources, which, strictly speaking, are not renewable but work with them in combination in distributed systems are considered in Chapter 5: photovoltaic systems, batteries, hydroelectric power systems, microturbines, and fuel elements. Chapter 6 is devoted completely to the simulation of wind generator systems. Diverse generator types and diverse structures are under consideration.

With the employment of SimPowerSystems, a very important problem is the compatibility of the developed models with diverse versions of SimPowerSystems, bearing in mind that every year two versions appear; at this, the new blocks appear, some blocks are declared obsolete, and some of these versions have new features that influence the model performance. It is natural that, bearing in mind the time that is necessary for writing and publishing the book, the author cannot keep up with the latest version. This book is written with the employment of versions R2012b and R2014a. The models can operate in both versions. Moreover, it has been verified that all models can start in version R2016a. Some of the models cannot work in version R2014a. For such models, special modifications are developed. The reader has to put *all* the appended files in the same folder in the MATLAB catalog; this folder has to be defined as the starting one.

This book can be used by engineers and investigators in the development of new electrical systems and investigations of the existing ones. It is very useful for students of higher educational institutions during the study of electrical fields, in graduation work and undergraduate's thesis.

Author

Viktor Perelmuter, DSc, earned a diploma in electrical engineering with honors from the National Technical University (Kharkov Polytechnic Institute) in 1958. Dr. Perelmuter earned a candidate degree in technical sciences (PhD) from the Electromechanical Institute Moscow, Soviet Union, in 1967, a senior scientific worker diploma (confirmation of the Supreme Promoting Committee by the Union of Soviet Socialist Republics Council of Ministers) in 1981, and a doctorate degree in technical sciences from the Electrical Power Institute Moscow/SU in 1991.

From 1958 to 2000, he worked in the Research Electrotechnical Institute, Kharkov, Ukraine, in the thyristor drive department, during which he also served as department chief (1988–2000). He repeatedly took part in putting power electric drives into operation at metallurgical plants. This work was commended with a number of honorary diplomas. Between the years 1993 and 2000, he was the director of the joint venture Elpriv, Kharkov. During 1965–1998, he was the supervisor of the graduation works at the Technical University, Kharkov. Dr. Perelmuter was a chairman of the State Examination Committee in the Ukrainian Correspondence Polytechnical Institute from 1975 to 1985. Simultaneously with his ongoing engineering activity, he led the scientific work in the fields of electrical drives, power electronics, and control systems.

He is the author or a coauthor of 10 books and approximately 75 articles and holds 19 patents in the Soviet Union and Ukraine. Since 2001, Dr. Perelmuter has been working as a scientific advisor in the National Technical University (Kharkov Polytechnic Institute) and in Ltd "Jugelectroproject," Kharkov. He is also a Life Member of the Institute of Electrical and Electronics Engineers.

1 SimPowerSystems™ Models

1.1 GENERAL CHARACTERISTICS

MATLAB® is a programming language that has been developed for resolving technical problems. The graphical programming language Simulink® is included in MATLAB and is intended for the simulation of dynamic systems. It uses MATLAB as a computational engine and therefore enables the use of MATLAB algorithms and sends simulation results to MATLAB for further analysis. Simulink contains the vast nomenclature of the blocks that model the blocks, devices, and units of the real dynamic systems; when the system model is developed, the proper blocks are carried from the Simulink library to the model block diagram and are connected according to their functions.

The great merit of MATLAB/Simulink is the possibility to utilize the toolboxes that are intended for certain branches of science and engineering. Their employment makes the investigation of the systems, which they are meant for, easier and faster.

Electrotechnical systems are very complex for full analysis of steady states and transients; therefore, simulation is the only reasonable method. The blocks, devices, and units that form the electrotechnical system have a very complex intrinsic structure; the employment of Simulink blocks for building an appropriate model demands from the staff developing this model in-depth knowledge about the device structure and processes in them. The set of the SimPowerSystems blocks [1] contains the models of rather complex but standard devices and units, whose fields of application are production, transmission, transformation, and utilization of the electric power, the electrical drives, and the power electronics. Evidently, renewable sources of electrical energy belong to these types of systems, and SimPowerSystems can be successfully used for their creation, simulation, and investigation.

The first step in the simulation is the creation of a model diagram consisting of the separate blocks of Simulink and SimPowerSystems. The terminals of Simulink blocks are designated as ">"; the connecting lines transmit the directional signals. The terminals of the blocks of SimPowerSystems are the electrical line; they transmit the electrical power in both directions and are designated as "□." The terminals of these two types cannot be connected with one another. For that, the blocks of SimPowerSystems that interact with the Simulink blocks have the special terminals, designated as ">" as well. Moreover, the special blocks are in the SimPowerSystems that convert the quantities of SimPowerSystems in the signals that are available for Simulink blocks and, on the contrary, convert the Simulink signals in the voltages and currents for SimPowerSystems.

Before the simulation start, the block parameters must be specified in the block dialog boxes. They can be given as numerical values or letters, which are determined either in the MATLAB program with extension *.m* (Model Parameters [MP] program) or in the option File/Model Properties/Callbacks/Init Fcn directly. The MP program must be executed before the simulation start. Its execution can be made manually, or the command "run <MPname>" has to be written down in the option mentioned earlier.

Before the simulation start, it is also necessary to decide if the model will be considered as a continuous or a discrete one; the choice is carried out in the block of user graphical interface **Powergui** (see the next paragraph). The authors of SimPowerSystems recommend the use of the continuous integration method, with inconstant integration step, for small systems, and the discrete method, with a constant step, for big systems, especially with nonlinear blocks. At this, a sampling time of 10–20 μs or more can be chosen for the systems without forced commutated electronic devices and 0.5–5 μs for the models with insulated-gate bipolar transistor, gate turn-off thyristor (GTO), and metal–oxide–semiconductor field-effect transistor. In most cases, the so-called stiff integration method **ode23 tb** turns out to be preferable. For the important tasks, it is recommended to experiment with the different integration methods, with different sampling times and with different integration tolerances.

The employment of the Phasor simulation method enables speeding up the simulation process essentially. Phasor is a time (unlike space) vector of the sinusoidal voltage or current that has a certain frequency (50 or 60 Hz in the systems under consideration). For the voltage $u(t) = U_m \sin(\omega t + \alpha)$, the phasor is $\mathbf{U} = U_m e^{j\alpha} = U_m \cos\alpha + jU_m \sin\alpha$, the same for the current phasor. All phasors in the system have the same frequency. The phasors are summed by the rules of vector algebra; two main laws of the circuit theory are valid for them: the sum of all voltage drop phasors around any closed network is zero, and the sum of all current phasors in the node is zero. If the voltage phasors $\mathbf{V}_a$ and $\mathbf{V}_b$ at the line ends are known, the current phasor is

$$\mathbf{I} = \frac{\mathbf{V}_a - \mathbf{V}_b}{\mathbf{Z}} = \frac{\mathbf{V}_a - \mathbf{V}_b}{R + j\omega L - j/\omega C} = (\mathbf{V}_a - \mathbf{V}_b)\mathbf{Y}, \tag{1.1}$$

where $\mathbf{Y} = G + jB$ is admittance, G is conductance, and B is susceptance.

The full circuit power is

$$\mathbf{S} = 0.5(\mathbf{VI}^*). \tag{1.2}$$

If not the maximal but the effective (root-mean-square) values are used for phasors, the multiplier 0.5 drops out. For more details about phasor employment, one can read, for example, in the work of Dorf and Svoboda [2].

When calculations are executed with the Phasor simulation method, it is not the instantaneous values of the voltages and the currents that are found, but their phasors that are coupled by the relationship in Equation 1.1. At this, the differential equations are replaced by the algebraic ones; the rapid proceeding processes are neglected. This method is used, mainly, under the study of electromechanical oscillations in the power systems, which contain many generators and motors, but gives the decision on one certain frequency. For simulation in this mode, the option Phasor is selected in **Powergui**, with the indication of the frequency (50 Hz for our systems).

The simple model **Example_Phasor** helps to better understand the peculiarity of this method. Two three-phase sources are connected to the ends of the three-phase circuit. The first one, with a voltage of 100 V and a frequency of 50 Hz, is connected permanently, and the second source is connected at t = 0.5 s with the help of the breaker, so that until this time the circuit is open. The second source, with a voltage of 50 V and a frequency of 50 Hz, fabricates additionally a voltage with a frequency of 250 Hz. The results of the simulation in the discrete mode and in the Phasor mode are shown in Figure 1.1. It is seen that the information about higher harmonics disappears completely.

The considerable speeding up of the simulation is obtained with the employment of the Acceleration mode. At this, the C-compiled code of the model is created. It is worth noting that in that mode, between the model start and the calculation beginning, some time elapses that depends on a number of factors and can be long sometimes.

When the simulation start is actuated, the Simulink model of the modeled system is created and saved in the block **Powergui**. Simulink employs the **State-Space** block for the modeling of the system linear part that is defined by the matrixes **A**, **B**, **C**, and **D**. The Simulink source blocks that are connected with the inputs of the block **State-Space** model the SimPowerSystems sources. Certain Simulink blocks model nonlinear elements. These blocks are placed in the feedback circuits between the voltage outputs and the current inputs of the linear part (Figure 1.2). When it is in demand, discretization is carried out. For the Phasor method, the **State-Space** model is replaced with the transfer matrix $\mathbf{H}(j\omega)$ that connects the voltage and the current phasors.

Many blocks of the SimPowerSystems are modeled as the current sources. Two such blocks cannot be connected in series; in an attempt to do that, a report about the error and an offer to set a resistor parallel to one of the blocks appear. The proper resistance of that resistor depends on many factors. Often its value can be chosen to be

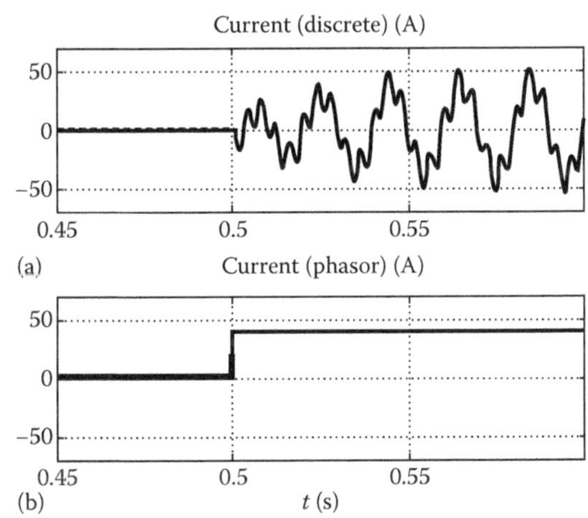

FIGURE 1.1 Explanation of the difference between (a) discrete and (b) Phasor modes.

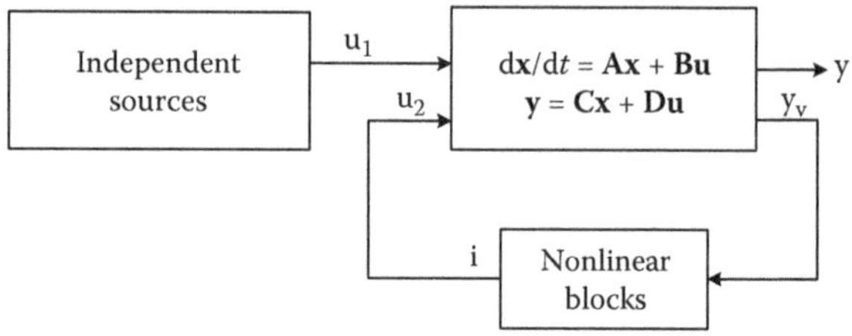

FIGURE 1.2 SimPowerSystems model structure.

very large and does not affect the simulation results. However, under the simulation of the alternating current generators, whose windings are connected in series with the inductors, cases have often occurred where the simulation process stopped when the resistance was not chosen to be sufficiently low. In order to make simulation possible, there are two ways: either to decrease the resistance or to decrease the sampling time. Under the first choice, the power loss in the resistor can distort the real power distribution, and under the second choice, the simulation time increases impermissibly. These problems will be considered in the following chapters when required.

A number of commands are included in the SimPowerSystems that are executed from the Command window. As a result, information appears that can be useful for experienced users to gain more knowledge about the model, but, in principle, to know them is not obligatory. Therefore, only one command is given here; its execution gives the opportunity to better understand the model structure; the matrixes **A**, **B**, **C**, and **D** of the linear part that are formed under this command can be used for subsequent analysis, for instance, with the employment of the Control System Toolbox.

The command appears as

```
[A,B,C,D,x0,electrical_states,inputs,outputs] =
power_analyze('sys'),
```

where `sys` is the name of the considered model. It computes the equivalent space system of the model and estimates the matrixes **A**, **B**, **C**, and **D** in the standard form:

$$\frac{d\mathbf{x}}{dt} = \mathbf{Ax} + \mathbf{Bu}, \mathbf{y} = \mathbf{Cx} + \mathbf{Du}. \tag{1.3}$$

The state variables in the vector **x** are the inductance currents and capacitor voltages. The nonlinear elements are the current sources that are controlled by voltages. The inputs in the vector **u** are the voltage and current sources and the currents from the nonlinear elements. The output vector **y** contains the measured voltages and currents and the voltages across the nonlinear elements. The breakers, switches, and the devices of power electronics are supposed to be open.

The simple model **Example2** is depicted in Figure 1.3. This model has three state variables: the currents in branches 1 and 2 and the voltage across the capacitor in branch 1. The outputs are the measured current and the voltage across the breaker.

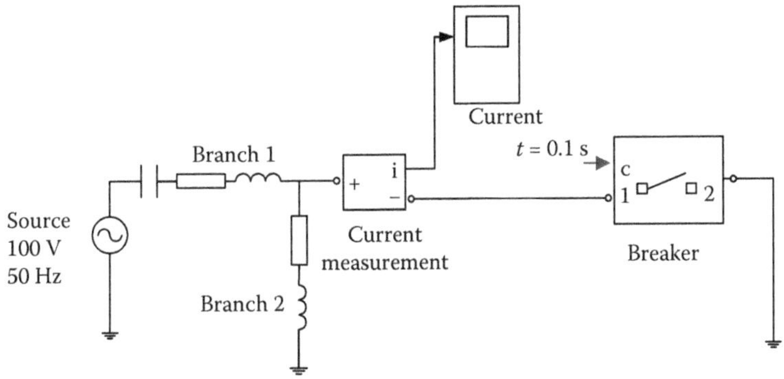

FIGURE 1.3 Simple model **Example2.**

The inputs are the source voltage and the breaker current. Executing the given command with `sys = example2` results in the following:

```
A = 1.0e+04 *
   -0.8375  -0.0008  0.8333
    1.0000        0       0
    1.0000        0  -1.0120

B = 1.0e+04 *
    0.8333   0.0008
    0             0
   -1.0000        0

C = 1000        0     -1000
       1        0        -1

D = -1000       0
        0       0

x0 = -1.3765
     -64.9571
     -1.4214

electrical_states = 'Il_Branch1' 'Uc_Branch1' 'Il_Branch2'

inputs = 'I_Breaker' 'U_Source'

outputs = U_Breaker I_Current Measurement
```

The simulation of the renewable electrical sources has a peculiarity that they almost always utilize the devices of the power electronics; it results in very little sampling time—not more than 5–10 μs. But the transient processes in such systems last minutes or dozens of minutes, demanding an enormous time for the full simulation. Therefore, it is necessary to take a number of simplifications under their investigations, including (not all simultaneously) neglecting the processes of the initial setting; neglecting the torsional elasticity; modeling by parts; decreasing the time constants of the mechanical, thermal, and chemical processes in comparison with actual values; and so on.

1.2 GRAPHICAL USER INTERFACE POWERGUI

The graphical user interface **Powergui** gives a number of opportunities to choose the wanted simulation conditions and to gain more information about the simulation results. This block is used for keeping the model data, so that its presence on the model diagram is obligatory. Only one such block may be placed on this diagram; its name cannot be changed.

On clicking the block picture, its dialog box opens as shown in Figure 1.4. When the item *Configure parameters* is selected, the new dialog window opens with three fields: *Solver, Load Flow*, and *Preferences.* The view of the first field depends on the chosen method of simulation (continuous, discrete, or Phasor). If the first method is selected, the opportunity appears to check the option *Enable use of ideal switching devices*; if it is the case, the following three fields, in which the details of the employed methods are made more precisely, appear, to turn off the snubbers, to take

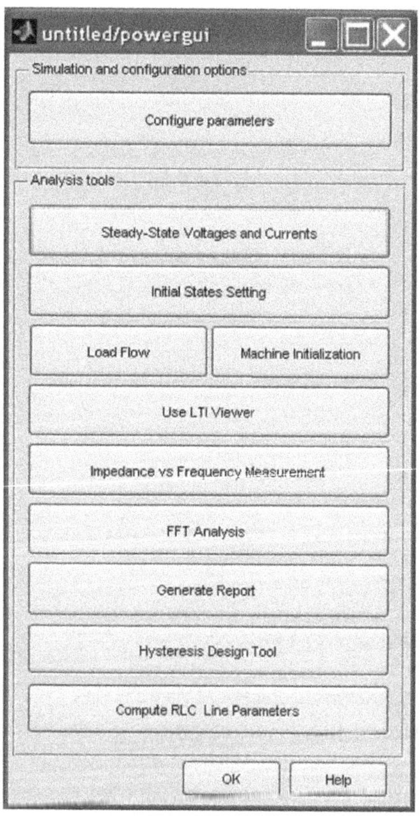

FIGURE 1.4 **Powergui** dialog box.

the inner resistances of the switches and the devices of the power electronics equal to zero, and to take the forward voltages of the power electronics devices equal to zero. Moreover, the opportunity appears to show the system differential equations in the Command window or in the window Diagnostic Viewer. When the discrete method of simulation or the Phasor method is selected, the sampling time or the frequency must be specified in the emerging lines, respectively.

The field *Load Flow* contains a number of lines for the specification of the parameters that can be used under the computation of the load flow for the model (see further): the frequency, the base power, the tolerance under the solution of the nonlinear equations with the approximating method, and so on. The field *Preferences* has four lines; the selection of three of them gives the opportunity to receive some messages about the model operation and to use the special function in the Acceleration mode (the utilization of this function often makes the simulation slower, and it is not considered here). In the last line, the initial conditions for the simulation are defined. There are three opportunities: *Blocks*, *Steady*, and *Zero*. In the first case, the initial conditions are defined in the model blocks; in the second one, they are taken equal to the steady-state values; and they are equal to zero in the third case.

Furthermore, the available analysis tools are listed. The item *Steady-State Voltages and Currents* gives the possibility to find the steady-state values of the variables in the vectors **x**, **u**, and **y** that were mentioned in Section 1.1. The window for the model depicted in Figure 1.3 is displayed in Figure 1.5. The opportunity is to

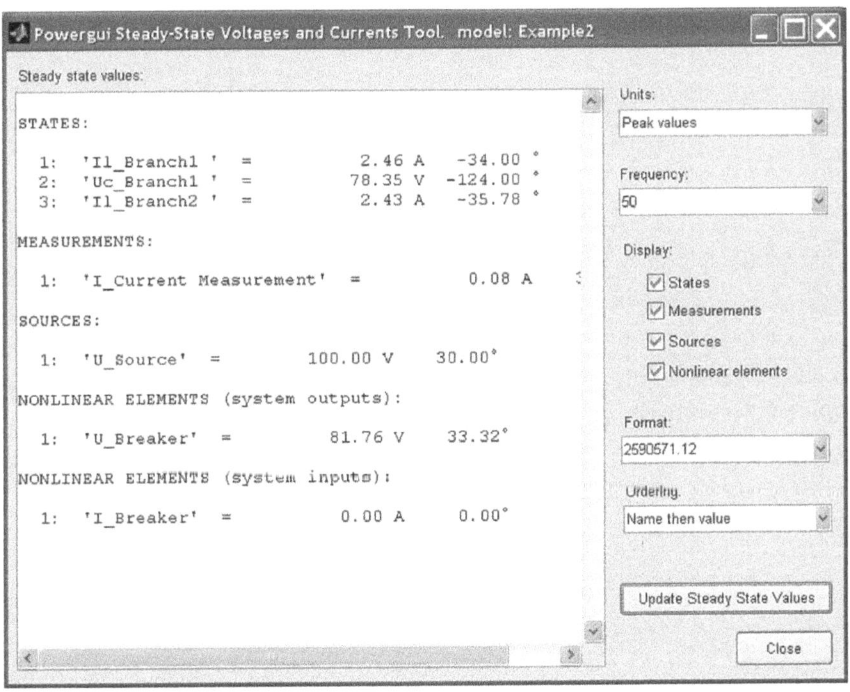

FIGURE 1.5 **Powergui** window under steady-state estimation.

select for the display of only the part of the variables: states, measurements, sources, inputs, and outputs of the nonlinear elements.

The item *Initial States Setting* defines the initial values of the state variables (they are the currents in the inductances and the voltages across the capacitors) for simulation. They can be equal to the values that were found in the previous item at $t = 0$, but they can be set equal to zero or, depending on model configuration, be set at will.

The next item *Load Flow* gives an opportunity to run the simulation, beginning from the steady state. The analysis of voltages and currents in the steady state is of great importance during the investigation of the electric power systems. Since the computation of a steady state is rather difficult, one begins the simulation simply with arbitrary initial values and waits to get the steady state, which often requires a lot of time. This item makes the simulation much quicker.

It is supposed that the system consists of n nodes, or collecting buses. The relationships that describe the system steady state can be expressed as follows. In the general case, to each ith node, the generator lines that deliver currents $\mathbf{I}_{gi}$ and have the complex powers $\mathbf{S}_{gi}$, the load lines consuming currents $\mathbf{I}_{li}$ and having the complex powers $\mathbf{S}_{li}$, and the lines that transfer in transit the complex powers $\mathbf{S}_{ti}$ with currents $\mathbf{I}_{ti}$ are connected, so that

$$\mathbf{I}_{gi} = \mathbf{I}_{li} + \mathbf{I}_{ti}. \tag{1.4}$$

If the node voltage vector is $\mathbf{V}_i$, then

$$\mathbf{V}_i\mathbf{I}_{gi}^* = \mathbf{V}_i\mathbf{I}_{li}^* + \mathbf{V}_i\mathbf{I}_{ti}^* \tag{1.5}$$

or

$$\mathbf{S}_{gi} = \mathbf{S}_{li} + \mathbf{S}_{ti}. \tag{1.6}$$

Because

$$\mathbf{S}_{gi} = P_{gi} + jQ_{gi}, \tag{1.7}$$

and it is the same for the other powers, it may be written for the active and the reactive power, respectively,

$$P_{gi} = P_{li} + P_{ti} \tag{1.8}$$

and

$$Q_{gi} = Q_{li} + Q_{ti}. \tag{1.9}$$

Therefore, with the knowledge of the vectors of the in-flow and out-flow currents, and also the voltage vectors, the active and reactive powers can be computed. It may be written for the currents $\mathbf{I}_{ti}$ as

$$\mathbf{I}_{\text{ü}}^{*} = \sum_{j=1}^{n} \mathbf{y}_{ij}\mathbf{V}_{j}, \quad i = 1,2,\ldots,n. \tag{1.10}$$

$\mathbf{y}_{ij}$ are the complex admittances between the nodes i and j, or

$$I_{\text{ü}}^{*} = \sum_{j=1}^{n} y_{ij}V_{j}\langle -\gamma_{ij} - \delta_{j}\rangle, \quad i = 1,2,\ldots,n, \tag{1.11}$$

where the expression in the braces means the vector angle, y_{ij} is the admittance module, γ_{ij} is its angle, V_j is the voltage module, and δ_j is its angle.

After substitution in Equation 1.5 and taking in mind Equations 1.6 through 1.9, we receive the following for each node:

$$P_{gi} = P_{li} + \sum_{j=1}^{n} y_{ij}V_iV_j \cos(\delta_i - \gamma_{ij} - \delta_j) \tag{1.12}$$

$$Q_{gi} = Q_{li} + \sum_{j=1}^{n} y_{ij}V_iV_j \sin(\delta_i - \gamma_{ij} - \delta_j) \tag{1.13}$$

The load powers P_{li} and Q_{li} are supposed to be known. Thus, for each node, there are two equations with four unknown quantities: P_{gi}, Q_{gi}, V_i, and δ_i; therefore, two quantities have to be assigned. Which of them depends on the selected bus type. Three available types exist: (1) The reference bus; in SimPowerSystems, it is called *Swing bus*. In this node (bus), it is accepted as $V_i = 1$, $\delta_i = 0$; the unknown quantities are P_{gi} and Q_{gi}. Such a node is one for the system. Usually, it is the generator of the largest power or the bus of the connection with the other system. (2) The load buses; they are buses with the known P_{gi} and Q_{gi}. Usually, they are the buses without the generator power, so that $P_{gi} = 0$ and $Q_{gi} = 0$; the unknown quantities are V_i and δ_i. (3) The generator buses; the system generator—prime mover—usually has the power and voltage regulators with fixed references, so it may be accepted: P_{gi} and V_i are known; Q_{gi} and δ_i have to be found.

The system of nonlinear equations that is received in this way is solved by an approximate method; subsequently, the computed values of voltages, currents, and powers are fed into the model automatically, and the simulation starts.

In order to better understand the **Powergui** operation, the model **Example3** is considered. Explanations provided here are brief, and the reader can return to this model later. G1 is the synchronous generator (SG) model of 250 MVA power, of 13.8 kV voltage. The step-up transformer with 13.8/230 kV is set at the SG output; via this transformer, the electrical energy is transferred to the network. The same unit SG—transformer (G2), but with the power of 200 MVA, is connected parallel to the first unit via the transmission line 100 km in length. The generators are driven with the hydraulic turbines. Each turbine has a governor that controls the turbine rotating speed (which is assigned as 1 pu); the governor characteristic has a droop that is assigned equal to 0.05. The excitation system maintains the SG voltage equal to the nominal value. The main load with the power of 100 MW is set at the high-voltage side of the second transformer. Moreover, there are the local loads with powers of 5 and 15 MW connected directly to the terminals of the G1 and the G2, respectively. The reference powers for the governors are chosen as 0.8 and 0.75 for SG1 and SG2, respectively, so that the power delivered to the network has to be $0.8 \times 250 + 0.75 \times 200 - 100 - 5 - 15 = 230$ kW.

If the simulation is run with the zero initial conditions (vector of the initial conditions [0 0 0 0 0 0 0 0 1.5]), it may be seen that the indicated powers are reached after rather long transients (about 200 s).

Now we will use the item *Load Flow*. Copy this model under the name **Example3a**. For the source of 230 kV, the Swing bus is selected, for the SG1 and the SG2 the PV buses, with the powers of $0.8 \times 250 = 200$ MW and $0.75 \times 200 = 150$ MW are selected, respectively; the mode with the constant impedance is determined for the loads. In a dialog box of **Powergui**, the item *Load Flow* is selected; after the execution of the command Compute, the item window receives the view displayed in Figure 1.6. Now to fulfill the command Apply, the calculated values are put into the corresponding fields of the model blocks that results in change in the model: the source voltage and its angle changed to 230×1.00176 kV, $\angle = -2.62°$. In the SG models, the vectors of the initial conditions change to nonzero; in the models of the governors, the values of the initial powers come to 0.80183 and 0.75161, respectively. To execute simulation, it will be seen that the assigned power values of SG1 (200 MW) and SG2 (150 MW) are obtained immediately, so that the investigations, beginning from the steady states (changes in the voltages and the loads, emergency operations), can begin on the simulation start.

The neighbor item *Machine Initialization* is the simplified version of the previous one (in the preceding versions of SimPowerSystems, only one item existed: *Load Flow and Machine Initialization*). In the model **Example3b**, the source 230 kV is replaced with the load of 100 MW. Let the SG2 produce a power of 125 MW (0.625 pu). Then the reference and the initial values of the power for the governor of SG2 are set to 0.625; the vector of the initial conditions of SG2 is chosen as [0 0 0 0 0 0 0 0 1.5]; and the bus type for the *Load Flow* is *PV* with a power of 125 MW. The swing bus is selected for SG1; the reference and the initial values of the power for the governor of SG1 are set arbitrarily as 0.5. The vector of the initial conditions is the same as for SG2. To run the simulation, it will be seen that after transients lasting for about 1 s, the SG1 and SG2 powers are equal to 110 and 111 MW, respectively. To choose the item *Machine Initialization* and to carry out the command Compute and Apply, one sees how the model parameters change: the references and the initial power values for SG1 and SG2 come to 0.345 and 0.675, respectively, and the nonzero vectors of the initial conditions appear in the SG models. The values of the powers that are obtained after the simulation are equal to 86 and 135 MW, respectively; at this, the transients are absent.

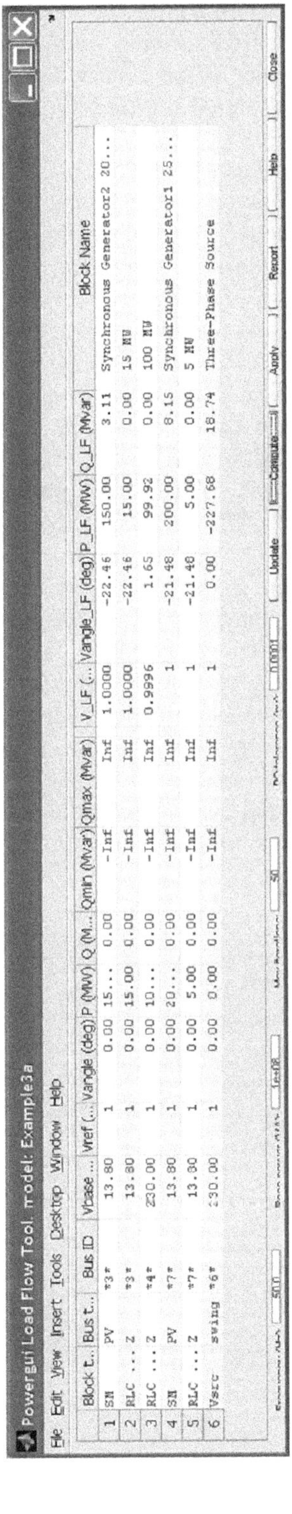

FIGURE 1.6 Window of the item *Load Flow*.

Carry on with the study of **Powergui**. The item *Use LTI Viewer* gives the opportunity to use the methods that are found in the MATLAB toolbox Control System Toolbox. Since the reader is not expected to be aware of this toolbox, this item is not described here. Interested readers can get to know this toolbox from Ref. [3].

The item *Impedance vs Frequency Measurement* gives the opportunity to build the frequency characteristic of the complex resistance of any model block. The model **Example3c** is a copy of model **Example3**. The block **Impedance Measurement** from the folder Measurement is added in the model; this block is connected in parallel to one phase of the transmission line. To actuate the aforesaid item, the frequency amplitude and the phase characteristics are plotted as shown in Figure 1.7.

The item *FFT Analysis* is intended for the frequency analysis with the help of the Fourier transformation of the recorded data. Under the data record, the scopes have to be in the following format: *Structure with Time*. For the demonstration of this item, we use the model **Three_Ph_Contr_Source**, which is considered in more detail in Chapter 2. The diagram of this model is depicted in Figure 2.1. In the controlled source, with the amplitude of the fundamental harmonic of $100 \times \sqrt{2} / \sqrt{3} = 81.6$ V at the frequency of 50 Hz, the 5th harmonic with the amplitude of 0.4 of fundamental and the 11th harmonic with the amplitude of 0.2 of fundamental are selected. After the simulation and the actuation of the item *FFT Analysis*, a window appears having a number of fields (Figure 1.8). In the upper right field, a list of the available structures that are formed in Workspace appears. There is only one structure in the case under consideration that is formed by ScopeData. The structure, the number of its input, and the number of the signal in this input are selected. Then the initial time of the processed signal, the number of the cycles of the fundamental frequency and the value of this frequency, the

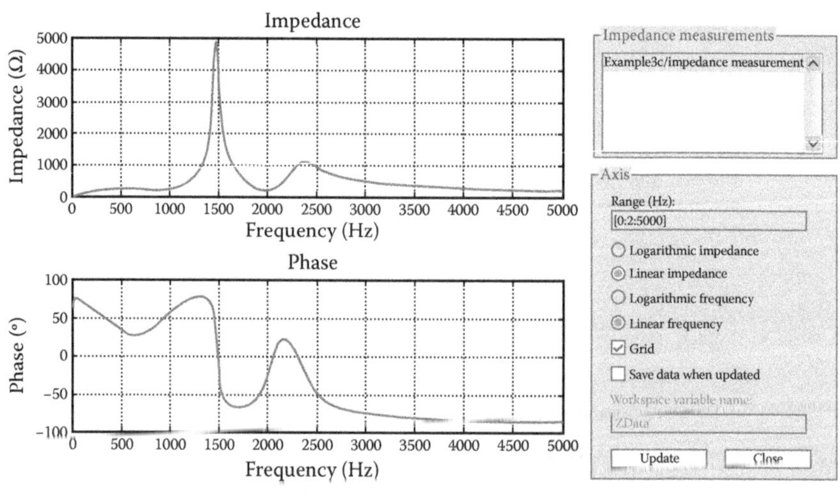

FIGURE 1.7 Amplitude–frequency characteristics of a phase of the transmission line model.

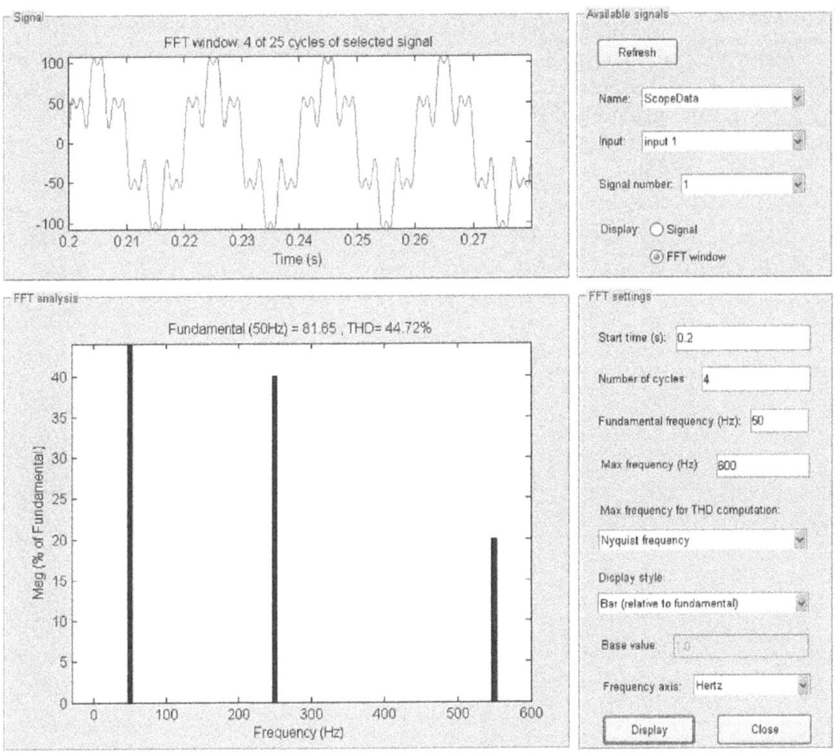

FIGURE 1.8 **Powergui** window during Fourier transformation computation.

maximal frequency for computation of Fourier transformation, the unit for axis x, and a style of result show—bar or list—are set.

The processed signal is displayed in the upper left window; at that point, a part of the signal (*Display selected signal*) or all signal (*Display FFT window*) can be chosen. With the command Display, the result of the Fourier transformation appears in the lower left field in the selected style of representation.

The next item *Generate Report* gives the opportunity to store the part of the computation results previously mentioned: the initial states, the steady states, the load flow, and machine initialization—as the text file with the model name and the extension .rep.

The item *Hysteresis Design Tool* is used with the transformer models while taking saturation into account and will be considered later. The last item *Compute RLC Line Parameters* is used for computing the active and reactive impedances of the electrical power transmission line by using the conductor data, the data of the line, and the support construction. This procedure is considered in Chapter 2 as well.

From the previous discussion, it follows that **Powergui** is a convenient tool that makes operations with SimPowerSystems models easier.

REFERENCES

1. MathWorks, *SimPowerSystems™, User's Guide*. MathWorks, Natick, MA, 2011–2016.
2. Dorf, R. C., and Svoboda, J. A. *Introduction to Electric Circuits* (9th ed.). John Wiley & Sons, Hoboken, NJ, 2013.
3. MathWorks, *Control System Toolbox™: User's Guide*. MathWorks, Natick, MA, R2013b—R2016b.

2 Models of the Elements and Devices

2.1 ELECTRICAL SOURCES

The following blocks that model the electrical sources are included in SimPowerSystems: (1) **DC Voltage Source**, (2) **AC Voltage Source**, (3) **AC Current Source**, (4) **Controlled Voltage Source**, (5) **Controlled Current Source**, (6) **Three-Phase Source** (with the inner impedance), (7) **Three-Phase Programmable Voltage Source**, and (8) **Battery**.

Before utilization, the block parameters must be put into the block dialog box. For the block of p.1, the direct voltage (V) is assigned. The blocks of p.2 and p.3 generate the sinusoidal voltage and the current, respectively; the voltage (V) or the current amplitude (A), the phase angle (°), the frequency (Hz), and also the sampling period are specified. There is an opportunity to measure the output block quantities without the employment of additional sensors if the option *Measurement* is selected; at this, the **Multimeter** is employed (see further). If the zero frequency is chosen, the phase angle must be equal to 90°.

Controlled Voltage Source and **Controlled Current Source** can operate in two modes: as the sources reproducing the Simulink signal that comes to their inputs or as the original sources. In the first case, the flag *Initialize* in the dialog box is reset. In the second case (flag *Initialize* is set), the parameters of the output voltage or current have to be put into the block dialog box.

The block **Three-Phase Source** models a three-phase voltage source that has a Y-connection and R-L internal impedance. The main block parameters are the phase-to-phase root-mean-square (rms) voltage (V), the phase angle (degree), and the frequency f (Hz). The block is employed often for the simplified modeling of the electric power grid. There are two ways to define R and L. If the option *Specify impedance using short-circuit level* is not checked, the lines appear, in which the values of R (Ω) and L (H) are entered. If this option is chosen, the three-phase short-circuit power (VA) at the base voltage, the value of the base voltage (V), and the ratio X/R are specified.

The short-circuit three-phase power P_{sc} at the given base voltage V_{base} is linked with the source inductance by the relationship

$$L = \left(\frac{1}{2\pi f}\right)\frac{V_{base}^2}{P_{sc}}. \tag{2.1}$$

The ratio X/R is linked with the inner resistance as

$$R = 2\pi f L/(X/R). \tag{2.2}$$

One of the parameters, L or R, can be taken as zero.

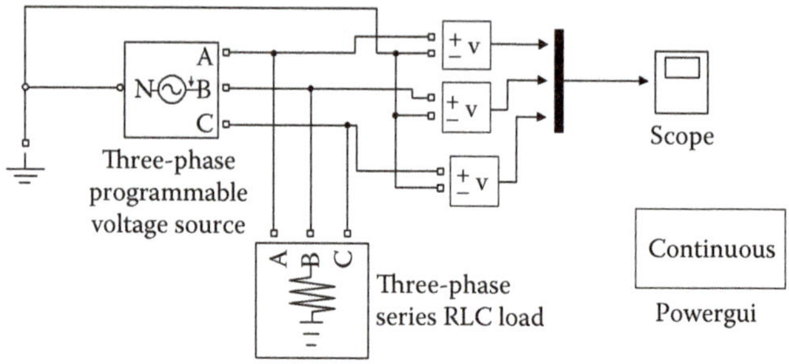

FIGURE 2.1 Diagram of the model **Three_Ph_Contr_Source**.

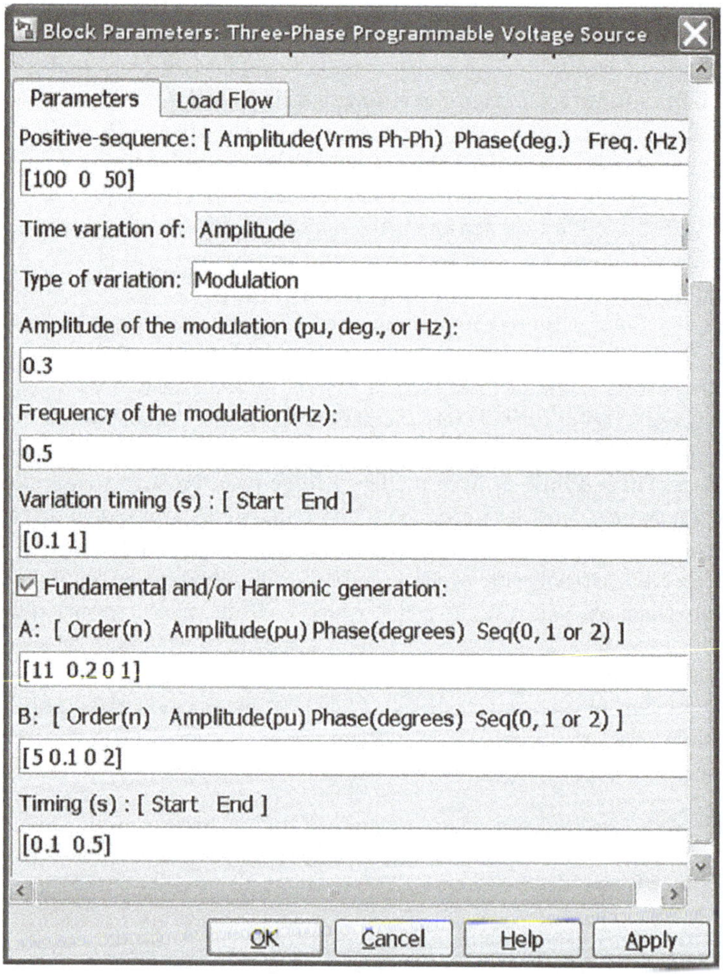

FIGURE 2.2 The main view of the dialog box of the block **Three-Phase Programmable Voltage Source**.

The **Three-Phase Programmable Voltage Source** models a nonstationary and nonsymmetric voltage source. For studying this block, the simple model **Three_Ph_Contr_Source** is used, whose diagram is depicted in Figure 2.1. The block has several possible configurations of the dialog box. The main one is shown in Figure 2.2. It is indicated in the line *Time variation of* which parameter of the supply voltage has to vary. In the drop-down menu, one can choose the following: *None*—no change, *Amplitude*, *Phase*, and *Frequency*. If some quantity (except *None*) is selected, in the line *Type of variation* of the drop-down menu, it is possible to choose the following: *Step*, *Ramp* (the linear changing with fixed rate), *Modulation*, and *Table of time–amplitude pairs*. Depending on the selected character of variations, the meaning of the next line can be changed, and the additional line can appear. In addition, the start and end times of actual changing are defined.

As an example, the phase A voltage is shown in Figure 2.3 under the modulation of the voltage amplitude during 0.1–1 s; the modulation depth is 0.3, and the frequency of modulation is 2 Hz.

If the item *Fundamental and/or Harmonic generation* is selected (as in Figure 2.2), the fields appear to assign parameters of two harmonics. For each harmonic, four parameters are defined: the harmonic order, its amplitude and its phase, and the type of the sequence (1—for the positive, 2—for negative, 0—for zero sequence). In Figure 2.2, the harmonic *A* of the positive sequence has the 11th order and a relative amplitude of 0.2; the harmonic *B* of the negative sequence has the 5th order and a relative amplitude of 0.1. After running simulation and activating the **Powergui** item *FFT Analysis*, the window shown in Figure 2.4 appears. In the upper area of the window, four periods of the phase *A* voltage are shown; in the lower area, the plot of the Fourier coefficients for this voltage is shown.

The item *Load Flow* is used when the block models a generator or the grid in the power system, and with the activation of the same function in **Powergui** (Chapter 1), it influences only the initial system state. There are three possibilities for the selection

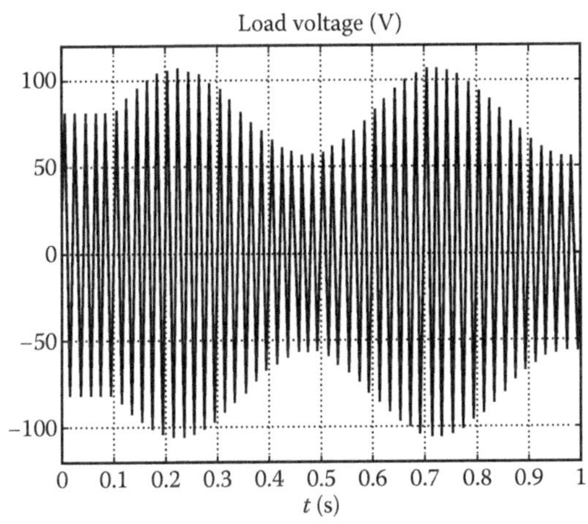

FIGURE 2.3 Source voltage under modulation of the amplitude.

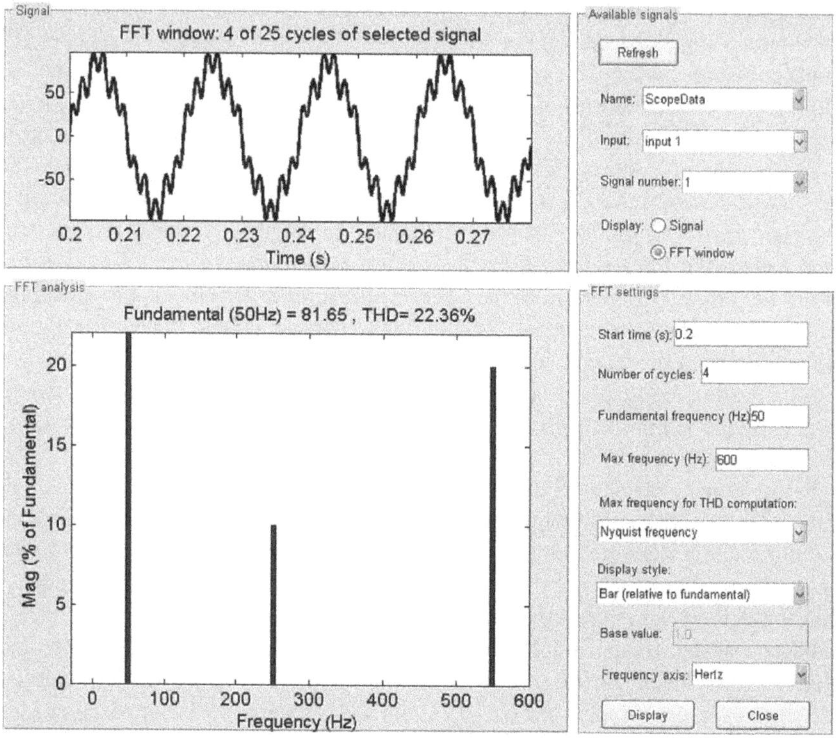

FIGURE 2.4 Fourier transformation window under the presence of the 5th and the 11th harmonics.

of the *Generator Type*: swing, *PV*, and *PQ*. The first two possibilities are considered in Chapter 1; in the third one, the generated active and reactive powers, which this source sends to the grid, are defined.

2.2 IMPEDANCES AND LOADS

The following blocks are included in SimPowerSystems that model various types of impedances and loads: (1) **Parallel RLC Branch**, (2) **Parallel RLC Load**, (3) **Series RLC Branch**, (4) **Series RLC Load**, (5) **Three-Phase Parallel RLC Branch**, (6) **Three-Phase Parallel RLC Load**, (7) **Three-Phase Series RLC Branch**, (8) **Three-Phase Series RLC Load**, (9) **Three-Phase Dynamic Load**, (10) **Three-Phase Harmonic Filter**, (11) **Mutual Inductance**, and (12) **Three-Phase Mutual Inductance Z1–Z0**.

The distinction of the branch models from the load models is that for the former, the values of *R* (Ω), *L* (H), and *C* (F) are defined, whereas for the latter, the active, reactive inductive, and reactive capacitive (W or var) powers are given under certain voltage and frequency. If some element (*R*, *L*, or *C*) is not needed or some power is equal to zero, the corresponding element is removed from the block picture.

The three-phase loads can have the configurations of Y or Delta, and in the first case, the neutral connection can be chosen: ground, floating, accessible neutral.

If it is necessary to measure the load voltage or (and) the current with the help of the block **Multimeter**, it is chosen in the falling window *Measurement*; what is to be measured are as follows: voltage, current, or both. The dialog boxes of the three-phase loads have page *Load Flow*, in which it is fixed, how the load behaves under computation of the load flow: as the constant impedance or as the consumer of the constant active and reactive powers. For the single-phase loads and branches, the initial values of the voltages across the capacitors and of the currents in the inductors may be specified.

Block **Three-Phase Dynamic Load** models a three-phase load, whose active P and reactive Q powers vary as a function of the positive sequence of the applied voltage V; this means that even under nonsymmetrical voltage, the current is symmetrical. The load impedance is constant, while $V < V_{\mathrm{min}}$. When $V > V_{\mathrm{min}}$, P and Q vary as

$$P(s) = P_0\left(\frac{V}{V_0}\right)^{n_{\mathrm{p}}}\frac{1+sT_{\mathrm{p1}}}{1+sT_{\mathrm{p2}}} \tag{2.3}$$

$$Q(s) = Q_0\left(\frac{V}{V_0}\right)^{n_{\mathrm{q}}}\frac{1+sT_{\mathrm{q1}}}{1+sT_{\mathrm{q2}}}, \tag{2.4}$$

where P_0 and Q_0 are the active and reactive powers under the initial voltage V_0 (the positive sequence), T_{p1} and T_{p2} are the time constants that control the dynamic of the active power, T_{q1} and T_{q2} are the same for the reactive power, n_{p} and n_{q} are the coefficients that usually are in the range of 1–3, and s is the Laplace transformation symbol.

Under the constant current load, $n_{\mathrm{p}} = 1$ and $n_{\mathrm{q}} = 1$, because in this case, the power is proportional to the voltage, and under constant load impedance, $n_{\mathrm{p}} = 2$ and $n_{\mathrm{q}} = 2$.

The employment of this block gives the opportunity to take into account the dependence of the load power consumption on the supply voltage more precisely. Some useful data are summarized [1] for the different types of consumers (in the order cos φ, n_{p}, n_{q})—residential consumers: 0.87–0.99, 0.9–1.7, 2.4–3.1; commercial consumers: 0.85–0.9, 0.5–0.8, 2.4–2.5; industrial consumers: 0.85, 0.1, 0.6; steel mills: 0.83, 0.6, 2; primary aluminum: 0.9, 1.8, 2.

As the simple example, the model **Var_load** is considered (Figure 2.5). The **Three-Phase Programmable Voltage Source** (380 V; 50 Hz) supplies the dynamic load that, with this voltage, has the active power of 100 kW and the reactive one of 50 kvar. The voltage source operates in the mode of the amplitude modulation with a modulation depth of 0.5 and with a frequency of 2 Hz; the modulation is active in the time interval of 1–5 s. The references for P and Q are proportional to the source voltage. It is seen by the scope that with the source voltage change, the load current is kept constant. The simulation is carried out in the Phasor mode, so that the scope fixes only the amplitudes of the harmonic quantities (Figure 2.6).

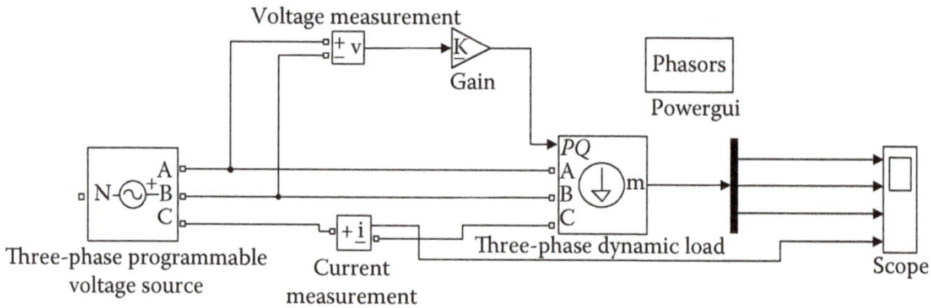

FIGURE 2.5 Diagram of the model **Var_load**.

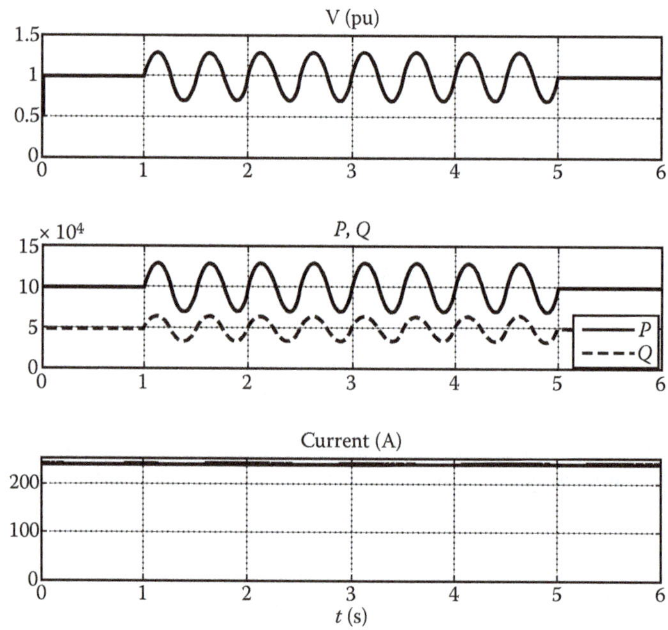

FIGURE 2.6 Processes in the model **Var_load**.

This block is useful under the simulation of the isolate operating renewable sources, when, with reduction of the external action (wind, illumination), the source power also decreases, and at this, it is necessary to reduce the load power.

A **Three-Phase Harmonic Filter** is intended for the simulation of the parallel filters in the power systems that are used for reducing the voltage distortions and for the power factor correction. These filters are capacitive at the fundamental frequency. Usually, the band-pass filters are used that are tuned for low-order harmonic filtering: the 5th, the 7th, the 11th, and so on. The band-pass filters can be tuned for filtering one frequency (single-tuned filter) or two frequencies (double-tuned filter). High-pass filters are used for filtering a wide range of frequencies. A special type of high-pass filter, the so-called C-type high-pass filter, has some merits over the other types of the high-pass filters [2]. In this filter, the reactor *L* is replaced with the series *LC* circuit that is tuned in resonance at the fundamental frequency. Therefore, the resistor is shunted with the resonance *LC* circuit at the fundamental frequency that results in nearly zero losses.

The filters contain *R*, *L*, and *C* elements. The values of these elements are determined by the following data: the filter type, the reactive power under the rated voltage, the tuned frequency or frequencies, and the quality factor. Four filter types that can be simulated in SimPowerSystems are shown in Figure 2.7.

The example of the filter parameter computations and obtained frequency characteristics are given in Ref. [3].

The block **Mutual Inductance** can operate in two modes: as a two- or three-winding reactor with equal mutual inductances or as a reactor with any winding number and with different mutual inductances. The choice is fulfilled in the dialog box. When the variant *Generalized mutual inductance* is chosen, the number of the windings *N* is specified as well as the matrix *N* × *N* of the self and mutual inductances. This model is used often for transmission line modeling.

The last block considered in this section is **Three-Phase Mutual Inductance Z1–Z0**. This block models a three-phase balanced inductive and resistive impedance with mutual coupling between phases. This block realizes the same function as

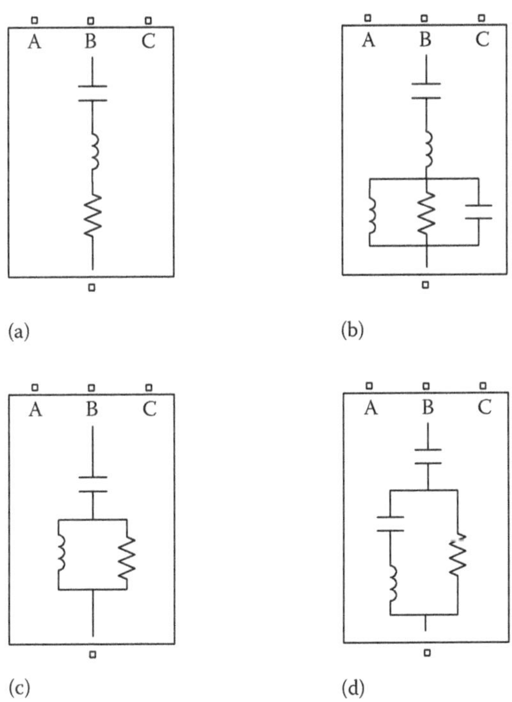

FIGURE 2.7 Pictures of the filter models: (a) single-tuned filter; (b) double-tuned filter; (c) high-pass filter; and (d) C-type high-pass filter.

the three-winding **Mutual Inductance** block, but it is more convenient for employment under the simulation of transmission lines. The block parameters are the active and reactive self and mutual impedances of the positive and zero sequences. These parameters are calculated usually during computation of the three-phase line parameters.

2.3 TRANSFORMERS

The following blocks that model the different types of the transformers are included in SimPowerSystems: (1) **Linear Transformer**, (2) **Saturable Transformer**, (3) **Multi-Winding Transformer**, (4) **Three-Phase Transformer 12 Terminals**, (5) **Three-Phase Transformer (Two Windings)**, (6) **Three-Phase Transformer (Three Windings)**, (7) **Zigzag Phase-Shifting Transformer**, (8) **Grounding Transformer**, (9) **Three-Phase Transformer Inductance Matrix Type (Two Windings)**, and (10) **Three-Phase Transformer Inductance Matrix Type (Three Windings)**.

The equivalent circuit of the single-phase transformer or of one phase of the three-phase transformer is shown in Figure 2.8. The leakage inductances L_1 and L_2 and the active resistances R_1 and R_2 of the windings, the magnetization inductance L_m, and the resistance R_m may be specified either in SI units or in per units, referring to the base resistance of the corresponding winding; for the magnetization circuit, the base values are the resistances of the primary winding. For the winding with the rated voltage U_i (V) of the transformer with power S (VA) and frequency f, the base values are

$$R_b = \frac{U_i^2}{S}, \; L_b = \frac{R_b}{2\pi f}. \tag{2.5}$$

If the option *Measurements* is checked, the opportunity appears to measure the winding voltages, the currents, the magnetizing currents, and the fluxes with the help of the block **Multimeter**. For some transformer types, there is the possibility to take the saturation of the core into account.

Consider the different types of the transformers in more detail. The single-phase transformer of p.1 can have two or three windings. The possibility to model the saturation both with and without taking hysteresis into account is intended for the single-phase transformer of p.2 that can have the two secondary windings as well.

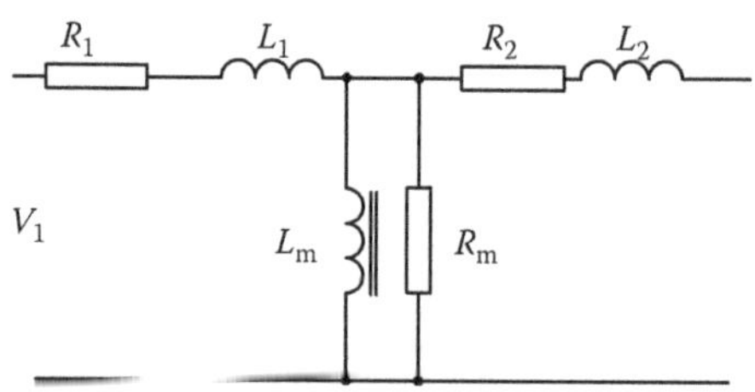

FIGURE 2.8 Equivalent circuit of the single-phase transformer.

Hysteresis modeling demands additional computation time; therefore, it has to be used only in specific cases.

With saturation modeling, the magnetization curve is defined as a collection of the pairs magnetization current—main flux in the earlier accepted units (pu or SI). The first point is the pair [0 0; …]. If the residual flux Φ_{res} is modeled, two pairs with zero abscissa are defined: [0 0; 0 Φ_{res}; …]. A piecewise linear approximation is carried out under simulation.

If hysteresis is modeled, the option *Simulate hysteresis* has to be marked; the main flux linkage of the transformer is computed by the integration of the voltage across the magnetization circuit. Both major and minor hysteresis loops are simulated using the function atan.

In order to create the file describing the hysteresis, it is reasonable to use the item *Hysteresis Design Tool* of **Powergui**. When this option is activated, the window that is shown in Figure 2.9 appears. In the right area of the window, the parameters of the curve are specified; in the left area, the curve is displayed corresponding to these parameters. We note that the option *Zoom around hysteresis* is marked, with which the loop takes up the entire screen that makes tuning of the curve easier. Without this option, all the workspace of the magnet system appears on the screen.

The built curve has to be saved on the disk under some name. The file receives the extension `.mat`. This name must be put into the line *Hysteresis Mat file* on the first page of the dialog box. It is worth noting that under the hysteresis simulation, the active resistance of the magnetization circuit takes into account only eddy current loss, without hysteresis loss.

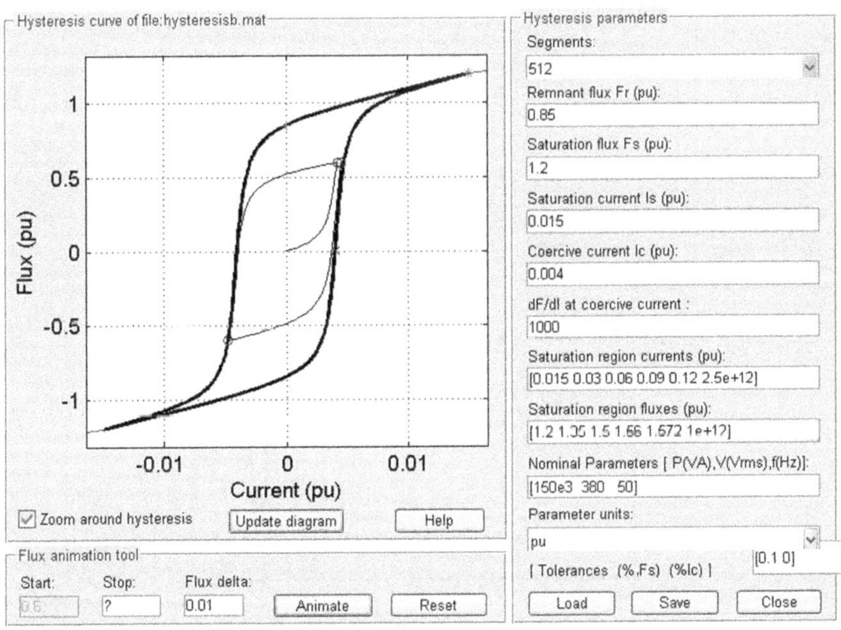

FIGURE 2.9 **Powergui** window under fabrication of the hysteresis curve.

There is one more interesting possibility for this item. There is an option *Flux animation tool*; the additional fields *Start*, *Stop*, and *Flux delta*; and two buttons *Animate* and *Reset*. With the help of these instruments, it is possible to observe the minor hysteresis loops. If, for example, to set *Start* = 0 and *Stop* = 0.6 and to fulfill the command Animate, and afterward, *Stop* = −0.6 and *Stop* = 0.6, the picture of the minor loop appears, which is shown in Figure 2.9 as well.

The model **Sat_trans** helps to study the characteristics of the **Saturable Transformer** with hysteresis. The hysteresis loop is displayed in Figure 2.9. It is saved in the file `hysteresisb.mat`. The transformer is supplied from the **Controlled Voltage Source** with the external control. The sinusoidal voltage $U_n = \sqrt{2} \times 380\,\text{V}$, 50 Hz is modulated by the amplitude with the **Timer** that assigns the following amplitude of the transformer primary voltage U: at $0 < t < 0.1$ s, $U = 0.8\ U_n$; at 0.1 s $< t <$ 0.2 s, $U = U_n$; and at 0.2 s $< t <$ 0.3 s, $U = 1.5\ U_n$. The transformer characteristics can be observed both with open and with closed secondary winding.

All variables of the transformer are measured with the block **Multimeter** and are fixed in the **Scope** in such an order: the transformer flux, the magnetization current without iron losses, the total excitation current including iron losses, the supply voltage, the primary transformer current; all quantities, except the voltage, are in per units. Moreover, there is another block **Multimeter1** that measures the voltage and the current of the primary transformer winding; these values in SI are used for the active and the reactive power computation. The graph plotter builds the magnetization curve. Firstly, the simulation is carried out with actuated option *Simulate hysteresis* and with the open secondary winding. One can observe as the graph plotter draws the hysteresis loops. There are three loops. The process in time is observed by **Scope**. The fraction of the magnetization current curve is given in Figure 2.10. Its essential increase at $t = 0.2$ s, when the supply voltage is drastically increasing, can be seen. The active power at $t = 0.15$ s (under rated voltage) is 935 W. Let us calculate the losses in the resistor R_m. The base value $R_b = 380^2/150{,}000 = 0.96\ \Omega$, then

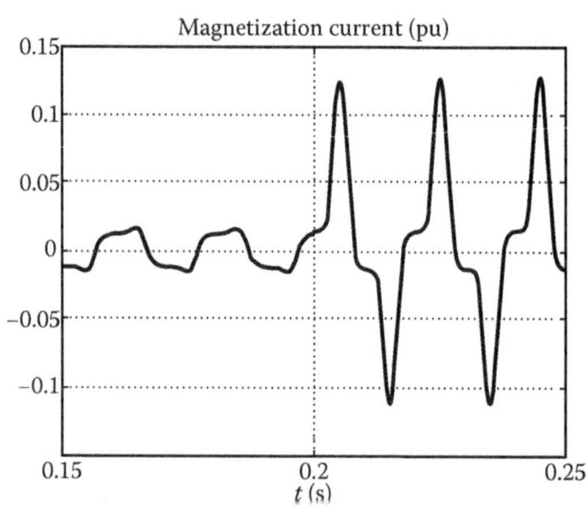

FIGURE 2.10 Changing of the transformer magnetization current in time.

$R_m = 0.96 \times 500 = 480\ \Omega$, and the losses are $P_l = 380^2/480 \approx 301$ W; hence, hysteresis losses are 634 W.

To repeat simulation without hysteresis modeling, the graph plotter displays the piecewise linear curve. **Scope** shows that until $t = 0.2$ s, the currents are sinusoidal. The power is 300 W at $t = 0.15$ s, which is in conformity with the calculation here. Thus, in order to appreciate the active losses in the magnetization circuit correctly, when the hysteresis is not modeled, it is necessary to set $R_m = (500/935) \times 300 = 160$ pu.

These experiments can be repeated when the **Breaker** in the secondary winding is closed. The magnetization currents do not change noticeably, but the influence of nonlinearity on the primary current reduces essentially.

The **Multiwinding Transformer** models a transformer with the changing number of windings both on the primary (on the left) and on the secondary (on the right) sides. The evenly spaced taps can be added to the first primary (upper on the left) or to the first secondary (upper on the right) windings. The voltage between the two next taps is equal to the winding total voltage divided by the number of taps plus 1. The total resistance and the leakage inductance of the winding are evenly spaced along the taps too. This transformer model is usually used for the simulation of the transformer that is controlled under a load change, with the aim to keep the bus voltage constant. The requested range of the voltage changing is ±(10–20%) of the rated value. The tap number is 6–10. Usually, the transformer has the main primary or secondary winding that is calculated for the rated voltage and the adjusting winding with taps that can be connected aiding or antiaiding to the main winding. Under the aiding connection to the primary winding, the output transformer voltage reduces, and under the antiaiding connection, it increases. Under the connection of adjusting winding in series to the secondary, the effect of switching over is reversed.

This type of the transformer unit has a drive for switching-over taps. The drive is controlled with the voltage controller. This unit is a rather complex device; there is the special model **Three-Phase OLTC Regulating Transformer** (OLTC is an abbreviation for On Load Tap Changer) in SimPowerSystems. This model is not included in the library of SimPowerSystems and is in the demonstration model **OLTC Regulating Transformer**. It is worth noting that in the library of SimPowerSystems, there is the model **Three-Phase OLTC Regulating Transformer** that operates in the Phasor mode only. The employment of this model can be reasonable in the complex systems with several controlled transformers, when simulation time can turn out too large.

The model **OLTC_trans** contains the **Three-Phase OLTC Regulating Transformer** with the following parameters: 80 MVA, 50 Hz, and 345/24.5 kV. The model diagram is depicted in Figure 2.11. To carry out the command Mask/Look Under Mask, it would be seen that the transformer model contains, for every phase, the block of the **Multiwinding Transformer** with taps, the block for tap switching-over, and the common control block. It is seven taps that give eight steps. Because the winding with taps can be connected as aiding or antiaiding to the primary one that, taking in mind the possibility to connect the latter directly to the supply, provides 17 voltage steps. The detailed description of the switching-over process and of the voltage control is given in Ref. [3].

In the model **OLTC_trans**, this transformer is utilized for the maintenance of load voltage under variations of the grid voltage and the load power. Firstly, it

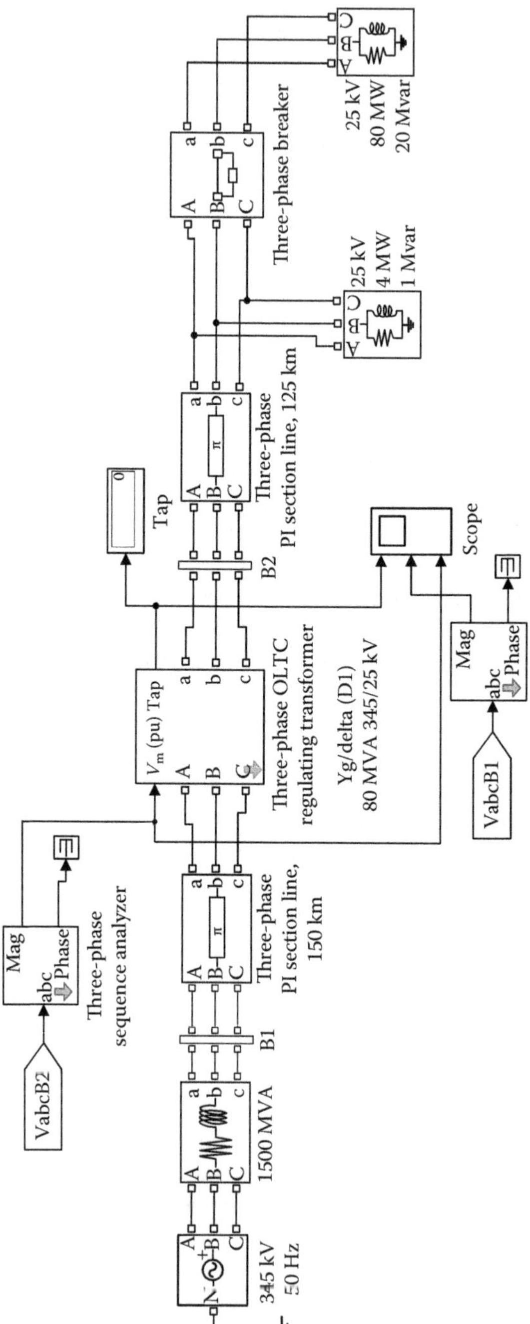

FIGURE 2.11 Diagram of the model **OLTC_trans**.

is necessary to specify the transformer parameters. Its dialog box has two pages, into which the transformer parameters and the controller parameters are entered. The reader can get to know the selected parameters by opening the model scheme. Specifically, the deadband is set as 0.01 pu.

In the considered model, the three-phase controlled source, with the nominal voltage of 345 kV, changes the output voltage from 1 to 0.9 and from 0.9 to 1.1 at t = 15 s and t = 40 s, respectively. For the interval [70 s, 95 s], the load power that was equal to 84 MW to 4 MW. The network and load voltages and the tap positions are recorded in Figure 2.12. It is seen that after transients, the load voltage is kept in the deadband boundary.

In systems with wind generators (WGs), OLTC transformers can be employed for matching WGs and network voltages in the cases, when the connection point is at a big distance from the WG location, and the essential voltage drops can be possible in the transmission lines, as is the case for the offshore WG sets.

The three-phase transformers are considered further. The first three types of such transformers consist of three single-phase transformers, whose equivalent circuit is shown in Figure 2.8. The nominal power and frequency are fixed in the transformer dialog box, and also the voltages and the parameters of the equivalent circuit for each winding. For **Three-Phase Transformer (Two Windings)** and **Three-Phase Transformer (Three Windings)**, winding connections are defined: star Y with isolated, grounded, or accessible neutral; delta lagging Y by 30° (D1); and delta leading Y by 30° (D11). For the latter two transformers, the opportunity to model saturation, both with and without hysteresis, is intended.

The **Zigzag Phase-Shifting Transformer** model uses three single-phase three winding transformers; its primary winding is formed by the connection of two windings

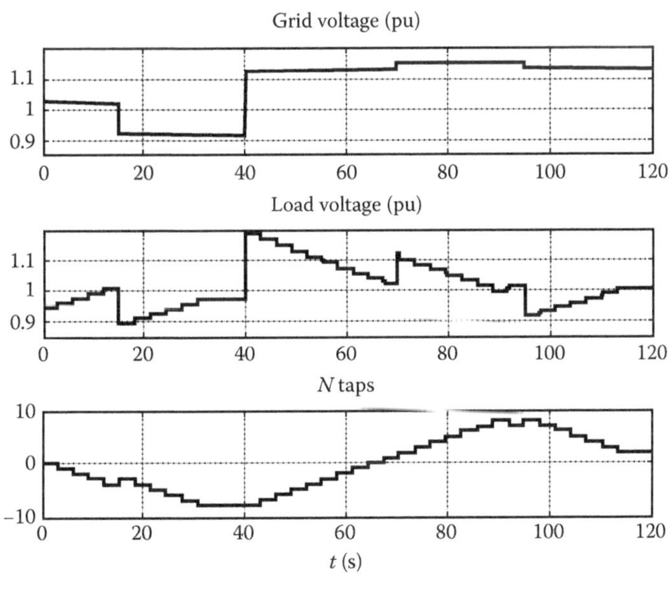

FIGURE 2.12 Variations of the network and the load voltages, the tap positions in the model **OLTC_trans**.

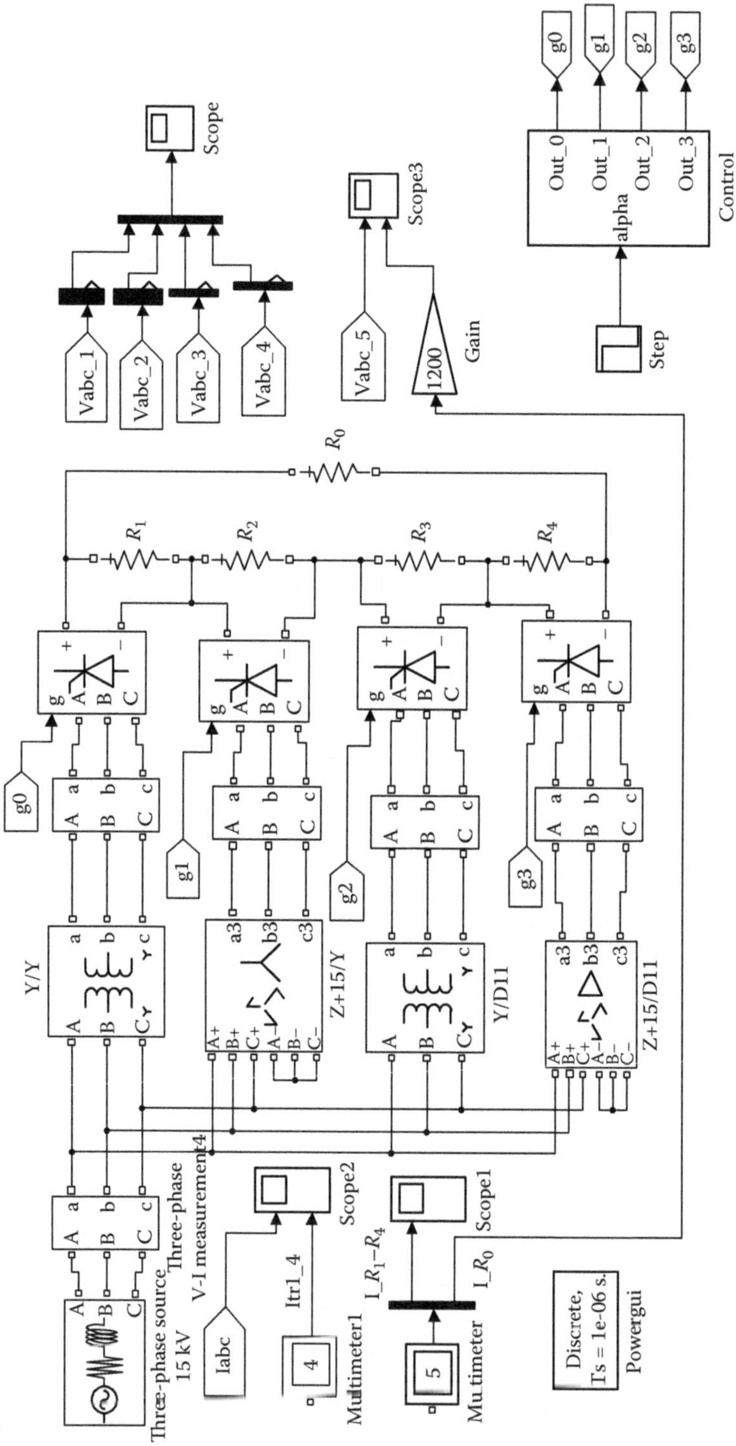

FIGURE 2.13 Thyristor rectifier with four phase-shifting transformers.

of the single-phase transformers in the zigzag configuration. On the second page of the dialog box, in addition to the usually defined parameters, the phase shift γ of the secondary voltage about the primary voltage is indicated.

The model **Zig_zag_trans** gives an example of this transformer employment. The model circuits are depicted in Figure 2.13. The model contains four transformers; each of them provides the shift γ about the preceding one. The first and the third transformers are modeled with the blocks **Three-Phase Transformer (Two Windings)** connected as Y/Y (or D1/D1) and Y/D11, respectively, and the second and the fourth transformers are modeled with the blocks **Zigzag Phase-Shifting Transformer**, whose secondary windings are connected as Y + 15° and D11 + 15°, respectively. The transformers have the power of 15 MW and the voltage of 15/50 kV. The transformers supply the thyristor bridges that are loaded on the 30 kΩ resistors R_1–R_4. The bridges are connected in series and supply the 1200 Ω load R_0. The firing pulses are fabricated with the blocks of Synchronized 6-Pulse Generator that have the same firing angle α. The synchronization voltages use the supply phase-to-phase voltages; at this, the first bridge employs this voltage directly, and for the rest of the bridges, instead of the voltages U_{ab}, U_{bc}, and U_{ca}, the voltages $-U_{bc}$, $-U_{ca}$, and $-U_{ab}$ are utilized, respectively. With the help of the first-order filters with the time constants of 0.85, 1.84, and 3.18 ms for the bridges of 2, 3, and 4, respectively, the requested phase shifts of 15°, 30°, and 45° at the frequency of 50 Hz are obtained.

The **Scope** fixes the A–B transformer voltages, the **Scope1** records the resistor currents, and the **Scope2** fixes the currents in the supply and in the phase A of the transformers. Under the simulation, the angle α changes from 30° to 15° at t = 0.1 s.

Some results of simulation are shown in Figure 2.14. It is seen that the pulsations of the voltages across the resistors R_1–R_4 are much more than the voltage pulsations

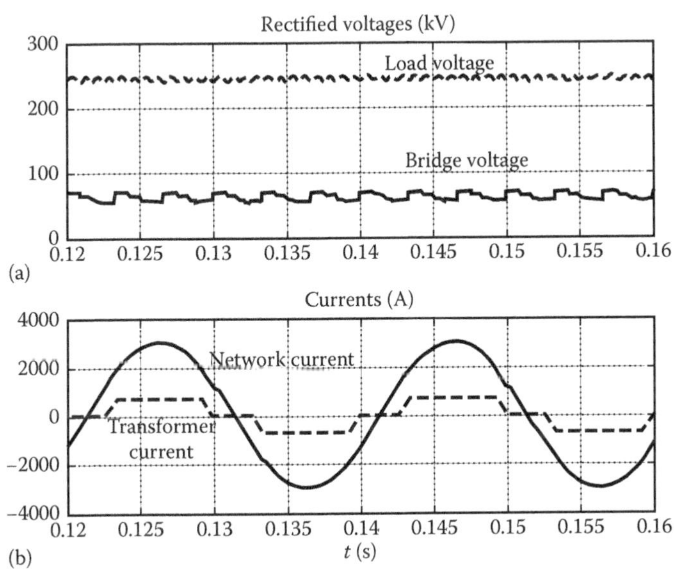

FIGURE 2.14 The (a) voltages and the (b) currents in the model **Zig_zag_trans**.

on the load resistor R_0. With the distorted currents in the primary transformer windings, the network current is practically sinusoidal.

The model **Zig_zag_transN** shows an example of how the new control blocks from R2013a version (see Section 2.6.2) can be used for the solution of the same task.

To use the option **Powergui/***FFT Analysis* for the records of ScopeData2 (Scope 2), it may be found that the transformer current has THD = 24%, whereas THD = 2% for the network current. Therefore, the employment of the transformers in zigzag configuration gives an opportunity to receive nearly perfect currents in the common load and in the supply without using the filters and the smoothing reactors.

The **Grounding Transformer** is used in distribution networks for providing a neutral point in a three-wire system, usually for supplying the single-phase loads connected to the ground. The transformer has three primary and three secondary windings connected in zigzag, and all six windings have the same number of the turns; in each phase, belonging to it are primary and foreign secondary windings connected with opposite polarity (antiaiding). The voltage of each of the six windings is $V/3$, where V is the rated network voltage.

The main characteristic of the **Grounding Transformer** is its zero sequence impedance $Z0 = R0 + jX0 = 2(R + jX)$, where R and X are the active and inductive resistances of one winding.

The blocks of the two- and three-winding transformers with inductances of the matrix type simulate these transformers taking into account the inductive coupling between windings of the different phases. The transformers can have three- or five-limb cores. The modeling of these transformers demands rather detailed information about them that can be received either by computation with known transformer construction or by testing the prototype.

2.4 TRANSMISSION LINE MODELS

The transmission line parameters are the active resistance R, the series inductance L, and the parallel capacity C. These parameters are given for a unit of the line length (meters or kilometers). The active resistance takes into account the skin effect, the stranding, and the environment temperature.

The inductance L is due to interaction of the alternating current (AC) in the line with the surrounding magnetic field. The phase inductance depends on the position of the phase conductors relative to the other phases and about the ground; that is, the inductances of the phases can be different. The transposition of the different phase conductors is used for inductance balancing.

The capacity C is owing to the potential difference between the line conductors, so that the conductors charge. The charge per unit of potential difference is C.

These RLC parameters are evenly distributed along the line length. Let $Z = R + j\omega L$ and $Y = j\omega C$, and ω is the network angular frequency. The transmission line is characterized by such parameters [4]:

$$\gamma = \sqrt{ZY} \tag{2.6}$$

is the propagation constant

$$\gamma = \alpha + j\beta = \frac{R}{2}\sqrt{\frac{C}{L}} + j\omega\sqrt{LC}, \tag{2.7}$$

where α is the attenuation constant and β is the phase constant.

The value

$$Z_c = \sqrt{\frac{Z}{Y}} \tag{2.8}$$

is the characteristic impedance. Usually, $R \ll \omega L$, then

$$Z_c = \sqrt{\frac{L}{C}} \tag{2.9}$$

is the wave impedance. The wavelength is defined as

$$\lambda = \frac{2\pi}{\omega\sqrt{LC}}. \tag{2.10}$$

The line voltages and the currents change as the electromagnetic waves: the voltage and current values in some point depend not only on the time but also on the position of this point. When the electromagnetic wave reaches the line end, it reflects and propagates in the opposite direction, so that the line voltages and currents are the results of superposition of the incident and reflected waves. If the line is loaded with the impedance Z_f, then under $Z_f = Z_c$, the reflected wave is absent and the voltage and the current have a constant amplitude along the line; such impedance is called *surge impedance.*

If the lossless line is supposed, when Z_c is defined by Equation 2.9, the surge impedance load (SIL) is defined as

$$\text{SIL} = \frac{V_{\text{nom}}^2}{Z_c}. \tag{2.11}$$

When such a power is transferred, the voltage along the line does not change. In Table 2.1, which is taken from Ref. [3], the typical parameters of overhead lines are given for the frequency of 50–60 Hz obtained from diverse sources.

The line parameters depend on many factors: Is the line overhead or underground? Which is the tower construction (a height, the distance between the phase conductors and the grounded wires)? What is the mutual placement of the phase conductors? How many lines does the tower carry? What is the construction of the phase

TABLE 2.1
Typical Parameters of Overhead Transmission Lines

Voltage (kV)	220–230	330–345	500	750–765	1100–1150
R (Ω/km) × 10^2	5–6	2.8–3.7	1.8–2.8	1.17–1.2	0.005
L (mH/km)	1.27–1.29	0.97–1	0.86–0.93	0.87	0.77
C (F/km) ×10^9	8.91–8.94	10.6–12	12.6–13.8	13–13.4	14.6
Z_c (Ω)	377–387	285–300	250–278	255–260	230–250
SIL (MW)	125–140	360–420	900–1000	2160–2280	5260–5300

conductors (the outside diameter, the thickness of conducting material, the number of the conductors in the bundle)? And so on. SimPowerSystems gives the possibility to calculate the line parameters that are necessary for the transmission line models. The matrixes of the resistances, inductances, and capacitances for the arbitrarily spaced overhead transmission line are computed; at this, for the system of N conductors, three $N \times N$ matrixes are computed: the matrix R of the series resistances, the matrix L of the series inductances, and the matrix C of the parallel capacitances (owing to the influence of the earth, the matrix R is not a diagonal one). For the three-phase line, the symmetric components are computed as well.

It is known [4], that the voltage phasors $\mathbf{V}_a$, $\mathbf{V}_b$, and $\mathbf{V}_c$ (or of other quantities) in any unbalanced three-phase system may be expressed as a sum of the three phasors $\mathbf{V}_0$, $\mathbf{V}_1$, and $\mathbf{V}_2$ in the symmetric three-phase systems that are called the *sequence* or *symmetrical components* as

$$\mathbf{V}_{012} = \mathbf{T}^{-1}\mathbf{V}_{abc}, \tag{2.12}$$

where $\mathbf{V}_{012}$ and $\mathbf{V}_{abc}$ contain the components $\mathbf{V}_0$, $\mathbf{V}_1$, $\mathbf{V}_2$ and $\mathbf{V}_a$, $\mathbf{V}_b$, $\mathbf{V}_c$, respectively; the reverse transformation is

$$\mathbf{V}_{abc} = \mathbf{T}\mathbf{V}_{012}. \tag{2.13}$$

Here

$$\mathbf{T} = \begin{bmatrix} 1 & 1 & 1 \\ 1 & a^2 & a \\ 1 & a & a^2 \end{bmatrix}, \tag{2.14}$$

$$\mathbf{T}^{-1} = \frac{1}{3}\begin{bmatrix} 1 & 1 & 1 \\ 1 & a & a^2 \\ 1 & a^2 & a \end{bmatrix}, \tag{2.15}$$

$$a = e^{j2\pi/3}.$$

If in the three-phase system the voltages and currents are linked with the impedance matrix $\mathbf{Z}_{abc}$

$$\mathbf{V}_{abc} = \mathbf{Z}_{abc}\mathbf{I}_{abc}, \tag{2.16}$$

then for the symmetrical components [4,5]

$$\mathbf{V}_{012} = \mathbf{Z}_{012}\mathbf{I}_{012}, \tag{2.17}$$

$$\mathbf{Z}_{012} = \mathbf{T}^{-1}\mathbf{Z}_{abc}\mathbf{T}. \tag{2.18}$$

For the totally symmetrical system $\mathbf{Z}_{abc}$ (all the diagonal elements are the same, and all the nondiagonal elements are the same as well), the matrix $\mathbf{Z}_{012}$ is diagonal and has only two different elements that are called *positive sequence impedance* and *zero sequence impedance.* For typical power systems, the matrix $\mathbf{Z}_{012}$ is nearly diagonal [4,5].

For line parameter computation, it is necessary either to execute the command `power_lineparam` or to use the **Powergui** item Compute *RLC Line Parameters.* At this, the window is open, in which the parameters of the line, of the tower, and of the conductors are entered. The detailed instructions as for filling out this window are given in Refs. [3,6]. After filling out the window and executing the command Compute *RLC line parameters*, the line parameters as the three $N \times N$ matrixes or the three vectors of the positive and zero sequence impedances appear in the new window. They can be sent to the selected model of the transmission line. There is also the opportunity to choose the line variant that had been computed in advance and is included in SimPowerSystems (the option *Load typical data*).

For power system analysis, instead of the full equations describing the performance of the transmission line, the models with lumped parameters are usually used. At this, the line model can be considered as a ladder circuit depicted in Figure 2.15. The number of the sections N depends on the requested maximum frequency of the model, which depends, in turn, on the aim of investigation. For the estimation of the maximum frequency, the following formula is recommended [6]:

$$f_{max} = \frac{N}{8d\sqrt{LC}}, \tag{2.19}$$

where d is the line length.

The f_{max} value of some hundred hertz may be sufficient for line representation under the analysis of the steady state and the transients but may be not sufficient under the investigation of the processes when the line is supplied from the devices of the power electronics.

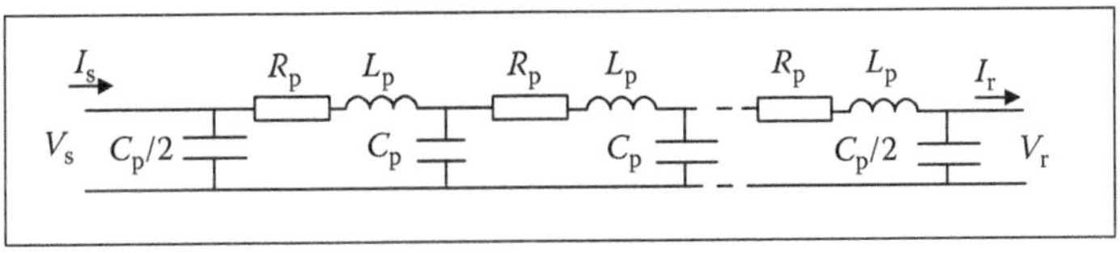

FIGURE 2.15 Ladder circuit presentation of the transmission line.

The block **PI Section Line** is included in SimPowerSystems that models the single-phase line. The block parameters are the frequency of the transmitted voltage, the line total length, the number of sections, and the values of R, L, and C per 1 km.

Each section is modeled with the use of the so-called equivalent π circuit depicted in Figure 2.16. Here [4]

$$Z_\pi = Z_c \sinh(\gamma d_s) = Z d_s \frac{\sinh(\gamma d_s)}{\gamma d_s}, \tag{2.20}$$

$$\frac{Y_\pi}{2} = \frac{1}{Z_c}\tanh\left(\frac{\gamma d_s}{2}\right) = \frac{Y d_s}{2}\tanh\left(\frac{\gamma d_s}{2}\right)\bigg/\left(\frac{\gamma d_s}{2}\right), \tag{2.21}$$

where d_s is the section length.

Evidently, for the short sections when $\gamma d_s \ll 1$, the π model turns into the nominal π equivalent model [4] with

$$Z_\pi = Z d_s = (R + j\omega L) d_s, \tag{2.22}$$

$$\frac{Y_\pi}{2} = \frac{Y d_s}{2} = \frac{j\omega C d_s}{2}. \tag{2.23}$$

The block **Three-Phase PI Section Line** models one section of the balanced three-phase line by using the equivalent π model. The specified parameters are the positive sequence impedances and zero sequence impedances per 1 km: r_1, r_0, l_1, l_0, c_1, and c_0. After multiplication by the line length d, the full impedance values are obtained: R_1, R_0, L_1, L_0, C_1, and C_0. Under simulation, these values of sequences are transformed into the self and mutual parameters of the model that consists of the elements described earlier. The line is supposed to be balanced (all the diagonal elements are the same, and all the nondiagonal elements are the same); it has been already said that in such a case, the impedance matrix is diagonal and has only two different elements that are called *positive sequence impedance* and *zero sequence impedance*. The matrix of the series impedance Z has the diagonal elements Z_s and nondiagonal elements Z_m. Then the sequence impedances are [7]

$$Z_1 = Z_s - Z_m,\ Z_0 = Z_s + 2Z_m, \tag{2.24}$$

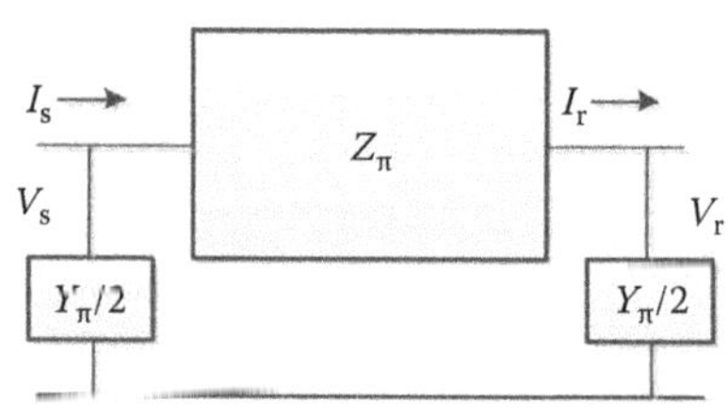

FIGURE 2.16 Equivalent π model of the transmission line.

from which it follows that

$$Z_s = \frac{1}{3}(2Z_1 + Z_0),\ Z_m = \frac{1}{3}(Z_0 - Z_1). \tag{2.25}$$

For consideration of the stray capacitances, the line model can be presented as it is shown in Figure 2.17a, where C_m is the interphase capacitance and C_s is the capacitance of the phase to the ground. This diagram can be transformed to the diagram shown in Figure 2.17b. For this scheme, the positive and zero components are $C_0 = C_s$ and $C_1 = C_s + 3C_m$ [7]. The **Three-Phase PI Section Line** is modeled as shown in Figure 2.17c. For this model, because it must be that $C_p = C_s + 3C_m$, it follows that $C_p = C_1$. Since it must be that

$$\frac{1}{C_s} = \frac{1}{C_p} + \frac{3}{C_g},$$

it follows that

$$C_g = \frac{3C_1C_0}{C_1 - C_0}. \tag{2.26}$$

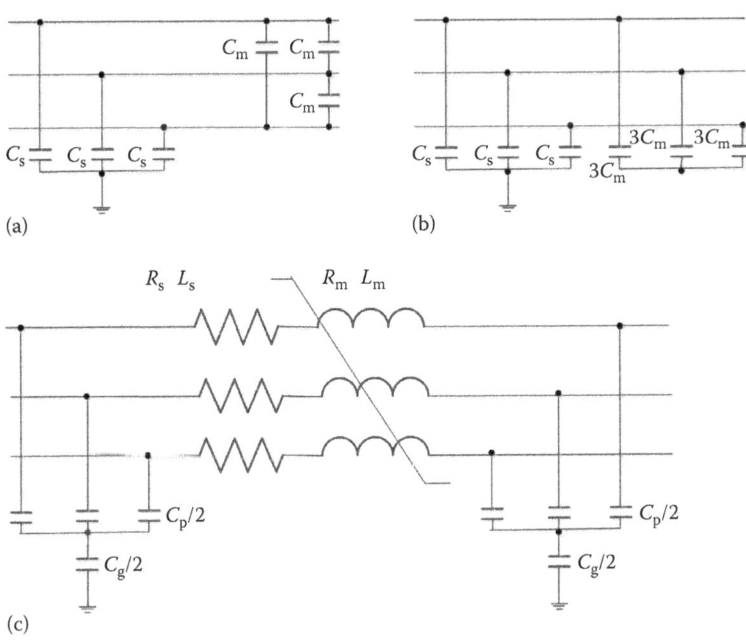

FIGURE 2.17 Equivalent π model of the three-phase transmission line: (a) diagram of the stray capacitances; (b) transformed diagram of the stray capacitances; and (c) full model diagram.

Before utilization in the given formulas, the values of the sequence impedances are corrected, using hyperbolic functions, which is how it was done in the previous model.

The **Distributed Parameter Line** models the *N*-phase line, whose phase parameters can be different. To neglect the losses, the line is characterized by three parameters: Z_c as Equation 2.9, the velocity of wave propagation $v = 1/LC$, and the length *d*. In the model, the fact is used that in the line without losses, the quantity $V + Z_cI$ in one line end reaches another end without changing, after a delay time of $\tau = d/v$. In order to take the losses into consideration, at the line ends, the resistors $R/4$ and in middle of line the resistor $R/2$ are set. The relationships that describe the model function are given in Ref. [6].

The model parameters, besides the frequency and the line length, are either sequence impedances, as in the previous model, or $N \times N$ matrixes of the resistances, inductances, and capacitances that are computed with the help of the **Powergui** option Compute *RLC Line Parameters*, as it was mentioned earlier.

In some cases, it is possible to neglect the capacitive elements and to use only *R–L* elements; the examples are given in some demonstration models.

The simple model **Trans_line** gives the opportunity to compare the performance of two models of the transmission lines. The model **Distributed Parameter Line** with $N = 1$ and the model **PI Section Line** transmit the same voltage and are loaded with the same load. The line parameters are the same as well; the line length is 200 km. The direct current (DC) voltage or the AC voltage can be transmitted. When it is the former case, one can see after the simulation that the processes in the line and in the load are essentially different in both models; at this, the voltages and the currents in the first model change stepwise, owing to the interaction of the incident and reflected waves. When the lines transmit the AC voltage, the processes in both models are close. It proves that the type of model and its parameters must be selected depending on the rate of change of the investigated variables.

Some simple models are given in Ref. [3] that demonstrate the employment of the described blocks, an origin of the incident and the reflected waves, their interaction, the load influence, and so on.

Together with the overhead transmission line, SimPowerSystems gives the possibility to calculate the parameters of the cable line with the command `power_cableparam`, which computes the inductances and capacitances of the system, consisting of *N*-shielded radial cables. It is supposed that the cable consists of the copper conductor with an outer screen and utilizes the polyethylene insulation.

Since the cable lines are usually not very long, the π model that is depicted in Figure 2.18 can be used for cable modeling. All conductors and screens are magnetic coupling; the capacitances of the conductors to the screen and of the screen to the ground are modeled with the lumped capacitors of the half capacitance that are placed at the cable ends. For the modeling of the self and mutual inductances of the conductors and screen, the block **Mutual Inductance** described earlier with the number of the windings $2N$ can be used, where *N* is the number of the conductors in the cable.

During the fulfillment of the command mentioned earlier, the window appears in which cable parameters that are determined by the cable construction are entered.

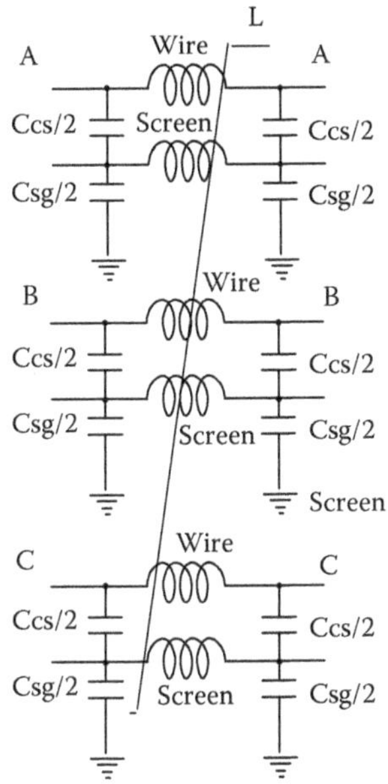

FIGURE 2.18 π cable model.

A number of options are presented at the low area of the window. The option *Load typical data* loads in this window the data that are fixed by the authors of SimPowerSystems: they are the parameters of a four-cable line that is utilized in the demonstration model **power_cable**. The option *Load user data* puts in this window the parameters that were developed by the user earlier and are saved on the disk under some name.

If the option Compute *RLC matrices* is selected, three matrixes $2N \times 2N$ appear in the new window: the R_matrix, the L_matrix, and the C_matrix. Before employment, these matrixes must be multiplied by the cable length. Moreover, for some rows of the inductance matrix, it may be that the mutual inductances turn out to be equal to the self inductances; it is impermissible, so that they have to be changed slightly. In the demonstration model mentioned earlier, it is made by the multiplication of the diagonal elements by 1.001. Therefore, before the simulation start, the command Send *RLC parameters to workspace* and afterward the program, as it follows, have to be executed:

```
Program of the cable parameter computation
Length = 10%km
R = R_matrix*Length;
L = L_matrix*Length;
C = C_matrix*Length;
nc = size (L,1);%row number
for ic = 1:nc;
L(ic,ic) = L(ic,ic)*1.001;
end;
```

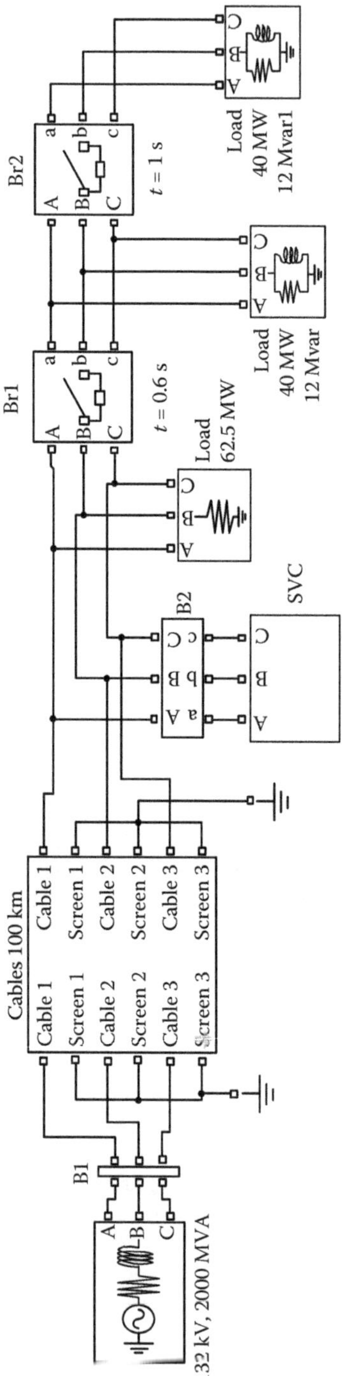

FIGURE 2.19 Diagram of the model **Ph3-Cable**.

The model **Ph3-Cable** shows the utilization of the described programs. The model diagram is depicted in Figure 2.19. The subsystem **Cables 100 km** is made according to the diagram in Figure 2.18. The cable parameters are given in Figure 2.20. The cable transfers the electric energy with the voltage of 132 kV at a distance of 100 km. The permanent load is 62.5 MW, and the turning on loads are 40 MW and 12 Mvar; they are connected at $t = 0.6$ s and $t = 1$ s, respectively.

For load voltage maintenance, static var compensator (SVC) is employed. Its design and operation are considered in Chapter 3; here we note only that it is a unit that controls the voltage at its terminals with a change of the amount of the reactive power, which is delivered to the grid or is consumed from the grid. When the voltage reduces, SVC generates the capacitive reactive power; during voltage rise, SVC takes

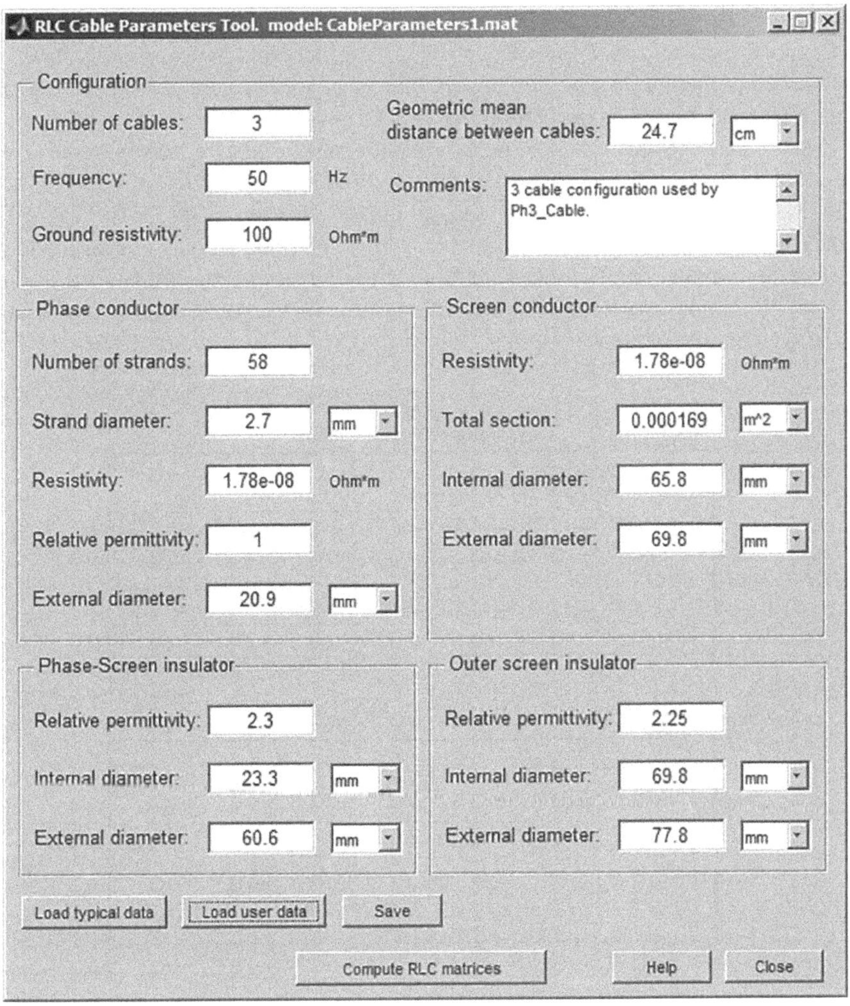

FIGURE 2.20 Cable parameters in the model **Ph3-Cable**.

off the inductive reactive power. The control of the reactive power flow is carried out with turning on and turning off of the capacitors and the inductors, by means of the thyristor switches, in the secondary winding of the matching transformer, whose primary winding is connected with the three-phase grid, or (and) with the smooth variation of the inductor equivalent inductance, by means of the firing angle change of the thyristors that are set in series with the inductor. When the inductor is fully turned off, the capacitive reactive power is $P_c = U^2B_{cmax}$, $B_{cmax} = 2\pi fC$ is the capacitive susceptance, and U is the phase-to-phase voltage. When the firing angle α decreases, the inductance increases, the SVC power becomes equal to $P = P_c - U^2B_{l\alpha}$, and $B_{l\alpha} = 1/2\pi fL_\alpha$, where L_α is the equivalent inductance at the given angle α; with $\alpha \approx 0$, $P = P_c - P_l$ and $P_l = U^2B_l$. The value P can be both positive and negative, depending on the choice L. The SVC control system consists of the device for the voltage measurement at the point of SVC connection with the grid, the voltage controller, and the Synchronized 6-Pulse Generator that fabricates the firing pulses.

In the considered model, the noncontrolled capacitor bank with power of 50 Mvar and the three-phase inductor with power of 60 Mvar are used; the equivalent inductance of the latter can change with the thyristor switches.

Firstly, it is necessary to open the window for cable parameter computation with the command `power_cableparam`, afterward with the activation of option *Load user data*, to enter in this window the parameters that are saved in the data file *`CableParameters1.mat`*, to fulfill the computation, and to send the results into workspace; and then the simulation can start. The results of the simulation without and with SVC are displayed in Figure 2.21. For the former case, the voltage steady deviation from the reference (1 pu) is 8%; for the latter case, it is only 1.5%.

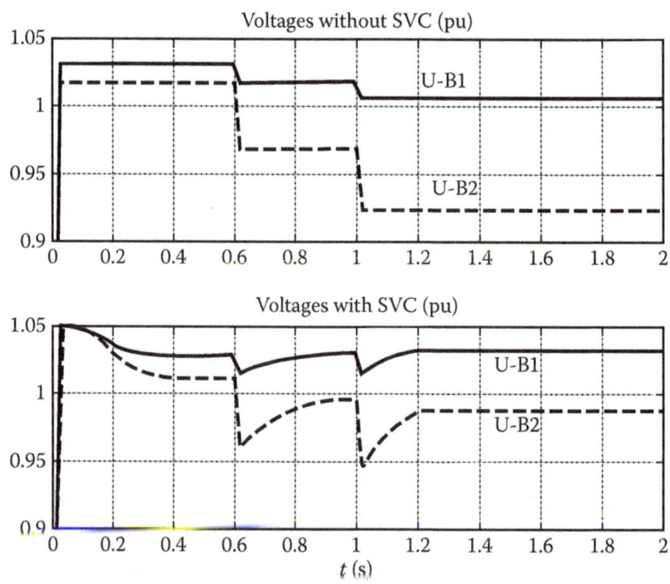

FIGURE 2.21 Voltage deviation in the model **Ph3-Cable**.

2.5 MISCELLANEOUS

The following elements from the folder *SimPowerSystems/Elements* that were not included in the previous three paragraphs are considered here briefly: **Breaker**, **Three-Phase Breaker**, **Three-Phase Fault**, **Surge Arrester**, and also the block **Ideal Switch** from the folder *Power Electronics*.

The **Breaker** models a switch that can switch over by the inner timer or by the outside signal Simulink. For the former option, the initial breaker state and the vector of the times when its state changes to the opposite are specified. For the latter case, the additional input for the signal Simulink appears at the block picture; the signal of the logical 1 turns on the block, and the signal of the logical 0 turns off. At this, the actual contact opening takes place after crossing of the conducted current zero level.

The **Three-Phase Breaker** consists of the three single-phase ones and has the same characteristics. Only part of the contacts can switch over; the rest of the contacts remain in the initial state.

The block **Three-Phase Fault** employs three blocks **Breaker**, which can be turned on or turned off individually, in order to model short-circuit phase–phase, phase–ground, or their combination. It is marked in the dialog box, to which phases of the given commutation times are related. The initial state is usually zero, but it can be set as 1, in order to begin simulation with the fault initially applied on the system. The block can be controlled by the external signal.

The block **Surge Arrester** is intended for the modeling of a nonlinear resistor that protects the line elements from the overvoltage. The nonlinear *V–I* characteristic of each column of the surge arrester is modeled by a combination of three exponential functions that are given in Ref. [6]. If it is necessary to dissipate the high power, several blocks are connected in parallel. The protection voltage and the reference current are specified in the dialog box; the latter is usually 500 or 1000 A. Examples of utilization of this block are given in Refs. [3,6].

The block **Ideal Switch** can be used for simplified modeling of the semiconductor devices (Chapter 3) and for the modeling of a switch in the main circuit that, unlike the **Breaker**, permits a commutation with current. The **Ideal Switch** is modeled as the resistor *Ron* in series with a switch that is controlled by the signal g. The block does not conduct current under $g = 0$ and conducts in any direction under $g > 0$. The switching over occurs instantly. The block contains the snubber that consists of R_s and C_s connected in series; this snubber can be connected in parallel to the switch contacts.

2.6 MEASURING AND CONTROL BLOCKS

2.6.1 Measurement of Main Circuit Variables

SimPowerSystems contains several blocks for measurement of the voltages, the currents, and some other variables in the power circuits and for conversion of these variables into the Simulink quantities. They are **Voltage Measurement**, **Current Measurement**, **Three-Phase V-I Measurement**, **Multimeter**, and **Impedance Measurement**.

The block **Voltage Measurement** measures the voltage instantaneous value between two points of the model. When the block is utilized in the model in the Phasor mode, an opportunity appears to select a mode of the voltage vector representation: the complex number, the real and imaginary parts as a two-dimensional vector, the module and the angle as a two-dimensional vector, and only the module.

The block **Current Measurement** measures the instantaneous value of the current that flows in the circuit. To use this block in the Phasor mode, see **Voltage Measurement**.

The block **Three-Phase V-I Measurement** is connected in series with three-phase devices.

Phase-to-phase or phase-to-neutral voltages and line currents can be measured in both SI and per units. In the last case, additional fields appear in the dialog box for the indication of the base values. As the voltage base value, the amplitude of the phase voltage that corresponds to the rms value of the phase-to-phase voltage *V* that is given in the dialog box is taken; as the current base voltage, the line current amplitude is taken that corresponds to the given values of the power *P* and of the voltage *V*. There are two options for the output of the measured quantities: as the labels with the assigned names or from the block output terminals. When the model operates in the Phasor mode, the form of presentation of the complex value has to be given, as described previously.

The block **Multimeter** is intended for the measurement of the voltages, the currents, and some other quantities of the blocks, in which the option *Measurements* is activated. To put **Multimeter** in the model, in which such blocks are present, and to open its window, it may be seen that this window consists of two parts. In the left part, a list appears containing all quantities that can be measured in this model. In order to select the quantities that have to be measured, they must be duplicated in the right part. When the model operates in the Phasor mode, the form of presentation of the complex values has to be given.

The block **Impedance Measurement** measures the impedance between two circuit points as a function of the frequency. It contains the current source of the variable frequency I_z connected to the block input/output and the meter of the voltage difference V_z between these points; the impedance is computed as V_z/I_z. The utilization of this block is demonstrated with the model **Example3c** in Chapter 1.

2.6.2 Meters with Employment of Simulink Blocks

SimPowerSystems has a number of blocks, which are used for processing both the main circuit quantities measured by the devices described in Section 2.6.1 and the signals fabricated with Simulink blocks. The outputs of these blocks are utilized for observation and control. These blocks are built by using Simulink blocks. In the previous versions of SimPowerSystems, these blocks were in the folders *SimPowerSystems/Extra Library/Measurements* or */Discrete Measurements*, but beginning from version R2013a, the continuous and discrete blocks were united and were placed in the folder *Control and Measurements Library/Measurements*. The nomenclature of the blocks and their circuits changes. Although the old blocks are not presented in the new library, they can be copied from the old SimPowerSystems versions or with use of the command *powerlib_extras*, and the model will still work.

Therefore, the most important and often-utilized blocks of the old and new versions are described briefly as follows. The former carry the additional designation R2012b; the latter—R2013a.

The block **Fourier** (R2012b, R2013a) performs Fourier analysis of the input signal over a running window of $T = 1/f$ width. The fundamental frequency f and the harmonic number, whose amplitude and phase are formed at the block outputs, are specified in the dialog box.

The block **Mean** (R2012b, R2013a) calculates the mean value of the input signal over a running window of $T = 1/f$ width. The block **Mean (Variable Frequency)** (R2012b, R2013a) has the additional input for the frequency of the measured signal; thus, the variable frequency of the input signal is possible. The possible minimal frequency has to be indicated in order to determine the buffer size of the Simulink block **Variable Transport Delay** that is used for building the considered block.

The block **RMS** (R2012b, R2013a) calculates the rms value of the input variable over a running window of $T = 1/f$ width. For **RMS** (R2013a), either the true value or the rms of the fundamental harmonic may be selected for computation.

The block **THD** (R2013a) or **Total Harmonics Distortion** (R2012b) computes the total value of distortion factor (THD) of the periodic signals as

$$\text{THD} = \frac{\sqrt{I_{\text{rms}}^2 - I_0^2 - I_1^2}}{I_1}, \tag{2.27}$$

where I_{rms} is the rms value of the total input signal, I_1 is the rms of its fundamental harmonic, and I_0 is the direct component.

The block **Power** (R2013a) or **Active & Reactive Power** (R2012b) computes the instantaneous values of the active and reactive powers at the fundamental frequency. The latter block also outputs the values of the voltage and current amplitudes.

The block **Power (3ph, Instantaneous)** (R2013a) or **Three-Phase Instantaneous Active & Reactive Power** (R2012b) uses the following relationships:

$$P = V_a I_a + V_b I_b + V_c I_c, \tag{2.28}$$

$$Q = \frac{1}{\sqrt{3}}(V_{bc} I_a + V_{ca} I_b + V_{ab} I_c). \tag{2.29}$$

The inputs are the vectors of the phase voltages and currents; the output is either the vector $[PQ]$ or the values P and Q separately.

Some additional blocks for power measurement are intended, including **Discrete Three-Phase Total Power** (R2012b). The **Discrete Three-Phase Total Power** block measures an active power of the three-phase system of voltages and currents that may contain harmonics. The power is computed by averaging the instantaneous power (that is equal to the sum of the products of voltages' and currents' instantaneous values of all phases) in a running window over one period of fundamental frequency width. The block inputs are the vectors $[V_a V_b V_c]$ and $[I_a I_b I_c]$, and the block outputs are the values of the instantaneous P_{inst} and mean P_{mean} powers.

The block **Discrete Three-Phase Positive-Sequence Active & Reactive Power** (R2012b) or **Power (Positive-Sequence)** (R2013a) calculates the active and reactive powers of the positive sequence fundamental harmonic in the unbalanced three-phase circuit, in which the voltages and the currents may have harmonics. The outputs are the fundamental harmonic amplitudes of the voltage and the current positive sequence and the vector $[PQ]$ for R2012b or the quantities P and Q for R2013a.

The block **dq-Based Active & Reactive Power** (R2012b) or **Power (dq0, Instantaneous)** (R2013a) uses the vectors $[V_d V_q V_0]$ and $[I_d I_q I_0]$ as the inputs:

$$P = 1.5(V_d I_d + V_q I_q + 2V_0 I_0), \tag{2.30}$$

$$Q = 1.5(V_q I_d - V_d I_q). \tag{2.31}$$

The block **Three-Phase Sequence Analyzer** (R2012b) or **Sequence Analyzer** (R2013a) calculates the positive, negative, and zero sequence components of the three-phase signal (Section 2.4). The frequency, the harmonic number n ($n = 1$ for fundamental), and the type of the sequence—positive, negative, 0, or all three simultaneously—are specified in the dialog box. The vector of three-phase quantities enters the block. The block has two outputs: for the amplitude of the selected sequence and for its phase angle.

It is reasonable to use the block as a voltage feedback sensor for the voltage regulation tuned with $n = 1$ for positive sequence, and for the study of the method symmetric components, see example in Ref. [3].

The blocks **1-phase PLL** (R2012b) or **PLL** (R2013a) and **3-phase PLL** (R2012b) or **PLL (3ph)** (R2013a) are used for the computation of the frequency and the phase of a harmonic signal. They are the devices with feedback. (PLL is an abbreviation for phase-locked loop.) The block diagram of **PLL (3ph)** is depicted in Figure 2.22. The q component of the voltage space vector V when the reference frame rotates with the angle frequency ω may be written as [8]

$$V_q = \frac{2}{3}\left[V_a \cos x + V_b \cos\left(x - \frac{2\pi}{3}\right) + V_c \cos\left(x + \frac{2\pi}{3}\right)\right]. \tag{2.32}$$

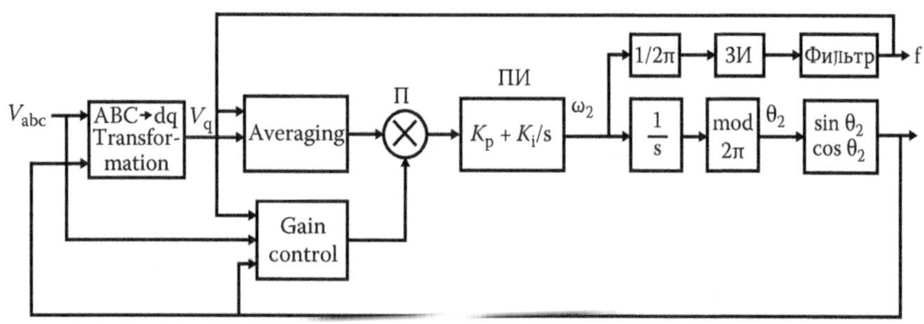

FIGURE 2.22 Block diagram of **3-phase PLL.**

If substituted in Equation 2.32 $V_a = V_m \sin\theta_1$, $V_b = V_m \sin(\theta_1 - 2\pi/3)$, $V_c = V_m \sin(\theta_1 + 2\pi/3)$, and $x = \theta_2$, where the last quantity is the estimation of the actual quantity θ_1, then $V_q = V_m \sin(\theta_1 - \theta_2)$; this value is equal to zero when the estimation is true ($\theta_2 = \theta_1$) and can be used for angle tuning. The block input quantities are the phase voltages V_a, V_b, and V_c; the outputs are the estimations of the frequency $f = \omega_2/2\pi$ and θ_2. The block **3-phase PLL** (R2012b) outputs, in addition to f and θ_2, the quantities of $\sin\theta_2$ and $\cos\theta_2$.

The **1-phase PLL** blocks have, in principle, the same structure; the input quantity $V = V_m \sin\theta_1$ is multiplied by $\cos\theta_2$; the output of multiplier $V_m \sin\theta_1 \cos\theta_2 = 0.5\, V_m [\sin(\theta_1 - \theta_2) + \sin(\theta_1 + \theta_2)]$ is filtered with the block **Discrete Variable Frequency Mean Value**, so that the second term disappears and the remaining quantity is processed, as in the three-phase blocks.

2.6.3 Transformation Blocks

In SimPowerSystems R2012b, the blocks for direct and reverse Park transformation were included: **abc_to_dq0 Transformation** and **dq0_to_abc Transformation**. The list of the transformation blocks is essentially increased in SimPowerSystems R2013a; it additionally includes **abc to Alpha-Beta-Zero**, **Alpha-Beta-Zero to abc**, **Alpha-Beta-Zero to dq0**, and **dq0 to Alpha-Beta-Zero**.

The first block of R2012b utilizes the relationships in Equation 2.32 and

$$V_d = \frac{2}{3}\left[V_a \sin x + V_b \sin\left(x - \frac{2\pi}{3}\right) + V_c \sin\left(x + \frac{2\pi}{3}\right)\right], \tag{2.33}$$

$$V_0 = \frac{1}{3}[V_a + V_b + V_c],\ x = \omega t. \tag{2.34}$$

in order to transform the space vector V from a stationary to a rotating, with speed ω, reference frame. The inputs are the vectors $[V_a V_b V_c]$ and $[\sin\omega t\ \cos\omega t]$; the output is the vector $[V_d V_q V_0]$.

The second block of R2012b carries out the reverse transformation by the following equations:

$$V_a = V_d \sin x + V_q \cos x + V_0, \tag{2.35}$$

$$V_b = V_d \sin\left(x - \frac{2\pi}{3}\right) + V_q \cos\left(x - \frac{2\pi}{3}\right) + V_0, \tag{2.36}$$

$$V_c = V_d \sin\left(x + \frac{2\pi}{3}\right) + V_q \cos\left(x + \frac{2\pi}{3}\right) + V_0. \tag{2.37}$$

The inputs are the vectors $[V_d V_q V_0]$ and $[\sin \omega t \cos \omega t]$; the output is the vector $[V_a V_b V_c]$.

The analogous blocks of R2013a fulfill, in principle, the same conversion, but with employment of the new blocks mentioned earlier, namely, for the direct conversion, **abc to Alpha-Beta-Zero** and **Alpha-Beta-Zero to dq0** connected in series; for the reverse conversion, **dq0 to Alpha-Beta-Zero** and **Alpha-Beta-Zero to abc** connected in series as well. Moreover, not quantities sin x and cos x come to the block input, but the value x direct. At this, the rotating frame may be aligned with the phase A axis or may be 90° behind.

The block **abc to Alpha-Beta-Zero** realizes the following relationships:

$$V_\alpha = \frac{2}{3}(V_a - 0.5V_b - 0.5V_c), \tag{2.38}$$

$$V_\beta = \frac{1}{\sqrt{3}}(V_b - V_c), \tag{2.39}$$

$$V_0 = \frac{1}{3}(V_a + V_b + V_c). \tag{2.40}$$

The reverse block **Alpha-Beta-Zero to abc** fulfills the following relationships:

$$V_a = V_\alpha + V_0, \tag{2.41}$$

$$V_b = -0.5V_\alpha + \frac{\sqrt{3}}{2}V_\beta + V_0, \tag{2.42}$$

$$V_c = -0.5V_\alpha - \frac{\sqrt{3}}{2}V_\beta + V_0. \tag{2.43}$$

The block **Alpha-Beta-Zero to dq0** carries out the transformation by the following equations:

$$V_d = V_\alpha \cos x + V_\beta \sin x, \tag{2.44}$$

$$V_q = -V_\alpha \sin x + V_\beta \cos x. \tag{2.45}$$

If the rotating frame is 90° behind the axis of phase A, the value $x - \pi/2$ must be used instead of x in Equations 2.44 and 2.45. The value V_0 remains the same.

The block **dq0 to Alpha-Beta-Zero** carries out the inverse transformation by the following equations:

$$V_\alpha = V_d \cos x - V_q \sin x, \tag{2.46}$$

$$V_\beta = V_d \sin + V_q \cos x. \tag{2.47}$$

Again, if the rotating frame is 90° behind the phase A axis, the value $x - \pi/2$ must be used instead of x in Equations 2.46 and 2.47.

2.6.4 Control Blocks

A number of the blocks that carry out the functions of control and regulation are in the folders *SimPowerSystems/Extra library/Control Blocks* and */Discrete Control Blocks* for version R2012b and in the folder *Control and Measurement Library* for version R2013a.

These blocks may be divided into two groups: the blocks for common applications and the blocks for special applications. Only the first group is considered in this chapter, and the rest are considered together with the systems for which they were developed.

The lists of these blocks for both versions coincide only partly, but, as said earlier, the blocks of R2012b version can work in the R2013a version; some blocks of the former version are mentioned here as well.

There are four models of the filters in version R2013a, namely, **First-Order Filter**, **Second-Order Filter**, **Lead-Lag Filter**, and **Second-Order Filter (Variable-Tuned)**. These blocks have the analogs in the version R2012b.

The **First-Order Filter** realizes the transfer function of the first-order element, whose form depends on the selected filter type: *Lowpass* or *Highpass*. In the first case when $\omega \to \infty$, the amplitude–frequency characteristic tends to be zero, and the phase–frequency characteristic, $-\pi/2$; in the second case, when $\omega \to \infty$, the amplitude–frequency characteristic tends to be 1, and the phase–frequency characteristic, 0. The time constant of the filter T is specified in the dialog box.

The block **Second-Order Filter** gives more possibilities for processing the signals. The filter types *Lowpass*, *Highpass*, *Bandpass*, and *Bandstop* may be selected. The filter natural frequency f_n and the damping factor ξ are specified in the dialog box. For *Lowpass* and *Highpass* filters, the amplitude–frequency characteristics are equal to $1/2\xi$ at f_n; this characteristic has a maximum at this frequency for *Bandpass* and a minimum for *Bandstop* filters.

The model **Filter_Ex** is used in order to better understand the filter operation. The signal *Xin* is the sum of the four signals of the frequencies 50 (fundamental harmonic), 350, 650, and 950 Hz. It inputs in the filter of the second order. The harmonic amplitudes both in *Xin* and at the filter outputs are measured (**Display1**, **Display3**).

The signal processing is carried out with the subsystem fast Fourier transform that is taken from the demonstration model **power_fft**. The brief block description is given in the block dialog box and in more detail in Ref. [3]. At the output $F(n)$, with the time interval that is equal to the specified value of *FFT Sampling time*, the harmonic amplitudes appear. At the output RMS, the vector of the rms values appears where the first value is the total signal rms; the second value is the rms of the positive component; the third value is the fundamental harmonic rms; the fourth value is the high-harmonic rms; and the fifth value is the total rms without positive component.

Let us suppose that it is necessary to select the fundamental frequency in *Xin*, which means that the filters must transmit the signal of 50 Hz with minimal distortion and must suppress, to the maximum possible extent, the signals with frequencies in the range of 350 Hz or higher. The natural (cutoff) filter frequency is accepted to be 100 Hz; the filter type is *Lowpass*.

To assign the harmonic amplitudes as 100, 40, 30, and 10 V for the 1st, 7th, 13th, and 19th harmonics, respectively, it is seen after simulation that the amplitudes of these harmonics at the filter output become 97, 3.2, 0.8, and 0.11 V. THD = 36 for *Xin*, and THD = 2.4 for the filter output. One can see effective smoothing with the second-order filter, looking at the curves on the **Scope** too.

Let the inverse problem be actual: an extraction of the high-frequency components. Since it is desirable that the signals with frequencies in the range of 350 Hz and higher are reproduced precisely, the natural frequency is taken as 150 Hz; the filter type is *Highpass*. After simulation, one can see that, with the amplitude of the input fundamental harmonics of 100 V, at the output of the filter, this amplitude decreases to 11 V. At this, the higher harmonic amplitudes are equal to 39, 30, and 10 V, respectively; that is, they are reproduced nearly ideally.

The filter operation as a band filter is demonstrated with help of model **Filter_Ex1**. The sinusoidal signal generators have the same amplitude of 40 V. The filter has the natural frequency of 350 Hz (it means that we try to pick out the seventh harmonic), type—*Bandpass*. After the simulation, the following harmonic amplitudes are seen (harmonic number/amplitude): 1/8, 7/40, 13/29, and 19/20. The seventh harmonic is reproduced precisely, but the values of rest of the harmonic amplitudes are essential too. It means that in the case under consideration, the second-order filter is inadequate. If the filter type is replaced with the *Bandstop* one, the harmonic amplitudes after simulation are as follows: 1/39.2, 7/0, 13/27, and 19/34.3. Thus, the seventh harmonic disappears, and the values of the rest harmonic amplitudes are close to the actual ones. So the second-order filter, as the *Bandpass* one, functions successfully.

For the **Second-Order Filter (Variable-Tuned)**, the natural frequency is defined with the external signal and can change. So this filter can be used for the filtration of the signals in the weak grids, in which the grid frequency can change.

The block **Lead-Lag Filter** realizes the transfer function

$$W(s) = \frac{1 + sT_1}{1 + sT_2}. \tag{2.48}$$

The time constants T_1 and T_2 are specified in the block dialog box. Depending on their relations, the block can be either a lead or a lag element.

Let us consider some other blocks. The block **Three-Phase Programmable Generator** (R2013a) or **3-phase Programmable Source** (R2012b) fulfills the same functions as the block **Three-Phase Programmable Voltage Source** discussed in Section 2.1. The difference is that the latter forms the voltages that are compatible with the block SimPowerSystems, and the former gives the three-phase signal as a vector [a, b, c] that is compatible with the block Simulink.

The **Three-Phase Sine Generator** (R2013a) fabricates the three-phase signal, whose amplitude, frequency, and phase are defined with the external input values. The block outputs, besides the three-phase signal *abc*, the signal *wt* that varies between 0 and 2π, synchronized with zero crossings of the fundamental of phase *A*. This block may be successfully employed in the models of isolated renewable electrical sources, whose output voltage is determined with the processes in the source (not the voltage of the connected grid).

The block **Timer** (R2012b) or **Stair Generator** (2013a) generates certain signals at fixed times. In the first line of the dialog box, the times are given when the output changes, and in the second line, the output values at these times.

The following blocks are in both versions: R2012b and R2013a.

For the block **Monostable**, the direction of the input change is specified for which the blocks respond (Edge: *Rising, Falling, or Either*). If this event takes place, the block outputs the logical 1 during the time that is fixed as *Pulse duration*.

The block **Edge Detector** continuously compares the current value of the input signal with the previous one. The block can be tuned for the generation of output logical signal 1 by increasing the input signal (Edge: *Rising*), by decreasing the input signal (Edge: *Falling*), or when there is any inequality between the current and previous values (Edge: *Either*). When the specified event takes place, the logical input signal 1 appears, whose duration is equal to one integration step.

The block **Bistable** is an *R–S* flip-flop that sets the output $Q = 1$ under $S = 1$ and $R = 0$, resets $Q = 0$ under $S = 0$ and $R = 1$, and holds Q under $S = 0$ and $R = 0$. The state under $S = 1$ and $R = 1$ is defined by the selected option *Priority to input*: If *Set* is selected, $Q = 1$, and if *Reset*, $Q = 0$. The opportunity is to fix the initial flip-flop state.

The output of the block **Sample and Hold** follows its input until at the block input *S*, the logic signal 1 holds ($S = 1$). When *S* is going to 0, the block output holds its input at this moment of time. The initial output value can be specified in the dialog box.

The block **On/Off Delay** is intended for the delay of the input logical signal for a specified time. In the dialog box, besides *Time Delay* and the input initial value before simulation start *Input at* $t = -eps$, the block operation mode is specified: *On delay* or *Off delay*. In the first case, the input = 1 appears at the output after *Time Delay* and is held in this state. When the input is going to 0, the output is going to 0 without delay. For the *Off delay* mode, in this description, 1 has to be replaced by 0 and 0 by 1.

The block **Discrete Shift Register** models a shift register with the serial input and the parallel output. If the vector signal of N_{sig} dimension comes at the input, the block stores the last $N \times N_{sig}$ inputs, and the number N is indicated in the dialog box. These signals appear at the block output as a vector of $N \times N_{sig}$ dimension that is arranged as follows: [$u1(k)$, $u1(k-1)$, $u1(k-(N-1))$, $u2(k)$, $u2(k-1)$, ...]. The sampling time and the initial register content are specified in the dialog box.

The blocks **Discrete PI Controller** and **Discrete PID Controller** were included in the folder *Discrete Control Blocks* of R2012b. The parameters of the controller, including the output limits and the sampling time, must be specified. This block was practically used in all developed models. But it is removed, beginning from the version R2013a; instead, the new developed block **PID Controller** from the folder *Simulink/Continuous* is recommended. This block gives some more opportunities during model development and investigation.

The block **Discrete Virtual PLL** was included in R2012b. This block has no input signals. It simulates the operation of the actual PLL by the use of parameters specified in the dialog box. This block may be used as a reference block in some cases.

REFERENCES

1. IEEE Task Force on Load Representation for Dynamic Performance. Load representation for dynamic performance analysis. *IEEE Transactions on Power Systems*, vol. 8, no. 2, pp. 472–482, May 1993.
2. Mohamed, I. F., Abdel Aleem, S. H. E., Ibrahim, A. M., and Zobaa, A. F. Optimal sizing of C-type passive filters under non-sinusoidal conditions. In *Energy Technology & Policy*, 1:1, pp. 35–44, Taylor & Francis Group, Abingdon, 2014.
3. Perelmuter, V. M. *Electrotechnical Systems, Simulation with Simulink and SimPowerSystems.* CRC Press, Boca Raton, FL, 2012.
4. Kundur, P. *Power System Stability and Control.* McGraw-Hill, New York, 1994.
5. Kothari, D. P., and Nagrath, I. J. *Modern Power System Analysis.* Tata McGraw-Hill, New York, 2003.
6. MathWorks, *SimPowerSystems™: User's Guide.* MathWorks, Natick, MA, 2011–2016.
7. Hase, Y. *Handbook of Power Systems Engineering with Power Electronics Applications.* John Wiley & Sons, Ltd, West Sussex, 2013.
8. Bose, B. K. *Modern Power Electronics and AC Drives.* Prentice Hall PTR, Upper Saddle River, NJ, 2002.

3 Simulation of Power Electronics

3.1 MODELS OF POWER SEMICONDUCTOR DEVICES

SimPowerSystems/Power Electronics has the following models of the semiconductor devices: (1) **Diode**; (2) **Thyristor**; (3) **Detailed Thyristor**; (4) **Gto**; (5) **Mosfet**; (6) **IGBT**; and (7) **IGBT/Diode**. The dialog boxes of these devices have many identical fields.

The parameters of the block **IGBT** are the inner resistance R_{on}, the inductance L_{on}, and the threshold voltage V_f. **IGBT** turns on when the voltage collector—emitter $U_{ke} > V_f$ and the control (gate) signal $g > 0$, and it turns off when $U_{ke} > 0$ and $g = 0$ and is in turning-off state when $U_{ke} < 0$. An antiparallel diode has to be used in order to make a path for the current in the last case, how it is made in the block **IGBT/Diode**. The block models the process turning-off: current decreasing from I_{max} to 0.1 I_{max} for the fall time T_f and further to zero for tail time T_t.

The block **IGBT/Diode** mentioned earlier is the simplified model of the pair insulated gate bipolar transistor (IGBT) (or GTO, or metal-oxide-semiconductor field-effect transistor [MOSFET])/diode that has only one parameter: inner resistance.

Mosfet is controlled by the signal $g > 0$ if the device current $I_d > 0$. The inner antiparallel diode conducts when the voltage across the transistor $V_{ds} < 0$. The block parameters are the transistor resistance R_{on}, the diode resistance R_d, and the inductance L_{on}. The transistor turns on when $V_{ds} > 0$ and $g > 0$. If $I_d > 0$, the transistor turns off when $g = 0$. If $I_d < 0$, the current flows via a diode. The device has the variable resistance: $R_t = R_{on}$ under $I_d > 0$, and $R_t = R_d$ under $I_d < 0$.

The block **Gto** models the GTO—thyristor that, unlike the usual thyristor, can be turned off by the control signal. It has the same parameters as IGBT. It turns on when the voltage anode–cathode $V_{ak} > V_f$ and $g > 0$. When g goes to 0, the process of thyristor blocking begins that is the same as for IGBT.

The device **Diode** conducts when $V_{ak} > V_f$ and stops to conduct when $V_{ak} < V_f$.

The device **Thyristor** is modeled as a GTO thyristor and turns on at the same condition but cannot be turned off by the control signal. When $g = 0$, it keeps in the conducting state until $I_{ak} > 0$. The times T_f and T_t are not specified.

For the block **Detailed Thyristor**, unlike the previous one, the latching current I_l and the turn-off time T_q have to be specified. When the thyristor is turning on, the duration of the firing pulse $g > 0$ must be sufficient for providing condition $I_{ak} > I_l$ (it is taken that $I_l = 0$ for the block **Thyristor**). If this condition was not fulfilled, after setting $g = 0$, the thyristor returns to the blocking state. During time T_q, the thyristor restores the ability to block the voltage anode–cathode after the current I_{ak} died (for the block **Thyristor**, $T_q = 0$ is accepted).

All these models give the opportunity to measure the current flowing in the device and the voltage drop across it when the option *Show measurement port* is activated.

The additional input *m* appears on the block pictures. The possibility to connect the snubber that consists of the connected in series R_s and C_s parallel to the blocks exists as well.

Together with the separated devices, the folder *SimPowerSystems/Power Electronics* contains the blocks that consist of some devices: **Universal Bridge** and **Three-Level Bridge**.

The first block can have one arm or model a single-phase or a three-phase bridge converter. The blocks **Diode, Thyristor, Gto** with antiparallel **Diode, Mosfet, IGBT/Diode**, and **Ideal Switch** can be utilized as the devices in this bridge. The vector of the gate pulses **g** comes to the block input *g*. When one develops his own pulse width modulation (PWM) system, one must know the correspondence of vector **g** components and the device position in the bridge; that is to say, one must know the device numeration in the bridge scheme. The numeration is different for the thyristors and for the bridge with the forced commutated devices and is depicted in Figure 3.1.

The snubbers can be set in the block that is indicated in the block dialog box. Their employment often improves the simulation process, especially in the discrete mode [1].

This block is the basis for building models of the rectifiers and the two-level inverters. In order to make simulation quicker, two additional modifications of this block are developed.

The two-level inverter fabricates six active and two zero states of the voltage space vector at the output. The certain combination of the gate pulses actuates each of these states. To neglect the transients and the voltage drop across the inverter devices, the output voltage space vector components are proportional to the voltage U_d across the capacitor in the inverter direct current (DC) link; at this, the proportional factors depend on the gate pulse combination. These factors are designated as the switching functions. The current that the inverter draws from the capacitor is determined by

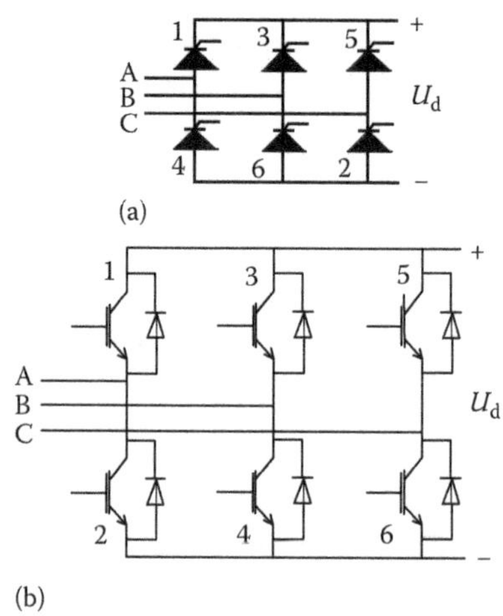

FIGURE 3.1 Device numeration in the block **Universal Bridge**: (a) thyristors and (b) IGBT.

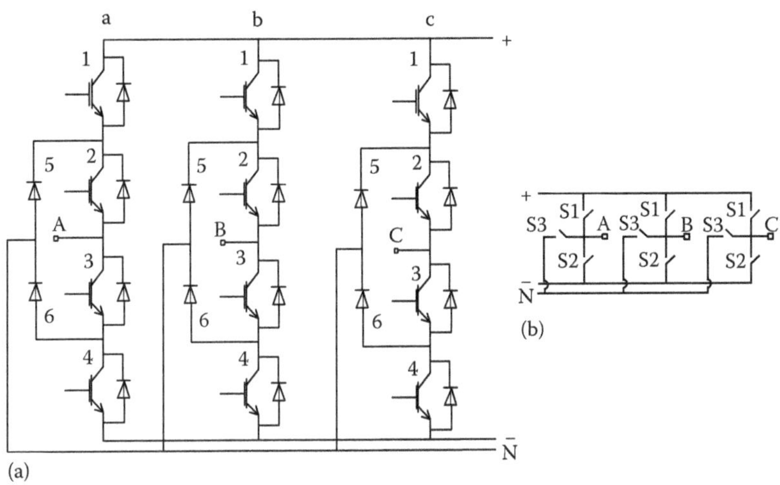

FIGURE 3.2 Device numeration in the block **Three-Level Bridge**: (a) for the forced commutated devices and (b) for the device **Ideal Switch**.

the phase currents and the switching functions too. When the structure *Switching-function based VSC* is selected, the block will contain two controlled voltage sources that fabricate the space vector and the controlled current source that forms the DC link current. The gate pulses come to this block as before, and the harmonic contents of the output voltage remain. When the second structure *Average-model based VSC* is chosen, the output voltage is proportional to the input signal U_{ref}; moreover, the proportional factor depends on the capacitor voltage. At this, the simulation runs faster, but the harmonic contents are not preserved. This possibility can be used for the simulation of the system that contains a lot of voltage source converters (VSC).

The device numeration for the block **Three-Level Bridge** is shown in Figure 3.2a. The forced commutated devices described earlier can be utilized in this unit, including the device **Ideal Switch**. The numeration for the last case is shown in Figure 3.2b.

When the block operates, two neighboring devices are turned on in each arm. The gate pulse vector comes to the input *g*; the vector dimension (4, 8, and 12) depends on the arm number. The vector component sequence is 1, 2, 3, and 4 for phase *A*, the same for phase *B* and phase *C*. When **Ideal Switch** is used, pulse 1 comes to switch S_1, pulse 4 to switch S_2, and the logic **AND** of pulses 2 and 3 to switch S_3.

3.2 CONTROL BLOCKS FOR POWER ELECTRONICS

A number of blocks are included in SimPowerSystems that fabricate the sequence of the pulses for controlling the blocks **Universal Bridge** and **Three-Level Bridge** with different devices. There are two groups of such blocks: the firing pulse generators for the units with the thyristors and for the units with the forced commutated devices—PWM blocks.

The first group contains **Synchronized 6-pulse Generator, Discrete Synchronized 6-pulse Generator, Synchronized 12-pulse Generator**, and **Discrete Synchronized 12-pulse Generator**.

The first two blocks control the converter whose diagram is shown in Figure 3.1a. Since the output voltage in the continuous current mode is

$$E_d = 1.35U_l \cos \alpha, \tag{3.1}$$

U_l is the rms phase-to-phase voltage and α is a firing angle, that is, the angle of delay, when the firing pulse comes, relatively the natural firing point, when the anode voltage is getting positive. Thus, the output voltage control is carried out by the change of the instants when the firing pulses come to the proper devices. The inputs are the three-phase system of the phase-to-phase voltages and the reference angle (grad). The pulse production fulfils only when the input *Block* = 0. The block output is the vector of sis pulses whose numeration corresponds to the thyristor numeration in the block.

The width is the main pulse parameter. Its duration must be sufficient that the thyristor current would exceed value I_l. This duration can be rather large sometimes, when the load inductance is big, as it takes place by supplying an excitation winding. In the bridge, every 60°, another thyristor is turned on. When the current is small or (and) the load inductance is also small, the duration of the thyristor conductance can be less than 60°; when the firing pulse comes to the next thyristor, the load current would be absent and the current will not appear because it must flow through two thyristors. In order to avoid such a situation, simultaneously with firing pulse fabrication for the next thyristor, the firing pulse for the previous thyristor must be provided. It can be done in two ways: by forming wide pulses up to 120° or simultaneously by sending a short pulse to the next thyristor, to send it to the previous one as well. Such a method is designated as a pulse doubling. The first method is simpler, but the devices that form the wide pulses can be rather bulky. The correct specification of the pulse width and the doubling necessity is especially important for multibridge converters with series bridge connection.

One can see how this block is built, if one carries on the command *Look under mask* (see Ref. [2] too).

The natural firing points are the instants when the phase-to-phase voltages cross the zero level. When the converter is supplied from the three-phase transformer, the primary voltages are used for synchronization, because they are less distorted than the secondary voltages. At this, attention must be paid for their correct connection. The phase-to-phase voltages that are indicated on the block picture are sent only if there is no shift between the primary and the secondary windings. For connection Y/D1 or D11/Y, the phase voltage U_c has to be sent to the input A, with a corresponding rearrangement for the other inputs, and for the connection D1/Y and Y/D11, the phase voltage $-U_a$ has to be sent to the input A.

One more possibility for synchronization should be taken in mind: to select the three-phase voltage system that leads the secondary voltages and to get the demanded phase shift with the help of the filters with corresponding time constants. Such a method was utilized in the model **Zig_zag_Trans**.

In the block discrete version, the frequency of the supplying voltage is not specified, and its value comes to the input either as a constant signal or as the PLL output. This option gives the opportunity to use the block under the simulation of the weak grids, in which the frequency can vary.

Synchronized 12-pulse Generator can control two thyristor bridges that are connected to two secondary windings of the three-winding transformer. At that, the primary and the first secondary windings are connected as star, and the second secondary as delta, whose phase voltages lag the star. The primary phase voltages are used for blocking synchronization. The firing pulses for the bridge that is connected to the star winding come to output PY and for the other bridge to output PD.

The discrete variant of the block has the input for the signal F_{req}.

The considered pulse generators use the linear reference; their drawback is nonlinear dependence of the output voltage on the input control signal. This drawback can be avoided with the help of arc cosine transformation of the control signal.

Beginning from the MATLAB version 2013a, the described blocks are excluded from the library browser. The block **Pulse Generator (Thyristor, 12-Pulse)** is developed instead. But the old blocks may also be used; they are available by typing `powerlib_extras` at the MATLAB command prompt. This new block can fabricate the firing pulses for one thyristor bridge or for two bridges, whose supplying voltages have the relative phase shift ±30°. The main difference of this block from the old ones is (that) not the supplying voltages, but the sawtooth waveform varying between 0 and 2π comes to the input *wt*, which is synchronized on zero crossings of phase *A* primary voltage of the converter transformer. This sawtooth waveform is usually obtained from a PLL. Which of the blocks is reasonable to use for simulation of a certain system depends on what method of synchronization is utilized in this system, so that the simulation results would be close to the reality.

The following pulse generators for the forced commutated devices were included in SimPowerSystems: **PWM Generator**, **Discrete PWM Generator**, **Discrete 3-phase PWM Generator**, and **Discrete SV PWM Generator**. In the new MATLAB versions, the blocks **PWM Generator (2-Level)**, **PWM Generator (3-Level)**, and **SVPWM Generator (2-Level)** may be used instead.

Since a VSC has only a limited number of states, its output voltage consists of rectangles that have various widths but only a limited number of heights. The fundamental harmonic of this voltage must have the wanted parameters that are determined by the generator input signal.

There are two main PWM methods: the comparison of the reference harmonic signal U_r with an isosceles triangle carrier waveform U_t (further—carrier) and a space vector modulation (SVM).

The first method is illustrated in Figure 3.3. Here U_{rabc} is a three-phase reference signal having the amplitude *m*, $m \leq 1$. Each phase is controlled separately. Take, for instance, phase *A*. When $U_{ra} < U_t$, the logic 0 comes to device 1 and it is turned off; at this, the inverse signal comes to device 2, and it is turned on. Thus, terminal *A* has the negative potential. The situation is contrary under $U_{ra} > U_t$, and the terminal has the positive potential. Since the frequency f_t of U_t is much more than the frequency of U_r, it can be proven that the average value of the phase voltage, relatively the middle point for time interval $1/f_t$, is U_{ra}.

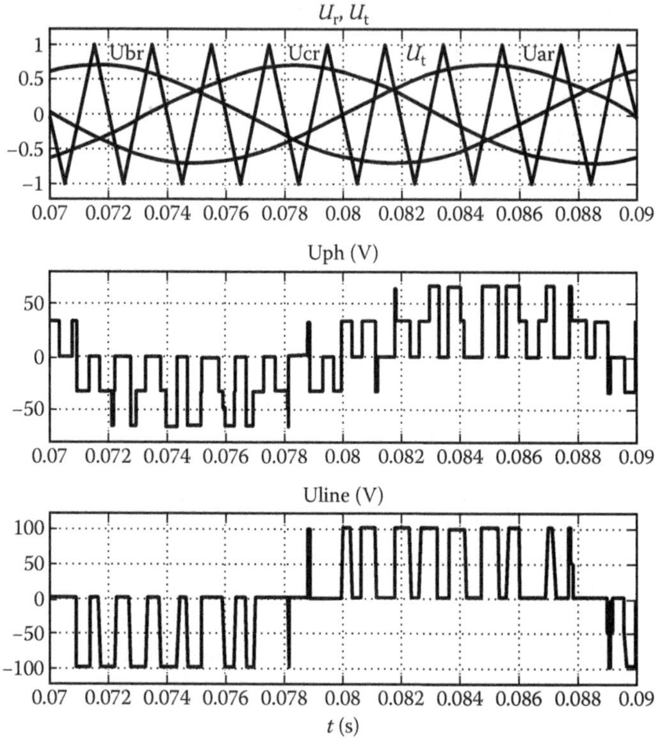

FIGURE 3.3 Output voltage fabrication by PWM with the triangle carrier waveform.

The effective value of the phase-to-phase voltage of the fundamental harmonic is

$$U_{11} = \frac{\sqrt{6}}{4} mU_{d} = 0.612mU_{d}. \tag{3.2}$$

The linear range can be extended to add to the three-phase reference signal zero sequence triple harmonic of the appropriate amplitude; it means that the same signal is added to all three phases. For example, this additional signal can be taken as

$$\Delta V = -0.5 \times [\max(V_{rabc}) + \min(V_{rabc})]. \tag{3.3}$$

Figure 3.4 illustrates this method.

In this case

$$U_{11} = \frac{1}{\sqrt{2}} mU_{d} = 0.707mU_{d}. \tag{3.4}$$

It means 1.155 more in comparison with Equation 3.2.

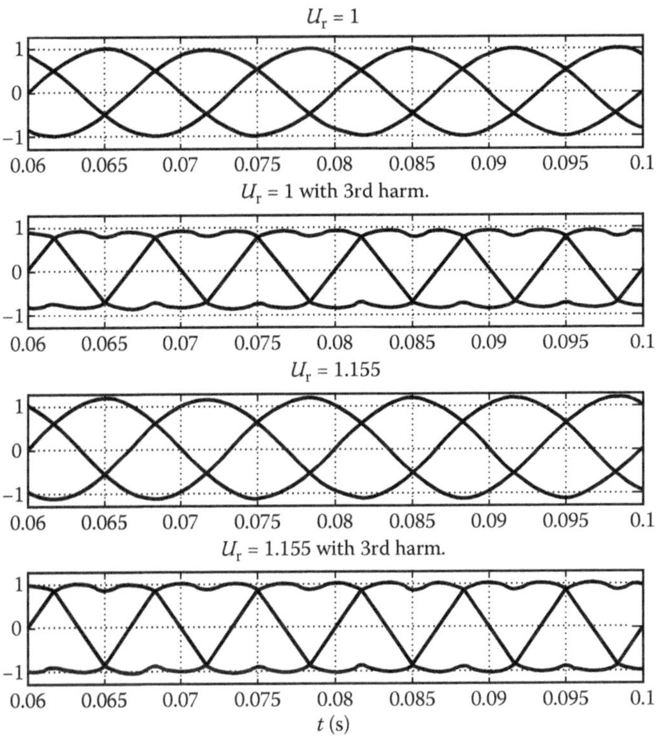

FIGURE 3.4 Addition of the 3rd harmonic to the reference signal.

SVM is a discrete method in principle. The system operates with a sample time T_c; during T_c, several inverter states are realized. Since the value T_c is small in comparison with the load time constant, the output voltage is supposed to be equal to the average value of the voltages used during T_c. When SVM is used, the control system fabricates the wanted vector of the output voltage **u** in the polar or the Cartesian coordinates. VSC can create at its output six nonzero space vectors and, in two ways, a zero vector. In order to produce the vector **u**, two neighboring inverter states are utilized and the zero vector as well. Figure 3.5 illustrates creation of the vector **u**.

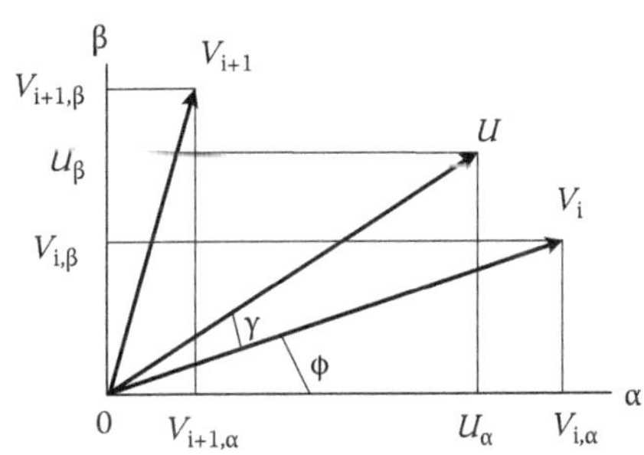

FIGURE 3.5 Fabrication of the space vector **u**.

Suppose that $\mathbf{V}_i$ is used for the time interval t_1 and vector $\mathbf{V}_{i+1}$ for t_2. Denote $\tau_1 = t_1/T_c$ and $\tau_2 = t_2/T_c$. Then the value τ_1 and τ_2 have to satisfy the following equations:

$$V_{i\alpha}\tau_1 + V_{i+1,\alpha}\tau_2 = u_\alpha$$

$$V_{i\beta}\tau_1 + V_{i+1,\beta}\tau_2 = u_\beta \tag{3.5}$$

Therefore,

$$\tau_1 = \frac{u_\alpha V_{i+1,\beta} - u_\beta V_{i+1,\alpha}}{V_{i\alpha}V_{i+1,\beta} - V_{i\beta}V_{i+1,\alpha}} \tag{3.6}$$

$$\tau_2 = \frac{u_\beta V_{i\alpha} - u_\alpha V_{i\beta}}{V_{i\alpha}V_{i+1,\beta} - V_{i\beta}V_{i+1\alpha}}. \tag{3.7}$$

During $T_c(1 - \tau_1 - \tau_2)$, the zero state is used. In the polar coordinates $V_{i\alpha} = U_0 \cos\phi$, $V_{i\beta} = U_0 \sin\phi, V_{i+1,\alpha} = U_0 \cos\left(\phi + \frac{\pi}{3}\right), V_{i+1,\beta} = U_0 \sin\left(\phi + \frac{\pi}{3}\right), u_\alpha = u\cos(\phi + \gamma)$,

$u_\beta = u\sin(\phi + \gamma), U_0 = \frac{2}{3}U_d$

After substitution in Equation 3.7, it follows that

$$\tau_1 = \frac{\sqrt{3}u}{U_d}\cos\left(\gamma + \frac{\pi}{6}\right),\ \tau_2 = \frac{\sqrt{3}u}{U_d}\sin\gamma. \tag{3.8}$$

The maximal rms of the phase-to-phase voltage is 0.707 U_d.

After the computation of t_1, t_2, and t_0, the problem of finding out the best way to arrange the corresponding inverter states in time appears. The criteria are contrary: less switching and less contents of the high harmonics in the inverter voltage. One method is to fix the state of one of the phases for every interval T_c. The inverter state is agreed to denote with a three-digit binary code (respectively *A B C* phases); at that, the digit is 1 when the upper device in the phase is turned on. Therefore, there are two zero states: 000 и 111. If, for instance, the state 001 corresponds to the vector $\mathbf{V}_i$ and 101 to the vector $\mathbf{V}_{i+1}$, the states change as (the time is indicated in the parentheses) 101(t_2/2), 111(t_0), 101(t_2/2), and 001(t_1); the devices do not switch over in phase *C*. However, it will have fewer harmonics, for example, under the following combination with more switching: 000(t_0/4), 001(t_1/2), 101(t_2/2), 111(t_0/2), 101(t_2/2), 001(t_1/2), and 000(t_0/4).

The block **PWM Generator** generates the gate pulses for two-level voltage source inverters (VSI), utilizing the triangle carrier waveform and can be used for the single-phase, two-phase, three-phase, and two three-phase schemes. In the last case, the pulses are in antiphase. There is an opportunity to fabricate the modulation signal in the block; the parameters of this signal are specified in the block dialog box. One can see in Refs. [1,2] how this block is built.

The block operation algorithm does not take into account a dead-time effect: the incoming device delays from the outgoing device of the same phase in order to prevent a shoot-through fault [3]. This delay tells on the output voltage. A number of methods have been developed to avoid this effect [3].

In the discrete variant of the block, the sample time has to be specified in the block dialog box.

The new block **PWM Generator (2-Level)** provides the same functions (excluding the control of two three-phase bridges) and can operate in continuous and discrete modes.

The block **Discrete 3-phase PWM Generator** can control two two- or three-level inverters; the output inverter voltages are in antiphase. The synchronization of the carrier signal with the modulation signal can be implied in this block. When the carrier signal has a constant frequency, and the frequency of the modulation signal changes, one period of the latter doesn't include, generally, the whole number of the periods of the former; it can cause the low frequency oscillations of the output voltage. With synchronization, it is not the carrier frequency, but the ratio to the modulation signal frequency that is specified as a whole number K_n. PLL must be employed.

When the block is used for a two-level inverter control, its operation does not differ essentially from the previous one. When this block is used for a three-level inverter control, its structure is shown in Figure 3.6.

The carrier signals C+ and C− are triangle waves that change respectively from 0 to 1 and from −1 to 0. If $U_r > 0$, the comparison element **COM1** operates, the output of **COM2** is the logic 0, and the output of **SM1** is 0 or 1. If $U_r < 0$, the comparison element **COM2** operates, the output of **COM1** is the logic 0, and the output of **SM1** is 0 or −1. The decoder converts these signals in the gate pulses by the following rule: when **SM1** = 1, the devices 1 and 2 are turned on, and 3 and 4 are turned off; when **SM1** = 0, devices 2 and 3 are turned on, and 1 and 4 are turned off; when **SM1** = −1, the devices 3 and 4 are turned on, and 1 and 2 are turned off. Pulse fabrication for the phase is shown in Figure 3.7.

The new block **PWM Generator (3-Level)** operates ditto, but can control only one three-level bridge.

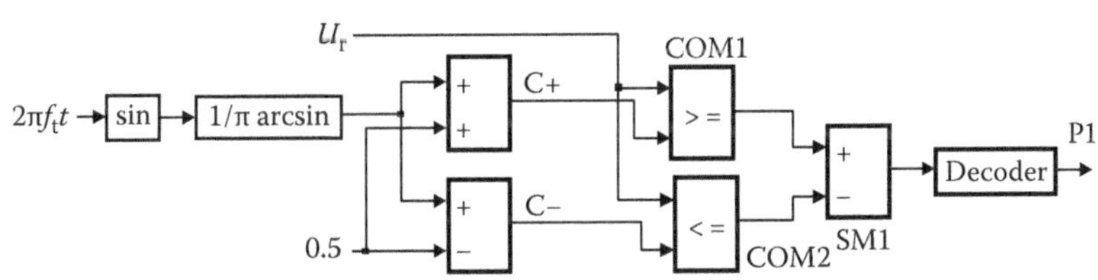

FIGURE 3.6 Block diagram of the block **Discrete 3-phase PWM Generator** under the control of a three-level inverter.

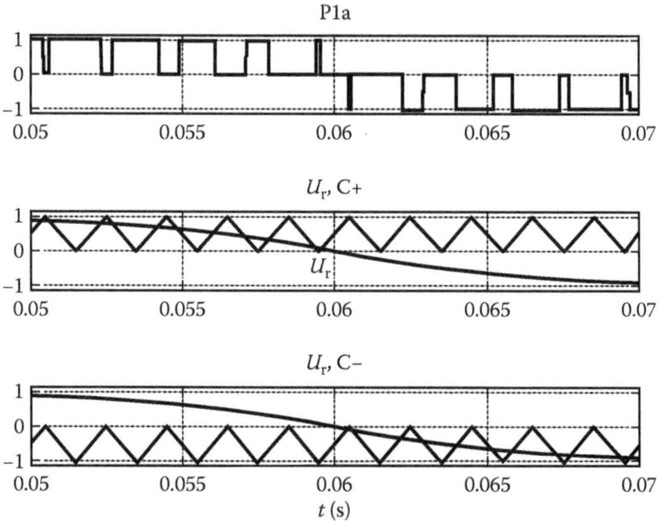

FIGURE 3.7 Gate pulse fabrication for the three-level inverter.

The block **Discrete SV PWM Generator** fabricates the gate pulses for a two-level inverter, using Equation 3.8. The wanted space vector **u** can be given as magnitude and angle or as α and β components; it can be formed inside the block too (the option *Internally generated*). The frequency of computations (*Chopping frequency*) that is equal to $1/T_c$ is specified in the block dialog box.

One of two types of state changes during one period T_c (the option *Switching pattern*) can be realized. These types were described earlier. For the option *Pattern #2*, the first state sequence is carried out, having less switching, and for *Pattern #1*, the second one, having more switching-over, but less harmonics. The new block **SVPWM Generator (2-Level)** operates in the same way.

The new block **Overmodulation** adds to the three-phase reference signal the third harmonic or triple harmonic zero sequence signal; it has been said earlier that, in so doing, the linear region of a three-phase PWM generator increases. The block output comes to the input of one of the PWM blocks. Some ways to fabricate the third harmonic are intended; one of them is according to Equation 3.3.

The new block **PWM Generator (DC-DC)** is intended for the control of the buck or the boost converter of the DC. The input signal is compared with the sawtooth waveform; the value of the input determines the percentage of the pulse period when the output is on. It depends on the circuit design, if the converter is buck or boost type. Such converters are used widespread in the renewable electric sources in Chapters 5 and 6.

When PWM is developed by the user, who cannot employ the standard PWM blocks described earlier, the sources of the sawtooth or the triangle waveforms are often needed. Such sources are included in the new version of SimPowerSystems: **Triangle Generator** and **Sawtooth Generator**.

3.3 SIMULATION OF CONVERTERS WITH THYRISTORS

The converters with thyristors are utilized in the wind-generator systems (WG) when a necessity appears to transfer the electric energy with the DC. Such a transfer is reasonable under a large distance from the place where WG is located to the point of its connection to the network—100 km and more. Such a situation appears for the offshore WG. The DC advantages in the case are the possibility to have different frequencies in the WG and in the network, avoidance of the capacitive currents, fewer losses in a cable, and big transferring power [4].

The example of such electrical transmission is shown in Figure 3.8. The WG farm has the total power of 500 MVA; the WGs are connected to the bus with a voltage of 30 kV. Since the WG output voltage can change under wind speed fluctuations, the synchronous compensator of the reactive power static synchronous compensator (STATCOM) is set; it will be considered further in this chapter. The thyristor bridges are connected in series and are supplied with voltages which are shifted one relative to another by 30° that compensates the 5th and 7th harmonics in the transformer primary current and the 6th harmonic in the rectifier voltage. The inverter is connected with the rectifier with underwater cable and has the same scheme.

The WG power changes randomly when the wind speed fluctuates; thus, the transmission line power has to conform to the present WG farm power. The transferred power is a product of the DC voltage by transferred current. The voltage value is determined, mainly, by the bridges voltage supply, whereas the current value is determined by the difference of the rectified voltages of the rectifier and the inverter. Thus, the system has to have the DC voltage and DC current controllers. The voltage control usually affects the inverter control pulses, and the current controller acts on rectifier control pulses. Because the inverter firing angle must be limited so that the thyristors can restore their blocking ability when the next firing pulse comes, the measurement and the limiting of the turn-off angle γ are carried out. If α is the firing angle, then $\beta = 180°—\alpha$ is the inverter angle, μ is the overlap angle, and $\gamma = \beta - \mu$. Instead of voltage control, the controller of the turn-off angle can be used.

Since the current pulsations lead to additional line losses, the smoothing reactors are set in the DC circuits. Since the higher current harmonics arise in the AC

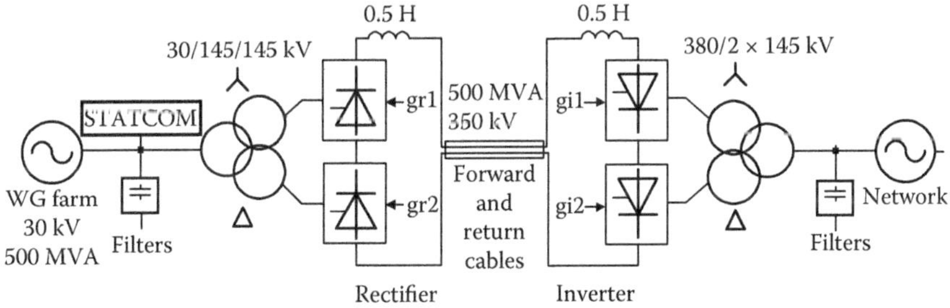

FIGURE 3.8 HVDC with thyristors.

network when it supplies the thyristor converters, the harmonic filters are set at the AC sides of the converters. The thyristor converters consume from the network the inductive reactive power whose value depends on the firing angle; the capacitor banks are set in parallel to the harmonic filters in order to compensate for this reactive power. When STATCOM is employed, the capacitance of these banks can be decreased.

The model **HVDC** simulates the DC transmission depicted in Figure 3.8; at this, some peculiarities that are caused by WG are not taken into account because they are considered in Chapter 6. There is the demonstration model **power_hvdc12pulse.mdl** in SimPowerSystems, which has a structure that is analogous to the model **HDVC**, for which the detailed control blocks are developed. The model **HVDC** is much simpler; it uses some blocks from the demonstration model, for example, the block for γ calculation and the block that fabricates the firing pulses for the rectifier and for the inverter.

The **Three-Phase Source** blocks are used for the source models. The subsystems **Rectifier** and **Inverter** include, each one, the three-phase three-winding transformer and two thyristor bridges connected in series. The transformer of the rectifier has the power of 350 MVA, 50 Hz, the primary voltage of 30 kV, and the secondary voltage of 2 × 150 kV; the transformer of the inverter has the power of 400 MVA, 50 Hz, the primary voltage of 380 kV, and the secondary voltage of 2 × 145 kV. The rectified current is measured and reduced to pu with the base of 1430A.

The forward and return cables are modeled as the circuits that consist of the resistor 1Ω and reactor 40 mH. Each subsystem has two filters: the filter with the high Q-factor ($Q = 100$) that is tuned for the suppression of 11th and 13th harmonics and the high-frequency filter with low Q-factor having the large bandwidth. The capacitor bank is set in addition. The total reactive power of the input filter is 200 Mvar, and 350 Mvar for the output one. The means for fault modeling are provided.

The rectifier is controlled by the system **R_Contr** that contains the controller of the direct current in line. The reference signal I_{dref} is compared with the actual current value *Id_R* and after the proportional plus integral regulator comes to the unit that fabricates firing pulses for a 12-pulse converter. The synchronizing signals for the unit are produced with the help of the block **PLL** and some other devices taking the transformer voltage that supplies the rectifier as a basis. The block **PLL** produces the value of the frequency too.

The short-circuit protection is realized in the considered subsystem when the DC voltage drops. If the voltage *VdL_R* decreases to some value that is fixed in the block **Relay2** (accepted 0.5 pu) during time interval of more than 40 ms, the signal appears at the regulator input that sets the firing angle as 165°; this angle puts the rectifier into inverter mode; the fault current drops down quickly and dies. The duration of this signal is determined by the monostable flip-flop **T2**. The time 50 ms is supposed to be sufficient in order to restore the normal operation. The block **Rate Limiter** assigns the reference current increasing with rather little rate and its decreasing with a big rate in order to quickly wipe out the fault current; the block **Rate Limiter1** forms an output signal increasing at a high rate in order to set the inversion mode quickly and decreasing of this signal with a lower rate for excluding excitation in the line.

If the duration of the fault signal is more than 90 ms, the pulse appears at the output of the **Delay** block **T3** that sets the flip-flop **T1**; at this, the current reference is turned off. After the current dies, the flip-flop **T2** resets, which blocks the pulse fabrication. The following line start begins with the signal *Reset* that resets the flip-flop **T1** giving the current reference and removing the pulse fabrication blocking.

The subsystem **I_Contr** controls the inverter. It contains the DC voltage regulator; its output comes to the system that fabricates firing pulses for a 12-pulse converter.

In the same subsystem, the subsystem **Discrete Gamma Measurement** is located, in which for each of the six thyristors of every rectifying bridge, the time interval is measured, from the instant when the thyristor current is getting to zero to the next moment of the natural firing. The signal of the frequency F_{req} is used in order to convert this time in the angle γ.

The null current threshold is specified in the dialog box (0.001 pu is advisable). The connection scheme of the secondary transformer winding (Y, D1, or D11) is defined in this dialog box too. The detailed subsystem description is given in Refs. [1,2].

When the mean value of γ becomes less than the assigned value, the reference signal of α decreases in order to keep the safe margin.

The flip-flop **T3** controls system starts. When both the voltage and the current are absent, the flip-flop is in reset state and blocks the pulse fabrication. On the origin of the rectifier current reference, the voltage signal appears firstly, the flip-flop comes to the set state, and the blocking is removed.

In the subsystem **Data Acquisition**, the scopes are collected that give the possibility to observe system operation.

Powergui defines the discrete simulation mode with the sample time of 50 μs (this value is put down in the option *File/Model Properties/Callbacks/Model preload function* as T_s). The plots of the rectifier current and its reference and the inverter voltage and its reference are displayed in Figure 3.9.

When the simulation of shorting in a DC circuit is carried out, the tumbler **Switch** is set in the low position, in the dialog box of the block **DC Fault**; the *Switching times* are defined as [0.7 0.74]. It is seen from Figure 3.10 that during the fault, the current, after briefly increasing, is limited, and after fault clearing, the normal line operation is restored. To repeat the simulation with *Switching times* as [0.7 0.88], it can be seen that after fault clearing, the line does not function. Only after appears the signal *Reset* in the subsystem **R_Contr** at $t = 2$ s does the line begin to work.

Under the simulation of the fault in the AC circuit, *Switching times* are defined as [0.7 0.8] in the dialog box of the block **AC Fault**. The voltages and the currents are shown in Figure 3.11.

It is seen that the short-circuit current is put out successfully, and after fault clearing, the line begins to work normally.

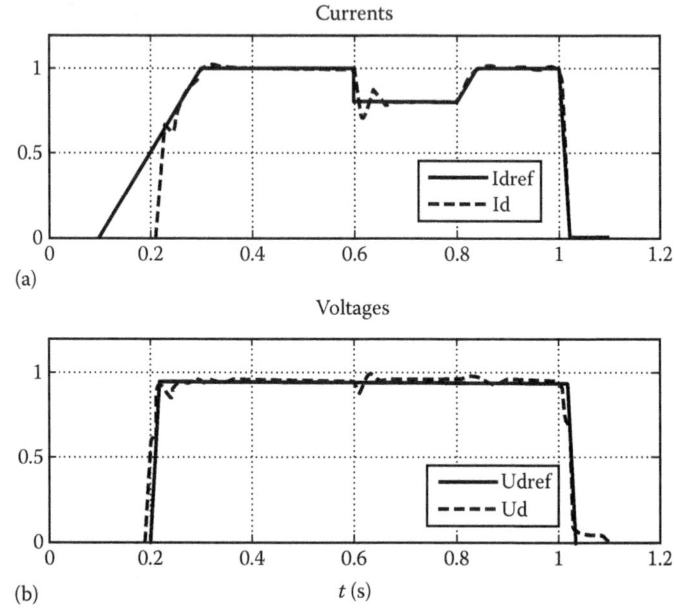

FIGURE 3.9 (a) Current and (b) voltage in HVDC under normal operation.

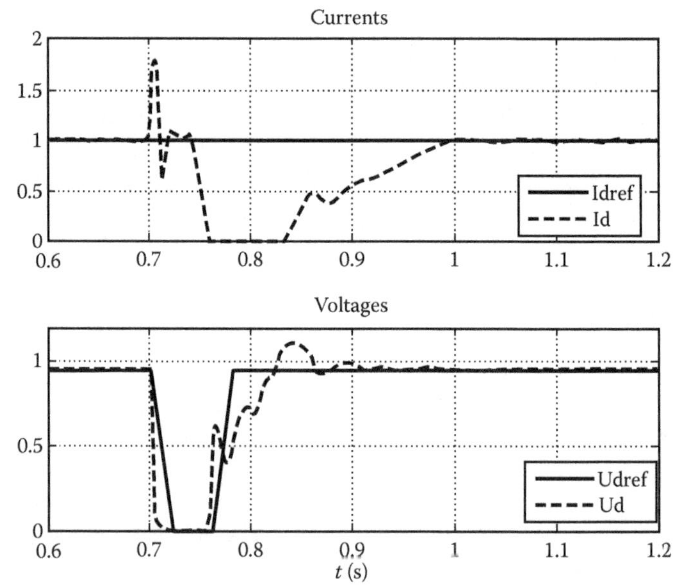

FIGURE 3.10 Processes in HVDC under the short-circuit in DC circuit.

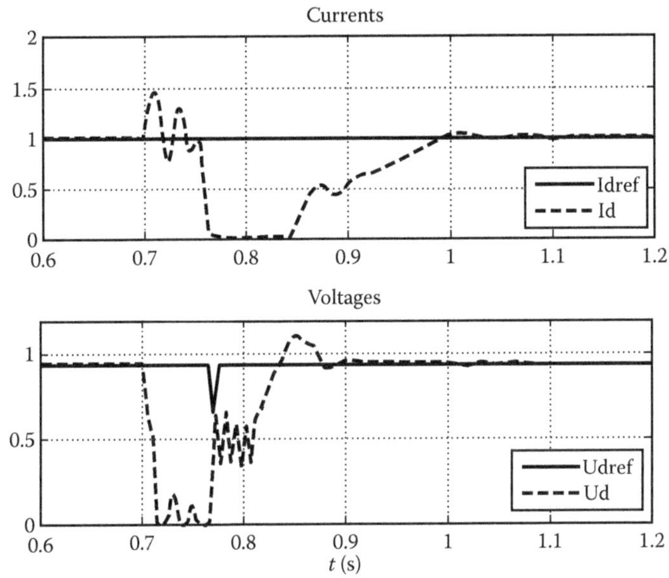

FIGURE 3.11 Processes in HVDC under the short-circuit in AC circuit.

3.4 SIMULATION OF CONVERTERS WITH FORCED COMMUTATED DEVICES

3.4.1 Models with Two-Level and Three-Level Inverters

The converters that utilize the forced commutated devices IGBT, GTO, or integrated gate-commutated thyristor (IGCT) are used practically in renewable electrical sources (RES). The relatively simple examples of the units with the forced commutated devices are considered in this section in order to better understand the special features of their operation.

The models **Two_level1** and **Two_level2** contain a two-level inverter, whose diagram is shown in Figure 3.1b. These models give the possibility to study the inverter operation with the various control systems, in a variety of modes. Specifically, such features can be investigated as the forms of the voltage and current curves in the various modes, the switching frequencies and the total harmonic distortion (THD) in the different operation modes, the effect of the third harmonic addition, an overmodulation mode when $m > 1$, an influence of the time interval arrangement on THD for SVM, the current distribution in the semiconductors that form a three-level inverter, the possibility to build the converter having very little harmonic content with using of several three-level inverters, and so on. These models are included here for completeness of interpretation because they are given and their utilization is described in detail in Ref. [2] as **model4_6** and **Model4_6a**. The same applies to the models **Three_level1** (the analog is **Model4_7**) and **Three-level2** (the analog is **Model4_8**). In the models developed for this book and considered further, these inverters are employed extensively.

3.4.2 Models with Four-Leg Voltage Source Inverter

In RES, especially in the isolated RES, the situation often appears when the main electric energy consumers are the single-phase loads that are connected to the phase and ground wires. Three-phase VSI supplies such loads. In order to create a four-wire system, the transform is set at the inverter output, whose secondary windings are connected in the star with the neutral wire. Such a transformer can be rather bulky.

For the creation of the neutral wire, the common point of two capacitors on the DC side of the inverter can be used; these capacitors are connected in series (Figure 3.12). The drawback of this scheme is that the capacitor voltages fluctuate very much when the load is unbalanced, so that the capacitors of great capacitance are necessary; the maximal amplitude of the phase voltage is $A_{max} = U_d/2$ [5,6].

The other opportunity that is utilized more often is to add the fourth arm as shown in Figure 3.13. The control of the new devices is different from the control of the main devices. The most popular method is to add the zero sequence signals to the modulating harmonic signals. It means that the signals $m \sin \omega t + V_0$, $m \sin[\omega t - (2\pi/3)] + V_0$, and $m \sin[\omega t + (2\pi/3)] + V_0$ are used for comparison with the triangle carrier waveform, and signal V_0 comes to the fourth arm (Figure 3.14). Here m is the amplitude of modulating harmonic signal divided by $U_d/2$.

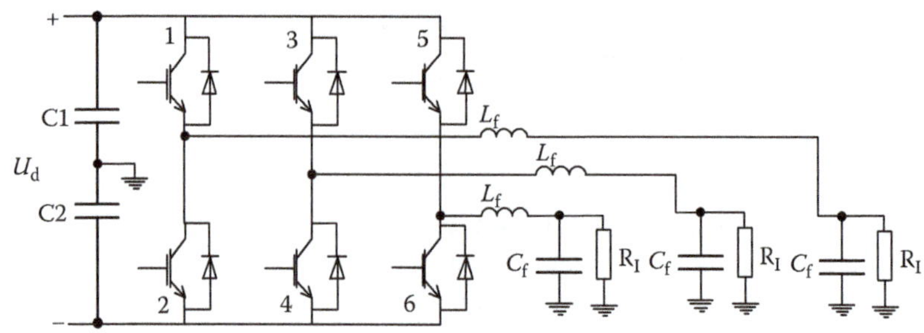

FIGURE 3.12 Connection of single-phase loads to the middle point of the capacitors.

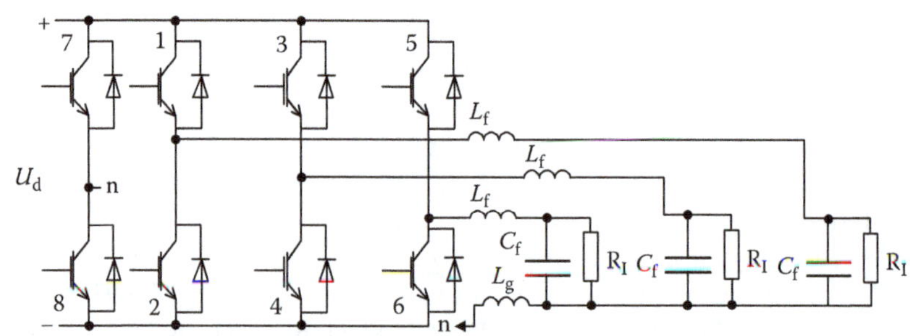

FIGURE 3.13 Diagram of the four-leg inverter.

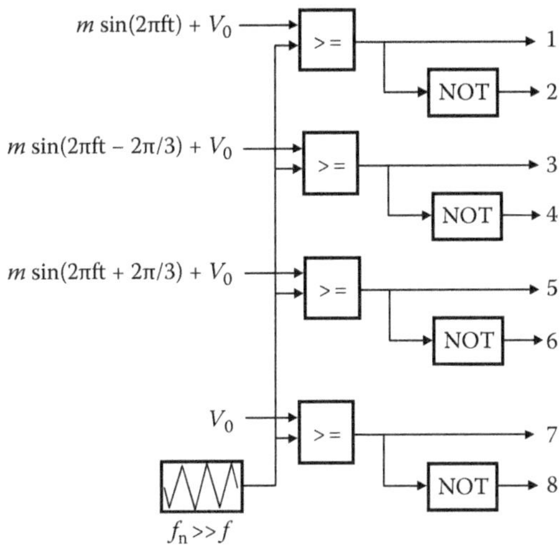

FIGURE 3.14 Gate pulse fabrication in the four-leg VSI.

It is recommended in Ref. [6] to take

$$V_0 = -\frac{\max(V_a, V_b, V_c) + \min(V_a, V_b, V_c)}{U_d}, \tag{3.9}$$

where V_a, V_b, V_c are the phase voltages or the harmonic modulating signals (in the latter case, the division by U_d is replaced with the division by 2). For such a scheme, $A_{max} = 0.577\ U_d$.

In the system under consideration, there are 16 inverter states, and 2 of them are zero states. They are given in Table 3.1.

To designate the device states with a four-digit binary code S_a, S_b, S_c, N, then $V_{an} = (S_a - N)U_d$ and so on.

In the models further considered, the inverter power is taken as 24 kW under the phase-to-ground voltage of 220 V; the rated current is 36 A; the load resistance is

TABLE 3.1
The Four-Wire Inverter States

Phase Voltages	1101	1100	0101	0100	0111	0110	0011	0010
V_{an}	0	U_d	$-U_d$	0	$-U_d$	0	$-U_d$	0
V_{bn}	0	U_d	0	U_d	0	U_d	$-U_d$	0
V_{cn}	$-U_d$	0	$-U_d$	0	0	U_d	0	U_d
	1011	**1010**	**1001**	**1000**	**1111**	**1110**	**0001**	**0000**
V_{an}	0	U_d	0	U_d	0	U_d	$-U_d$	0
V_{bn}	$-U_d$	0	$-U_d$	0	0	U_d	$-U_d$	0
V_{cn}	0	U_d	$-U_d$	0	0	U_d	$-U_d$	0

6 Ω; U_d = 600 V; the filter capacitance C_f is 50 μF; the inductor resistance is 0.1 Ω; and its inductance can change under investigation.

In the model **Four_legs1**, the inductor inductance is 2 mH. The diagram shown in Figure 3.14 is realized in the subsystem **Control**. To turn off the **Breaker** and to fix the three-phase reference as 1.05, it will be seen after simulation both when **Breaker1** is turned on and when it is turned off that the load voltage is nearly constant (Figure 3.15); its rms value is 221 V under THD = 1.2–1.7%; the rms value of the current of the fourth arm (scope **Zero-Current**) increases from 1.5 to 6 A when **Breaker1** is turned on.

When the loads are the single-phase ones, the phase currents are usually different; it means that the inverter load is unbalanced. To set the phase load resistances as 6, 12, and 24 Ω, the load voltages turn out to be rather balanced, with **Breaker1** turned on, and equal to 221–226 V for the different phases, but the rms current value of the fourth arm increases to 29 A. When **Breaker1** is turned off, the large voltage asymmetry appears: 210, 239, and 222 V that is caused by the common inductance of the neutral wire, but the rms current value of the fourth arm decreases to 24 A.

The essential potential difference between the source zero point and the ground exists in the considered four-wire systems. The source middle point that is formed with two capacitors connected in series often grounds that decreases the potential difference between the system elements and the earth to $U_d/2$. This potential difference can disturb the normal operation of the units connected to the system, for instance, of the computers, so that it is reasonable to look for the possibility to decrease this potential difference. The simplest way is to set the capacitor of greater capacitance between the zero point and the earth. Such a scheme is realized in the

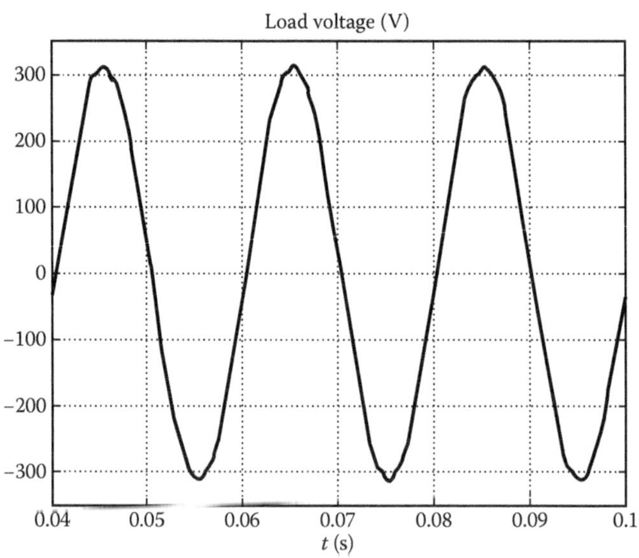

FIGURE 3.15 Load voltage in the four-leg VSI under balanced load.

model **Four_legs1** when the **Breaker** is turned on. One can make sure after simulation that, with the grounding capacitor of 50 μF, the maximal potential difference between the zero wire and the common capacitor point (U_{0g}) decreases from 300 to 100 V; at this, the load voltage THD increases about twice as large and the current in the fourth arm increases about 30–40% under balanced load.

The possibility to use the load voltage controller is considered further.

A combination of the direct reference and the feedback controller is used in the model **Four_legs2**. Two voltage control modes are provided in the subsystem **Control**: the single-loop controller **Voltage_Control1** and the cascaded controller with the inverter current inner loop (**Voltage_Control2**). A choice is carried out with the constant **Mode**: 0—the first option; 1—the second option. The possibility to simulate the nonlinear loads (**Br1** is closed) and the loads that actuate the DC component (**Br** is closed) in the phase current is provided in this model.

The plots of the output voltages for the case of the unbalanced loads (6 Ω, 12 Ω, and 24 Ω) mentioned earlier, with **Br** and **Br1** open, are shown in Figure 3.16 when the first control system option is employed. THD is <3% under the filter inductance of 1.2 mH. To repeat simulation when the breaker **Br** is closed, the current with a DC component of 4 A flows in phase *C*; at that, the rms of the fundamental harmonic is 14 A and THD = 15%. The curves of the load phase voltages do not noticeably change. To repeat simulation when **Br** is open and **Br1** is closed, it will be seen that phase *C* current has rms of the fundamental harmonic as 19 A and THD = 19%. The phase *C* voltage keeps as 220 V. About the same are the system features when **Mode** = 1.

The model **Four_legs3** is intended for the study of the system dynamic characteristics. The DC source of 600 V is replaced with the regulated source, whose

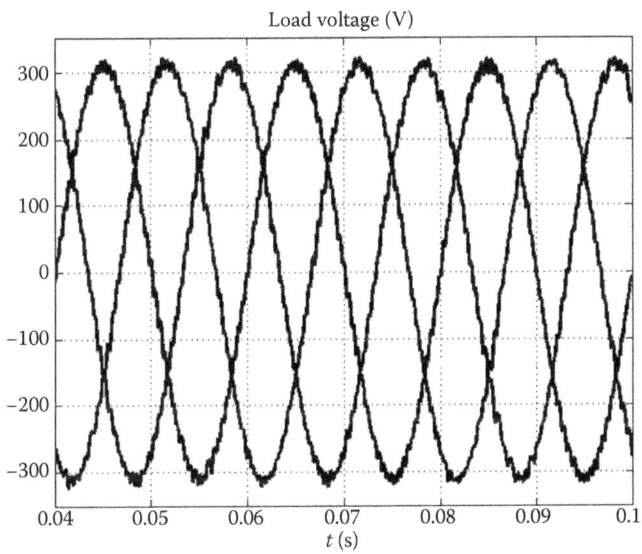

FIGURE 3.16 Load voltage in the four-leg VSI under unbalanced load and voltage control.

voltage at $t = 0.1$ s increases by step to 660 V (by 10%). One can see by the scope **Load_Voltages** that the voltage amplitudes of the load phases increase by 1–1.6%.

Thus, the described models give the opportunity to estimate the influence of the various parameters on the system quantities and to choose the necessary parameters for the wanted results to be achieved.

3.4.3 Cascaded H-Bridge Multilevel Inverter Simulation

The cascaded multilevel inverters that consist of some single-phase bridges connected in series are widespread at present. Its diagram is shown in Figure 3.17. Depending on the conducting semiconductors, each bridge can fabricate the voltage $\pm E$ or 0.

Thus, the phase voltage can change from $-nE$ to nE where n is the number of the bridges in a phase; in this way, $2n + 1$ level, including 0, can be received. The rms value of the output phase-to-phase voltage is

$$U_1 = 2kmnE, \tag{3.10}$$

where m is the modulation factor and $k = 0.612$. This scheme gives an opportunity to reach the high output voltage with the limited voltage of one bridge; the harmonic contents in the output voltage are less; the modular design makes production and maintenance lighter. The drawback is the necessity to have a lot of DC sources that are isolated from one another; as such sources, the three-phase diode rectifiers are used, which are connected to the secondary windings of the multiwinding transformer. In such a scheme, recuperation is impossible; in order to attain the recuperation, the active rectifiers with IGBT must be employed.

Nonsymmetrical inverters with different voltages of the bridges in the same phase give the possibility to increase the number of the attainable levels. If, for example,

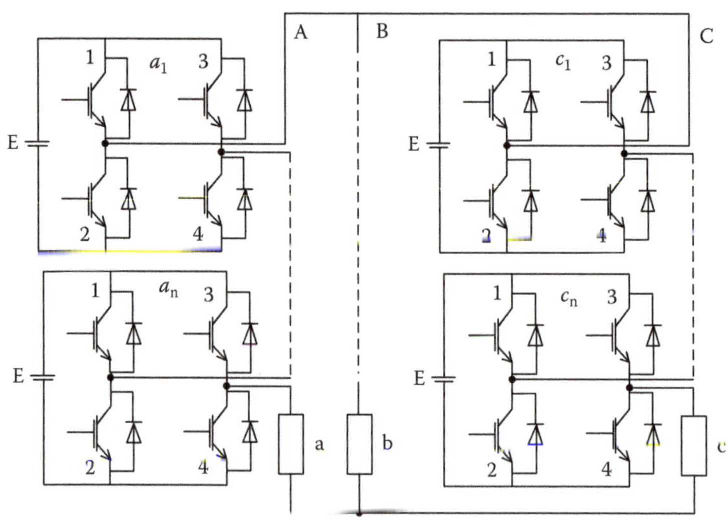

FIGURE 3.17 Cascaded H-bridge multilevel inverter.

DC in the bridge a1 is E, in the bridge a2 is 3 E, and in the bridge a3 is 9 E (in the common case, $3^k E$, $k = 0, 1, \ldots, n - 1$), then the phase voltage can change from −13 to 13 E with the step E, that is, 27 voltage levels exist (3^n in the common case) with the maximal value of $(3^n - 1)/2\ E$. But this unit does not have the modular design, and the bridges have to implement recuperation even in the two-quadrant converters.

If the output frequency changes within small bounds, the output transformers can be utilized, which have the primary windings with the same number of turns, whereas the numbers of the turns of the secondary windings relate as w^3. All the blocks are the same. Voltage levels of 3^n can be achieved in this system; any value of the maximal inverter voltage can be chosen by the turn number change. What is more, the single DC source can be used for all three phases.

In the converters under consideration, a comparison of the reference harmonic signal having the demanded frequency with the triangle carrier waveform that has the frequency F_c can be used for the fabrication of the control pulses for the inverter. Both unipolar and bipolar carrier waveforms can be utilized. At that, each bridge is controlled independently, and the carrier waveforms of the bridges are shifted one of the other by $2\pi/n$, in order to achieve the minimal THD value in the inverter phase-to-phase voltage. Pulse fabrication for these two cases is displayed in Figures 3.18 and 3.19. In the first case, the high harmonics are located around the frequency nF_c,

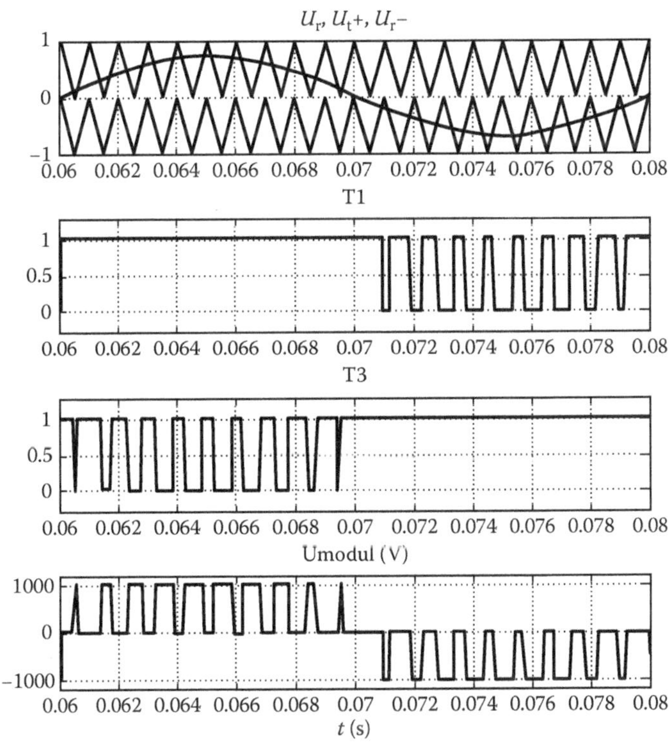

FIGURE 3.18 Gate pulse generation by unipolar PWM.

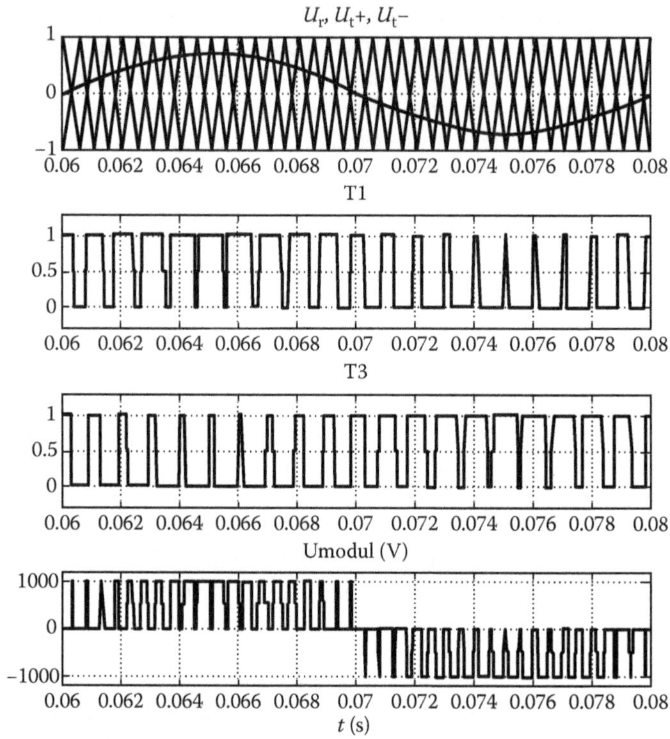

FIGURE 3.19 Gate pulse generation by bipolar PWM.

and in the second case, around the frequency $2nF_c$ that makes filtering easier, but the average frequency of switching increases twofold.

The demonstration model of the cascaded inverter with five blocks in a phase is given in SimPowerSystems as **Five-Cell Multi-Level Converter**. The inverter is supplied from the grid of 6.6 kV and supplies the load of $R = 10\ \Omega$, $L = 1$ mH in a phase. The inverter model consists of three identical subsystems-phases; each of them has five blocks according to the diagram in Figure 3.17 that are supplied from the phase-shifting transformers with the relative shift of 12°. The transformer secondary voltage is 1320 V so that $E \approx 1.35 \times 1320 = 1780$ V (simulation shows that this voltage is in fact ≈1700 V because of the voltage drops on the system components). The source of the input modulation signal defines the frequency of 60 Hz under the modulation factor 0.8 so that one can expect $U_1 = 2 \times 0.612 \times 0.8 \times 5 \times 1700 = 8323$ V. Every power block has the PWM system that carries out the bipolar modulation with $F_c = 600$ Hz. To apply the tool *Powergui/FFT* to the output voltage plot, it will be found that the rms of this voltage is 11,730/1.41 = 8320 V and that harmonics are around the frequency $2n \times 600 = 6$ kHz. It is worth mentioning that the simulation runs very slow even for the passive load: the ratio of the simulation time to the process time is about 500 even with the employment of the Accelerator mode.

In RES, the use of the cascaded inverters is most reasonable in the STATCOM, whose modeling is considered further in this chapter, or in photovoltaic systems, in which the corresponding parts of all the systems can be used as the isolated DC sources.

3.4.4 Simulation of Z-Source Converters

The output voltage of the RES can change to a considerable extent; therefore, in order to have the demanded voltage at the inverter output, the DC/DC boost converter is set in the inverter DC link. The inverter that has the intricate circuit of impedances in its DC link (Z-source converter—Figure 3.20) gives the possibility to achieve the output voltage that exceeds the input voltage very much and, in this way, to exclude the DC/DC boost converter.

The same eight inverter states are used as in the usual inverter. The difference is that the additional zero state is utilized when both devices in a phase (or in all phases) conduct. The shorting would take place in the usual inverter in such a case because the capacitor holds the constant voltage. In the considered scheme, however, the rate of the current rise in the conducting devices is limited by the inductors, and this current can be turned off with these semiconductors. When these devices are turned on, the current rises in the inductor under the influence of the capacitor voltages, the voltage across the reactors is increasing, the diode D cuts off, $U_L = U_c$ and $U_i = 0$; when the short-circuit gets open, $U_L = U_d - U_c$ and $U_i = 2U_c - U_d$. If the device turn-on time is T_0 and turn-off time is T_1, then, bearing in mind that the average voltage drop across the inductor during a period must be null,

$$U_c = \frac{T_1 U_d}{T_1 - T_0}. \tag{3.11}$$

The mean value of the inverter input voltage is

$$U_{iav} = \left(\frac{2T_1}{T_1 - T_0} - 1\right)\frac{U_d T_1}{T_1 + T_0} = \frac{U_d T_1}{T_1 + T_0} = U_c. \tag{3.12}$$

Because the short-circuit mode must not affect the inverter operation when the inverter active states are used, this mode is carried out when the zero states are used;

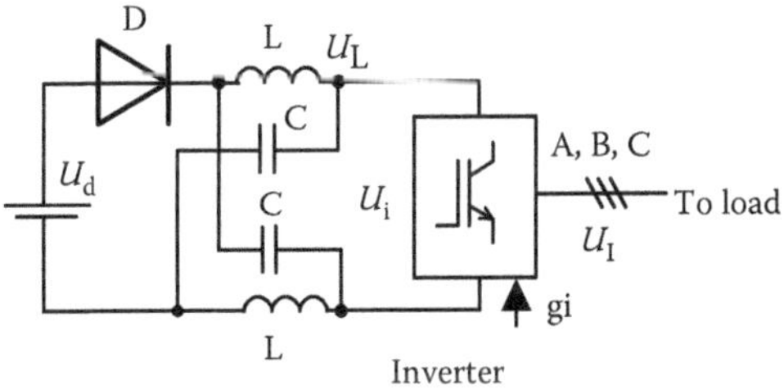

FIGURE 3.20 Diagram of the Z-source converters.

it means that the fact that during the interval T_0, $U_i = 0$ does not affect the output voltage. Then, if $T_1 = T - T_0$, where T is the period of PWM [7]:

$$U_{iekv} = 2U_c - U_d = \left(\frac{2T_1}{T_1 - T_0} - 1\right)U_d = \frac{U_d}{1 - 2t_0} = BU_d \tag{3.13}$$

$$t_0 = T_0/T,\ B = \frac{1}{1 - 2t_0} > 1 \tag{3.14}$$

Thus, with PWM with the triangle carrier waveform U_t

$$U_{1leff} = 0.612\ BMU_d \tag{3.15}$$

$0 < M < 1$ is the modulation factor.

There are different ways to realize this scheme. For instance, take $M_{max} < 1$ and choose $D \geq M_{max}$. When $|U_t| > D$, the pulses come to all devices making the short-circuit mode. Then

$$B = \frac{1}{1 - 2(1 - D)} = \frac{1}{2D - 1}. \tag{3.16}$$

The maximal inverter voltage is

$$U_{1leff} = \frac{0.612M_{max}U_d}{2M_{max} - 1}. \tag{3.17}$$

Let, for example, $U_d = 200$ V, and the wanted rms phase-to-phase output voltage is $U_{1leff} = 250$ V. The solution of Equation 3.17 gives $M_{max} = 0.66$. In the usual inverter, it was only $U_{1leff} = 122.4$ V under $M = 1$.

When this method to receive the short-circuit mode is utilized, the number of the switching-over of one device per second (turn-on–turn-off) is $N = 4F_c$, where F_c is the carrier waveform frequency. To carry out the short-circuit mode only with devices of one phase (for example, whose input signal is the greatest at the moment), the value N decreases to $N = 2.67F_c$.

The model **Z_inv** helps to understand a function of such an inverter. Each reactor has an inductance of 160 μH; the capacitance of the each capacitor is 1 mF. The load power is 50 kW under the voltage of 400 V. The gate pulses come to the devices either when the modulation signals cross the triangle carrier waveform that has the frequency $F_c = 10$ kHz or when the absolute value of the latter crosses the level D (subsystem **Control**).

The subsystem **Control1** is built more intricately. It has the subsystem **Max**, which with the help of the comparison blocks and the logic elements, fabricates the logic 1 at the output of the **AND** gate that corresponds to the modulation signal that has the maximum absolute value. The upper part of the scheme picks out the maximum of the positive signals, and the low part picks out the negative signal that has the maximal absolute value. Depending on the sign of the carrier waveform in the given moment, the signal that participates in the fabrication of the pulses in this moment is selected. Thus, the pulses that produce the short circuit are formed only in one inverter arm.

In the model, the block **Counter** counts the number of switching-over of the device +A during the time that is determined by the block **Timer1**. There is the unit to produce the third harmonic in the control system according to Equation 3.3.

Let us carry out the simulation with AC voltage of 220 V, under $M_{max} = D = 0.64$, without the third harmonic. With help of the scope **Scope** and the tool *Powergui/FFT Analysis*, the rms of the output voltage can be found as 387 V. Since DC voltage $U_d = 284$ V, from Equation 3.17, it follows that $U_{1leff} = 397$ V, which is sufficiently near to the real value. The voltage at the Z-scheme input fluctuates in the range of 200–1270 V. The voltage at the inverter input has a form of the rectangle oscillations 0–1050 V. The current and voltage of the reactors, voltages across the capacitors, U_d, and at the Z-scheme input are shown in Figure 3.21. At moment *A*, the inverter devices are shorted, the inductor current rises, the voltage across inductor increases

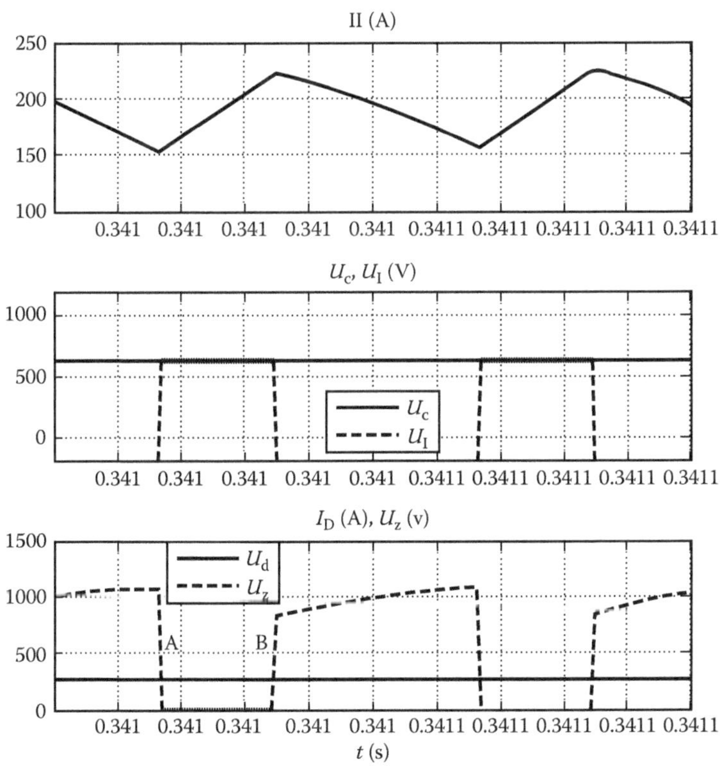

FIGURE 3.21 Current and voltage of the inductors, voltages across the capacitors, U_d, and at the Z-scheme input.

sharply, and the diode cuts off. At moment *B*, the voltage across the inductor begin to drop off and the diode restores the conduction. The voltage across the capacitors does not noticeably change and is 625–635 V.

The utilization of the third harmonic decreases the amplitude of the modulation signal by 1.155 at the same inverter voltage and, with this way, gives the possibility to decrease the *D* value; it means to increase the gain *B*. In order to receive the wanted voltage, the factor *M* is found from the equation

$$U_{\text{lleff}} = 0.612\frac{MU_{\text{d}}}{1.732M-1} \tag{3.18}$$

and $D = M/1.155$. For instance, in order to have $U_{\text{lleff}} = 390$ V under $U_{\text{d}} = 280$ V, it must be $M = 0.774$ and $D = 0.67$.

To carry out the simulation with the utilization of the third harmonic when $M = 0.775$ and $D = 0.67$, it can be found that $U_{\text{lleff}} = 387$ V under $U_{\text{d}} = 284$ V. The voltage across the capacitor is 540–550 V, and the maximum of the voltage at the inverter input is 900 V, which is less than without the utilization of the third harmonic.

As for the number of the switching over *N*, it can be found that in the case when all arms take part in the shorting, for the computation time of 0.1 s, $N = 4000/3820$, and when only one arm makes shorting, $N = 2670/2505$ (the numerator without the third harmonic, the dominator with its utilization), which is sufficiently near to the values given earlier.

Under the development of the output voltage controller, it is necessary to have in mind that the modulation factor must change under U_{d} changing as shown in Figure 3.22. If it is necessary to keep the output voltage U_{leff} (rms phase-to-phase)

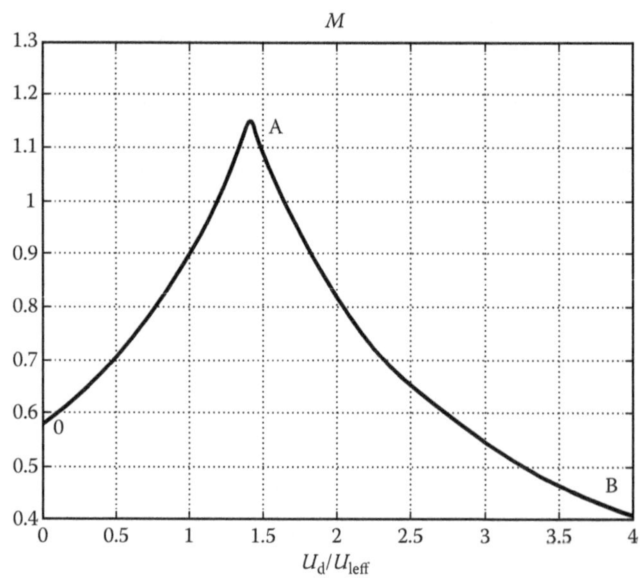

FIGURE 3.22 Dependence of the modulation factor on DC voltage under constant output voltage.

up when U_d rises, beginning from the small values, firstly, M has to increase until the value $M = 1.155$ is reached (sector 0A), and afterward, M has to decrease (sector AB). Moreover, there are two ways to affect the output voltage: M and D; at that, $D > M/1.155$ in sector 0A, $D > 1$ in sector AB.

It is possible to combine the feedforward and the feedback control. If U_d can be measured or estimated by some way, then the values M and D can be calculated by the formulas given earlier; since these computations are not very precise, the system is provided with the output voltage controller that affects the value D and operates in the close range $\pm a$. In order not to violate the condition $D > M/1.155$, the value M that was calculated from the value U_d decreases by a.

The corresponding block diagram is represented in Figure 3.23. The three-phase source that has the variable amplitude (and, perhaps, frequency) models, for instance, a synchronous generator with permanent magnets PMSG that is set in rotation either with WG or with the system for power taking off from the some main motion drive (for traffic or propulsion). It is supposed that the source output voltage can be measured or calculated. It is also possible to measure the voltage across the diode rectifier output at the times when the short-circuit mode is not used. If U_s is the rms of the source output voltage, then $U_d = 1.35U_s$. By this value, the values M and D are calculated.

The computed value D is corrected by the integral controller of the output voltage **I-reg**, whose output is limited by $\pm a$, and the value a is subtracted from the calculated value M. The gain of the D-controller changes as a function of M for its linearization.

If the voltage U_d when it increases gets more than $1.41U_{leff}$, the control system structure changes, $D = 1$, and the PWM input M is defined by the output of the PI voltage control (Figure 3.23).

The described system is simulated in the model **Z_invV**. The power circuits are taken from the previous model.

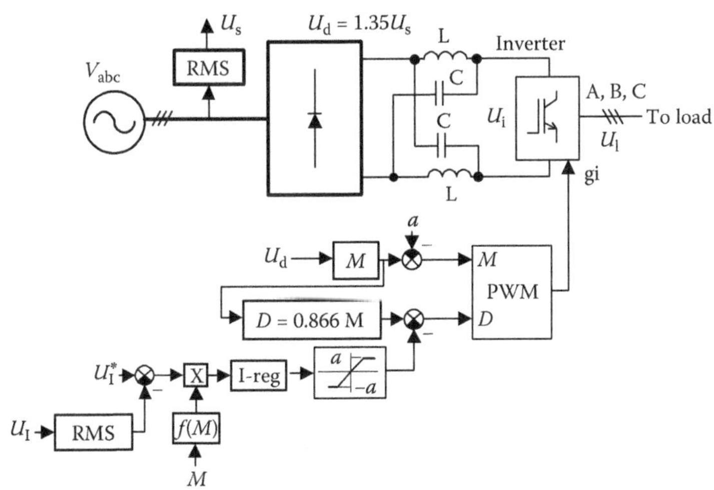

FIGURE 3.23 Z-source inverter voltage control (the first option).

The block **Control** fabricates signals for PWM. The block **Fcn** computes M that, after division by 1.155, gives D reference value. This value is adjusted by the controller that has limitations ±0.05. The value 0.06 is subtracted from the calculated value M. This new M value is multiplied by the three-phase signal, whose amplitude is equal to 1, and comes to PWM.

In order to become a more stable simulation, the sampling time is taken sufficiently small, and the deadband is set in the control error circuit.

When U_d reaches 550 V, the **Switch** switches in the upper position; at this, the output of the PI voltage regulator comes to PWM input.

To define the voltage amplitude as 480 V and its rate of decreasing as 0.05 nom/s (–24 V/s) in the block **Three-phase Programmable Voltage Source**, the process that is displayed in Figure 3.24 will be received. At the time interval [0.1–12 s], the voltage U_d changes from 640 to 250 V; at this, the load voltage is constant at 320 V (phase-to-ground amplitude). The value M is increasing firstly, and afterward decreasing; at this, the value D that was equal to 1 firstly begins to decrease too. The slight oscillations of the amplitude of the output voltage that can be seen in Figure 3.24

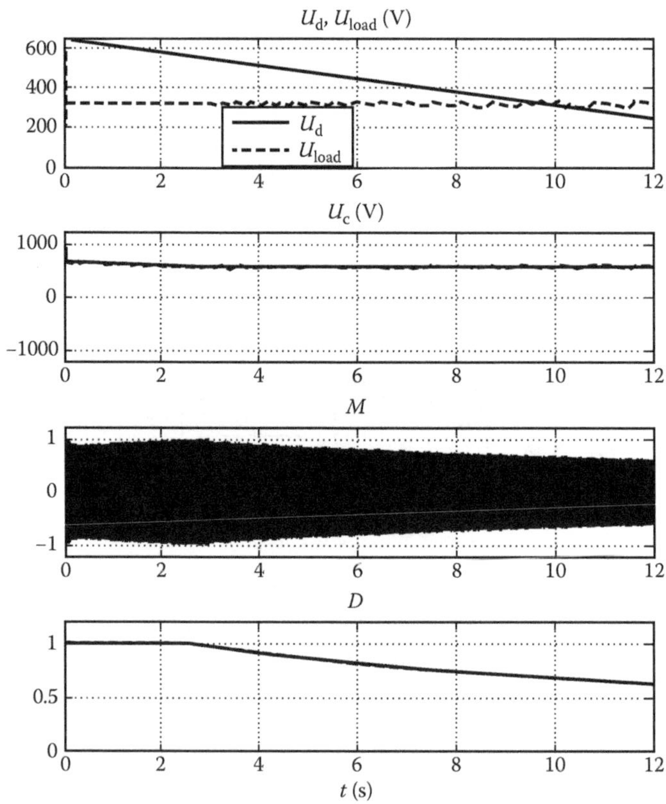

FIGURE 3.24 Variation of the quantities in Z-source inverter under voltage control (the first option).

are caused by the high speed of the source voltage changing accepted during simulation; they disappear in the steady state.

Together with the considered control system, the feedback system with two controllers can be developed; the first one controls the load voltage affecting M. The second controller affects D; it can be either the regulator that controls the inverter input maximum voltage U_i [8] or the capacitor voltage regulator in Z-circuit U_c [9]. In the first case, the quantity is controlled that is in direct proportion to the inverter output voltage under the given modulation factor and that is equivalent to the U_d voltage in the DC link of the usual inverter. However, it is difficult to measure, because this voltage fluctuates from 0 to U_{imax}; moreover, the pulsations of the rectified supply voltage tell on. The voltage U_c changes in the much closer bounds, but when it is constant, the voltage U_{imax} varies inversely as D. The second possibility is investigated in the model **Z_invV**.

The control system diagram is shown in Figure 3.25. The control of the shorting duration is carried out with **PI_regD**; the phase lead unit **Wk** is set at its input [8].

The output voltage control is implemented with **PI_regM**; its output is limited by the current value D with the factor $K = 1$ if the third harmonic is not in use, and $K = 1.155$, if it is used.

In order to get the considered system, the tumbler **Switch1** is set in the upper position. The reference capacitor voltage is 543 V. In Figure 3.26, the same process as in Figure 3.24 is fixed. The load voltage keeps constant. The voltage across the inverter input reaches 1050 V.

The control system with the measurement of the inverter input voltage U_i is considered in the model **Z_invV1**. The block **Sample & Hold** is used to fix the maximum U_i of the inverter input voltage. This voltage reaches its maximum just before the fabrication of the boost signal. This voltage comes to the block input; this value is stored at the moment when the carrier waveform crosses the D level. The block output, after the filter with the time constant of 1 ms, is the feedback signal and is used in the structure that is shown in Figure 3.25, instead of the signal U_c. The reference value for the voltage maximum is taken as 900 V.

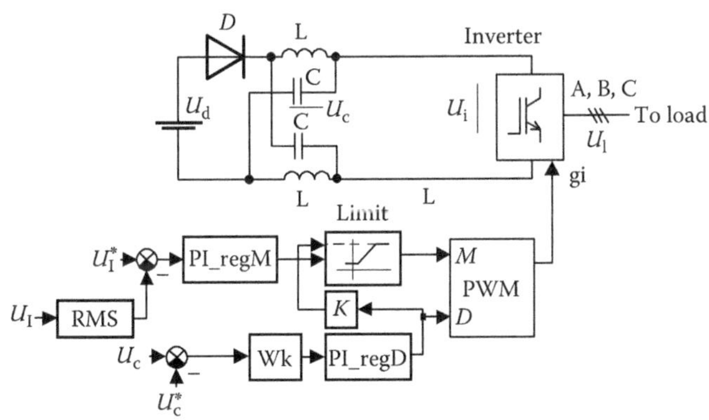

FIGURE 3.25 Z-source inverter voltage control (the second option).

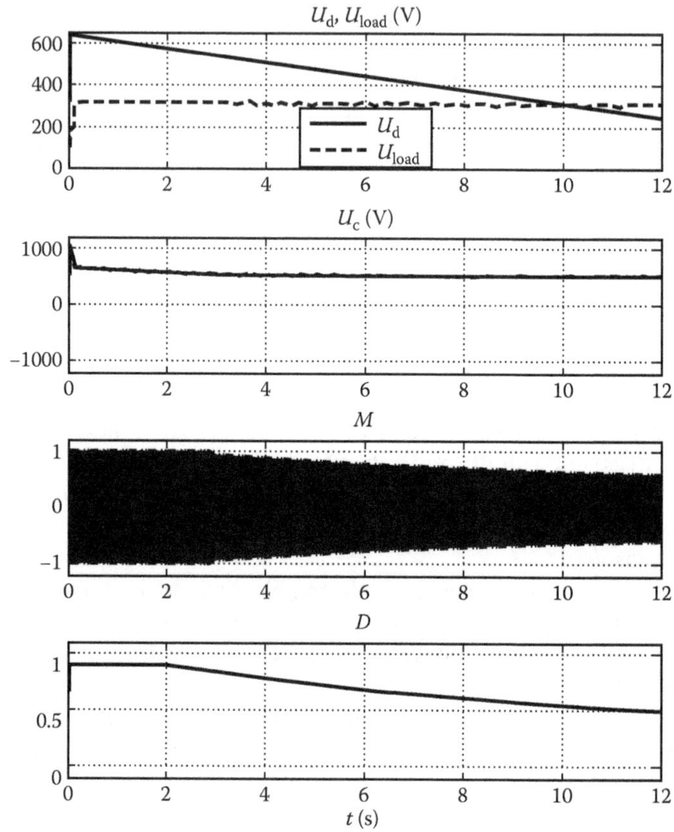

FIGURE 3.26 Processes in the model **Z_invV**.

The process of variation of the system quantities is shown in Figure 3.27. As for the previous model, the load voltage is constant in all ranges of DC voltage change. Since the U_i maximum is constant, the modulation factor M is constant as well and equal to $390/(0.612 \times 1.155 \times 900) = 0.61$.

The voltage across the capacitor decreases from 750 to 550 V as U_d drops.

3.4.5 Simulation of Modular Multilevel Converters

Modular multilevel converters (MMC or M2C) are relatively a new type of semiconductor converters. The diagram of such three-phase converters is depicted in Figure 3.28. It consists of three same arms (phases) that are connected in parallel and are supplied from the DC source V_d. Each phase contains $2n$ same blocks (modules); the middle points of the arms are the converter outputs. Every module consists of two IGBT with antiparallel diodes and the capacitor that is charged to the voltage V_c. The IGBT are controlled in antiphase. When T1 is turned on and T2 is turned off, the voltage between the points X and Y is V_c; when T2 is turned on and T1 is turned off,

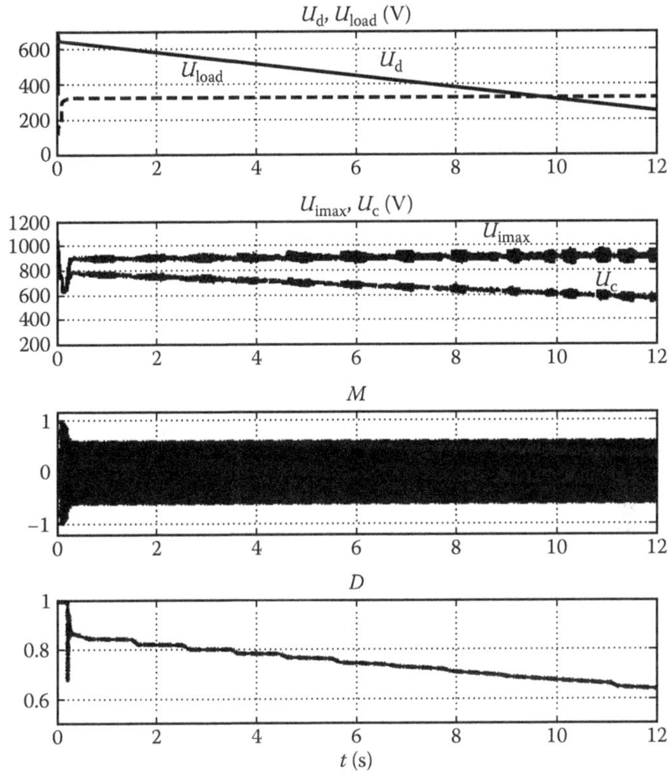

FIGURE 3.27 Processes in the model **Z_invV1**.

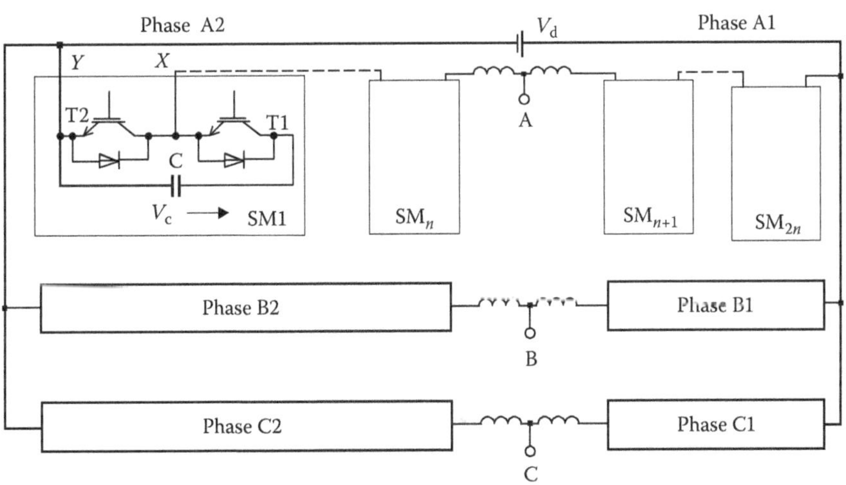

FIGURE 3.28 Modular multilevel converter.

this voltage is zero. Therefore, the voltage of a semiphase V_{a1} or V_{a2} changes from 0 to nV_c. The value V_c is taken as V_d/n; the modules are controlled in such a way that at every moment, n modules are in the active state. Therefore, the semiphase voltage changes from 0 to V_d. If sinusoidal modulation is used, the reference voltages for the right and left semiphases of the phase A are

$$V_{ref1,2} = V_d/2 \pm mV_d/2 \sin \omega t, \tag{3.19}$$

where $0 \leq m \leq 1$ is the modulation factor; for phases B and C, the reference voltages are shifted by 120° and 240°, respectively. With the third harmonic, the value m can be increased to 1.155. The phase-to-phase voltage is

$$V_{ab} = \frac{V_d}{2} + m\frac{V_d}{2}\sin \omega t - \left[\frac{V_d}{2} + m\frac{V_d}{2}\sin\left(\omega t - \frac{2\pi}{3}\right)\right] = 0.866 m V_d \sin\left(\omega t + \frac{\pi}{6}\right). \tag{3.20}$$

These converters consist of many same blocks that give an opportunity to build the high-voltage converter units, which often allows dispensing with the transformers for connection to the high-voltage grid. It is not difficult to set the additional blocks and, with this way, to increase the system reliability and survivability.

The large number of the modules makes it, practically, impossible to have PWM for each module. The modulation signal as in Equation 3.19 is usually used; after division by V_c, the result is rounded to the whole number, and in this way, the number of the modules k is found, which must be in the active state. Such an operation repeats every $T_{me} = 1/F_c$ s.

Because of the most part of the modulation signal period $k < n$, the existence of the redundancy gives the possibility to use it for voltage balancing across the module capacitors. It is carried out with the following way.

Firstly, all the capacitors have to be charged to the voltage V_c. It can be made with the help of the low-power source, having this voltage, which connects instead of the source V_d. The possible algorithm and corresponding model are considered in Ref. [2]. Afterward, if the phase current is positive, the capacitors of the modules that are in active state (their transistors T1 are turned on; thus, they take part in building of the phase voltage) increase their voltage, and if the phase current is negative, they decrease their voltage. Therefore, if the current is positive, k modules are set in the active state that have the least voltage; if this current is negative, k modules are set in the active state that have the largest voltage. Duty balancing of the modules takes place as well.

Such a control method is realized in the model **M2CDCR**.

The inverter is supplied with the DC 10 kV. Each inverter leg (semiphase) has 10 modules, which are charged beforehand to 1 kV. The inverter load is 4 Ω, 1 mH in each phase. The sampling frequency is $F_c = 1000$ Hz. The capacitance of the capacitors is 6 mF (it is specified in the option *Callbacks*); the inductance of the semiphase inductor is 2.5 mH. The starting charge is not modeled.

The block diagram of the choice of the active modules is displayed in Figure 3.29. The capacitor voltage vector $\underline{\mathbf{V}}_c$ and the value $\underline{k}$ are stored in the units for sampling and

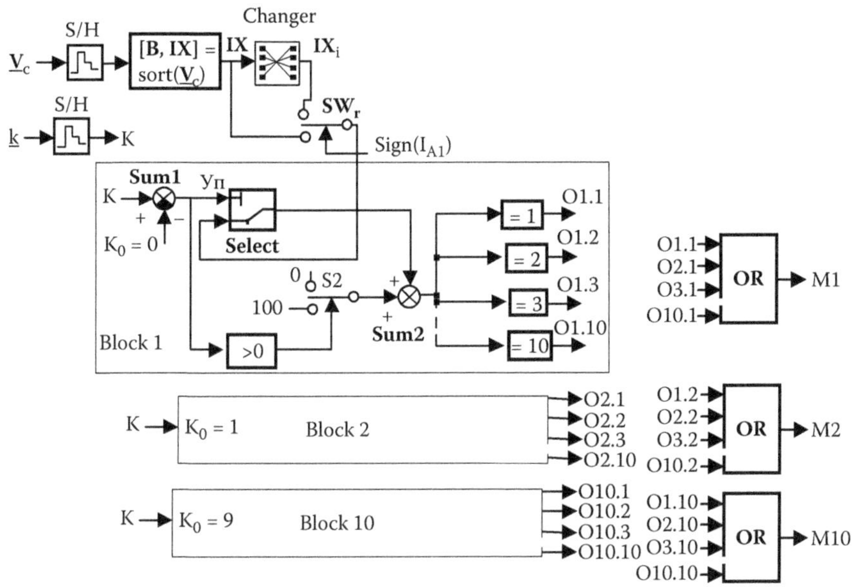

FIGURE 3.29 Control of the modular multilevel converter.

storage (**Sample&Hold**) during the time interval T_{me}. The subsystem **SORT** carries out the choice of the active modulus. The value k, the sign of the semiphase current, and the vector $\underline{\mathbf{V}}_c$ come to the subsystem inputs. The MATLAB function **[B, IX] = sort (V$_c$)** is utilized to order the vector components. This function builds the vector **B** that consists of the vector $\mathbf{V}_c$ components in increasing order and the vector **IX** that contains the component indexes as they follow in the vector **B**. If, for example, $\mathbf{V}_c$ = [1060, 900, 950, 980], then **IX** = [2 3 4 1]. The switch **SWr** changes the index order for the reverse one; either **IX** or **IX**$_i$ are used depending on the current sign in the semiphase.

The block **Select** selects, from the input vector components, one that corresponds to the whole number that is at the input CI. When CI = 1, the first component is chosen, when CI = 2, the second one, and so on. Let, for instance, $k = 4$ and the least capacitor voltages are in the modules 5, 3, 9, and 7. In block 1, CI = 4, so that the output **Select** is 7; in block 2, CI = 3, so that the output **Select** is 9; in block 3, CI = 2, and the output **Select** is 3, the output **Select** of block 4 is 5. Because the outputs **Select** for blocks 5–10 are < 1, the adder **Sum2** is set in order to prevent the failure: this adder sends, in this case, to the comparison elements some number that is ineffective for comparison, for instance, 100.

Therefore, the logic 1 will be at the outputs O1.7, O2.9, O3.3, and O4.5 of the blocks 1, 2, 3, and 4, respectively, and the logic 0 at the outputs of the other blocks. With the help of the elements **OR**, the resulting signals are fabricated that control the modules M1–M10. It follows from what was said earlier that the gate pulses will be sent to the modules 3, 5, 7, and 9 having the least capacitor voltages.

Under the simulation, the modulation signal of 4 kV decreases down to 3600 V at t = 0.5 c. Both the amplitude of the load voltage and the instantaneous values are shown in Figure 3.30. THD = 7.5%, but it is necessary to have in mind that for HVDC, the number of M2C modules can be a few dozens and more, so that the THD value will be much less. On the same figure, the capacitor voltages of some modules are shown; the voltage balancing is seen to be provided.

With certain control system complication, the less THD value can be received even with a small module number. The integral part k = floor (V_{ref}/V_c) and the fractional part k_{pwm} = rem(V_{ref}/V_c) are calculated; the value k determines how many modules must be activated in the next interval T_m. During interval $k_{pwm}T_{me}$, the module $k + 1$ is activated as well; it means that this module operates in PWM mode.

This method is employed in the model **M2CDCR1**. The values k and k_{pwm} are computed in the semiphases A1, B1, and C1. The system sampling time is $T_{me} = 1/F_c$ and $F_c = 1000$ Hz.

Figure 3.31 depicts how the module $k + 1$ realizes PWM. On the fabrication of the next pulse T_{me}, the integrator is reset and afterward begins to integrate the signal

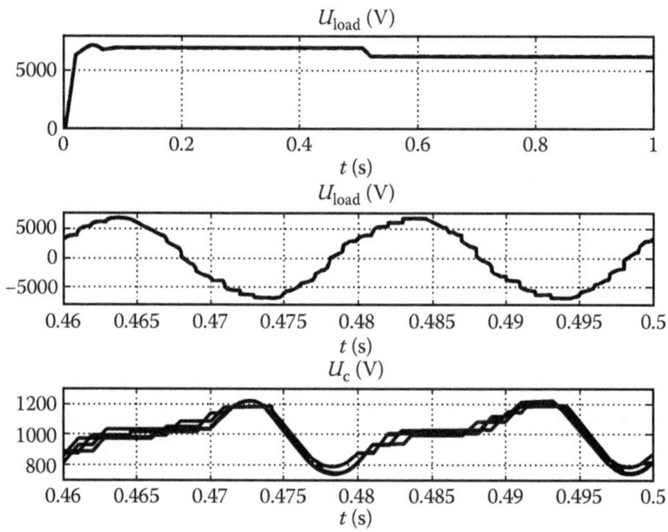

FIGURE 3.30 Load and capacitors voltages in the model **M2CDCR**.

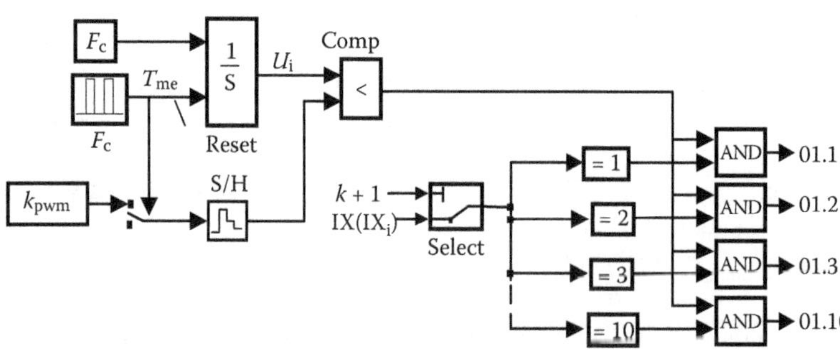

FIGURE 3.31 PWM for $k + 1$ module.

that is equal to F_c. Until the integrator output $U_i < k_{pwm}$, the output of the comparison block Comp = 1, and the module $k + 1$, is in the active state. When $U_i > k_{pwm}$, Comp = 0, the module is deactivated. For semiphase A2, the value k is a complement with respect to nine of the value k in semiphase A1, and PWM is carried out after PWM in semiphase A1 was over; in this way the condition is kept that the number of the active modules in a phase must be n always.

The load voltage controller is used in the considered model. Subsystem **Control** produces a three-phase harmonic signal with amplitude <1.15 and frequency of 50 Hz with the addition of the third harmonic. This signal is modulated in amplitude by the voltage regulator output. This reference signal is multiplied by the base phase voltage of 4 kV.

The block **Step** assigns the phase-to-phase voltage amplitude as 6 kV, with increasing to 6.9 kV at $t = 0.5$ s. Some simulation results are given in Figure 3.32. The load voltage has slightly less THD in comparison with the previous model, and the great thing is that the harmonics are shifted to the higher frequencies that make their filtration easier. The capacitor voltage oscillations cause the second harmonic.

In HVDC, the inverter load is a step-up transformer, with the aid of which the electric energy is transferred to the network. Such a system is investigated in the model **M2CAC**. The 4.8/35 kV step-up transformer is used. Of course, the transformers with the higher voltages are utilized in practice, but in order to simulate such a system, the module number has to be increased, which would make duration of simulation impossibly long.

This model is a revision of the model **M2CDCR**. Because it follows from Equations 3.19 and 3.20 that the phase-to-phase voltage surpasses the modulation signal by $\sqrt{3}$, in the case that it is 4.8/1.732 × 1.414 = 3.92 кВ ≈ 4 кВ. The transfer of

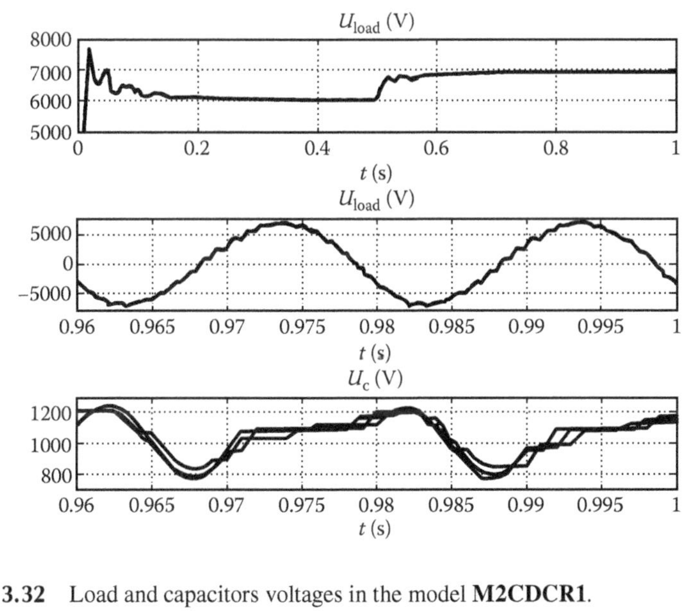

FIGURE 3.32 Load and capacitors voltages in the model **M2CDCR1**.

the active power through reactive impedance (through the transformer in the case) is carried out by the phase difference of the space voltage vectors at the ends of this reactance, 4 kV in our case, by the shift of the phase of the inverter voltage space vector relative to the network space voltage vector. Such a shift β is fulfilled in the subsystem **Controller** that contains the block PLL and the additional elements, including ones for building of the third harmonic. The angle β can be defined and changed with the block **Step** or with the power controller.

The transients under β changing by 4° are fixed in Figure 3.33. The power transferred in the network decreases, and the current is nearly sinusoidal (THD < 5%). The oscillations of the capacitor voltages decrease when the power diminishes. It is interesting to note that, with the same β, the value and even the direction of the transferred power depend on the capacitance of the capacitors. So, to increase the capacitances from 6 to 9 mF, the power will transfer from the network to the DC circuit, and in order to reach the former power value, $\beta_0 \approx 15°$ has to be set.

This fact makes it reasonable to use the power controller; for that, the tumbler **Switch** is set in the low position. The process under power reference changing from 12 to 10 MW at $t = 2$ s is shown in Figure 3.34.

The considered models operate in the inverter mode, converting DC voltage to AC voltage. In HVDC, one converter operates as a rectifier. The rectifier is considered in the model **M2CACR**. The converter is supplied from the transformer with the secondary voltage of 6 kV, and the load resistance is 10 Ω. The model gives an opportunity to investigate the influence of various parameters (the AC voltage, the

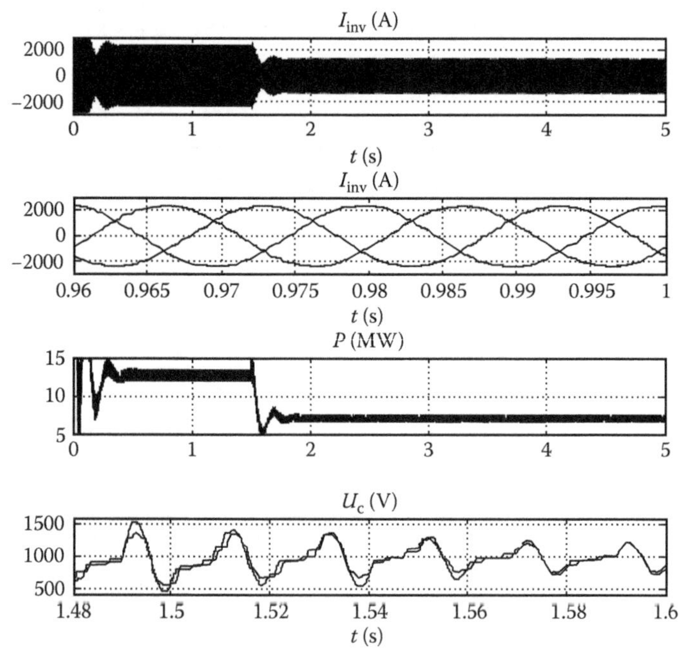

FIGURE 3.33 Processes in the model **M2CAC** under β changing.

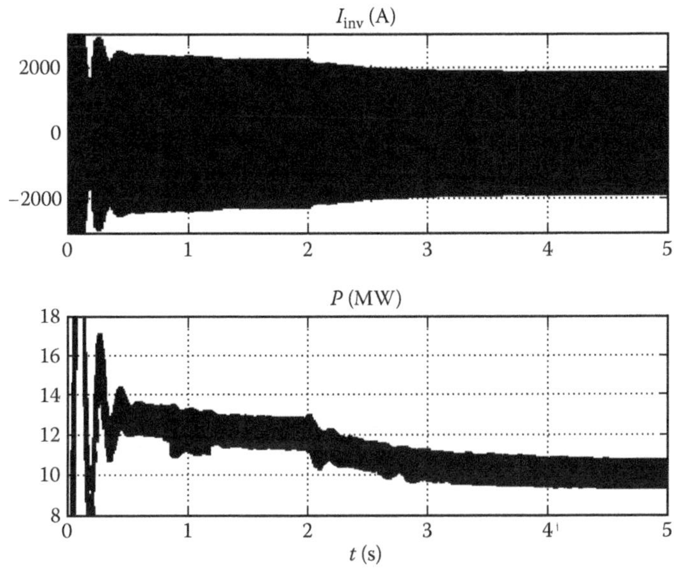

FIGURE 3.34 Processes in the model **M2CAC** with power controller.

direct component *KUd*, the variable component *m* of the modulation signal, the module capacitances C_v, and the rated value of the capacitor voltages V_{cn} that is taken into account under the number of the active modules *k* computation) on the DC output voltage, the consumption of the active and reactive power from AC source, and the capacitor voltage oscillations. So, for example, under $C_v = 6$ mF, $KUd = 4.5$ kV, $m = 4$ kV, and $V_{cn} = 1200$ V, the output voltage is 11.3 kV; the active and reactive network powers are 13 MW and 3.5 Mvar, respectively; the capacitor voltages change in the range of 1.3–1.7 kV; under $KUd = 5$ kV and $m = 4.5$ kV, the output voltage increases to 13.7 kV under the active network power of 18 MW, but the reactive power increases to 27 Mvar, and the capacitor voltages change in the range of 1–2 kV.

The DC power transmission is modeled in the model **M2CHVDC**. The two models described earlier are utilized for the rectifier and the inverter. The transformer voltages are 35/6 and 4.8/35 kV. The power transfer is controlled by the action on the angles β. When the tumblers **Switch** in the models of the **Rectifier** and the **Inverter** are in the low positions, the rectifier power controller and the DC voltage controller are turned on. The following mode is simulated: the initial values of the power and the voltage are 20 MW and 11.2 kV; power is decreasing to 15 MW at $t = 1.5$ s; and voltage is decreasing to 10 kV at $t = 3$ s. The reference and measured values of the power and the voltage, and also the capacitor voltages of one rectifier and one inverter module are shown in Figure 3.35. The values of the transferred power and the voltage are seen to be controlled independently; only during a short interval after the reference change that the mutual influence can be seen.

The curves of the currents in the transformer low-voltage windings of the inverter and the rectifier are displayed in Figure 3.36 for the condition 15 MW, 10 kV.

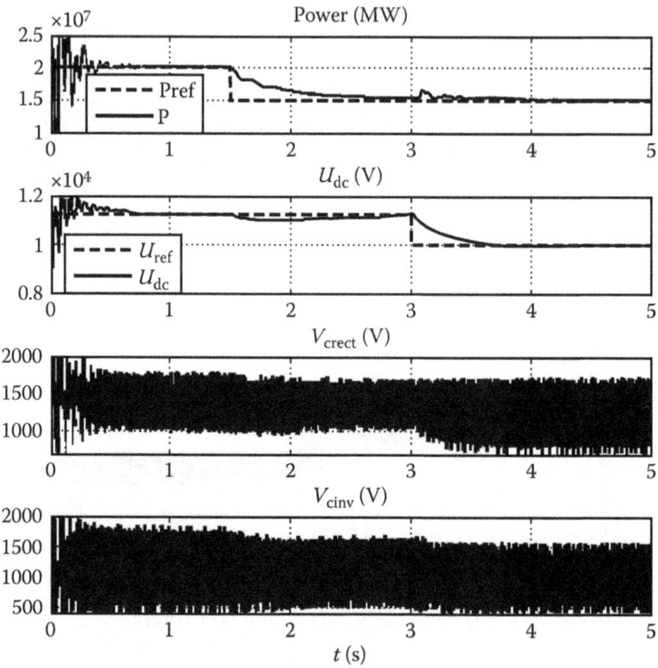

FIGURE 3.35 Reference and measured values of the power and voltage and the capacitor voltages in HVDC.

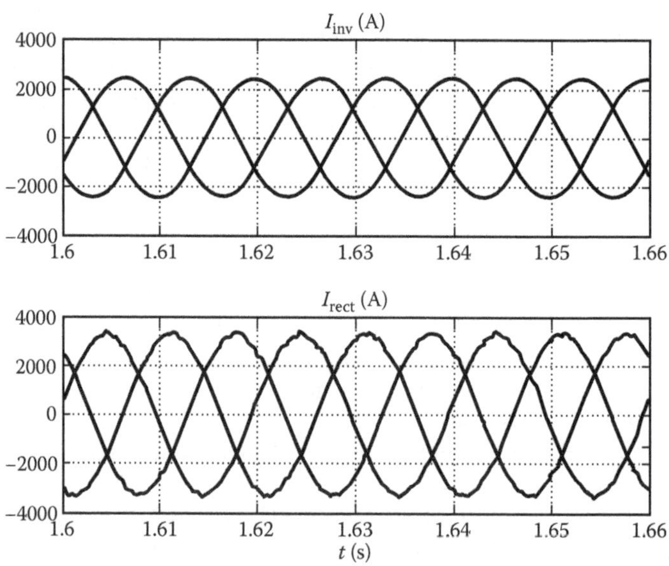

FIGURE 3.36 Currents in the transformer low-voltage windings of the inverter and the rectifier in HVDC.

The contents of the higher harmonics are not great: THD = 2.1% and 1.5% for the rectifier and inverter respectively.

In conclusion, it can be noted that the creation of the perfect control system for the converters with the module inverters is complicated and, up to present, still not completely a decided task, despite a lot of published articles. The readers, taking an interest in this problem, can develop and investigate their own control systems, using the given models.

3.4.6 Simulation of Five-Level Inverter 5L-HNPC

The diagram of one phase of a 5-level H-inverter is displayed in Figure 3.37. It contains two of three arms of the three-level inverter (Figure 3.2). The admissible device states, the output voltages fabricated under these states, and the active capacitors are given in Table 3.2 that is taken from Ref. [2].

Each phase provides five voltage levels: 0, $\pm V_d/2$, $\pm V_d$. To connect three such units to the three-phase star load, the phase-to-phase voltage will have nine levels: 0, $\pm V_d/2$, $\pm V_d$, $\pm 3V_d/2$, $\pm 2V_d$. In this unit, every device has to withstand the voltage $V_d/2$, under the output voltage maximum $2V_d$, whereas in the three-level inverter, the output voltage maximum is only V_d. The existence of the states that produce the same voltage but utilize different capacitors can be used for voltage balancing across them.

Three converter phases must be supplied from three isolated DC sources, or the AC three-phase winding must have the isolated phases. The first option is used in the electrical drives of a great power. The second option is employed in WG [10].

From Table 3.2, it follows that the states of the devices 1 and 3, and also 2 and 4, are opposite; thus, it is sufficient to control only four devices, for example, *a*1, *a*2, *b*1, and *b*2. The comparison of the modulation signals U_{ref+} and the antiphase U_{ref-} with

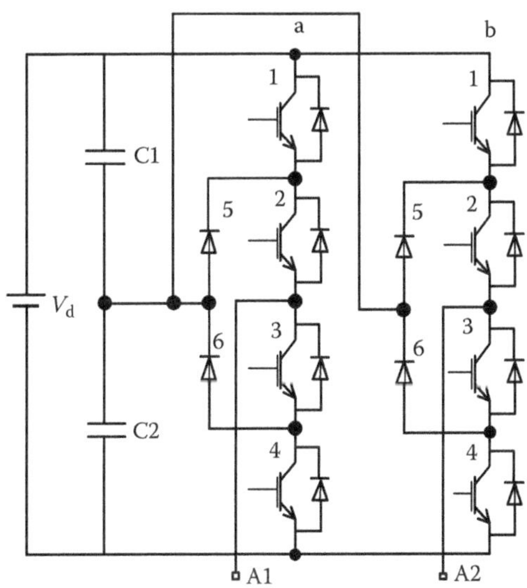

FIGURE 3.37 A phase diagram of **5L-HNPC** inverter.

TABLE 3.2
Voltages of a Phase of 5L-HNPC Inverter

1a	2a	3a	4a	1b	2b	3b	4b	V_{out}	C
0	0	1	1	0	0	1	1	0	–
0	0	1	1	0	1	1	0	$-V_d/2$	C_2
0	0	1	1	1	1	0	0	$-V_d$	$C_1 + C$
0	1	1	0	1	1	0	0	$-V_d/2$	C_1
0	1	1	0	0	1	1	0	0	–
0	1	1	0	0	0	1	1	$V_d/2$	C_2
1	1	0	0	0	0	1	1	V_d	$C_1 + C$
1	1	0	0	0	1	1	0	$V_d/2$	C_1
1	1	0	0	1	1	0	0	0	–

two in-phase triangle carrier waveforms is carried out in PWM; they have the frequency F_c and the change in the ranges of 0–1 and −1–0, respectively. The gate pulse fabrication is shown in Figures 3.38 and 3.39. The average switching frequency of one device is $F_c/2$, and the higher harmonics are around $2F_c$ and around multiple values. However, such a control is possible when each capacitor is supplied from the DC

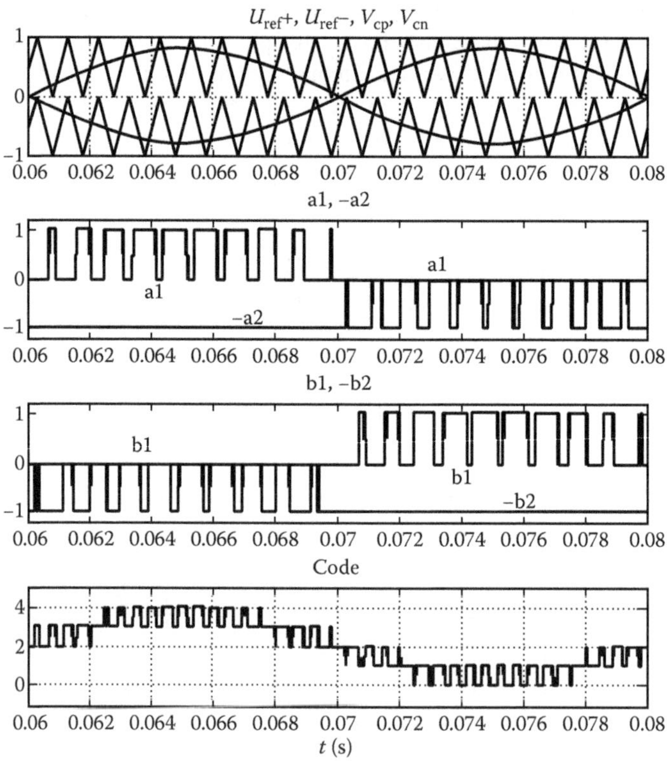

FIGURE 3.38 Pulse fabrication in **5L-HNPC** inverter.

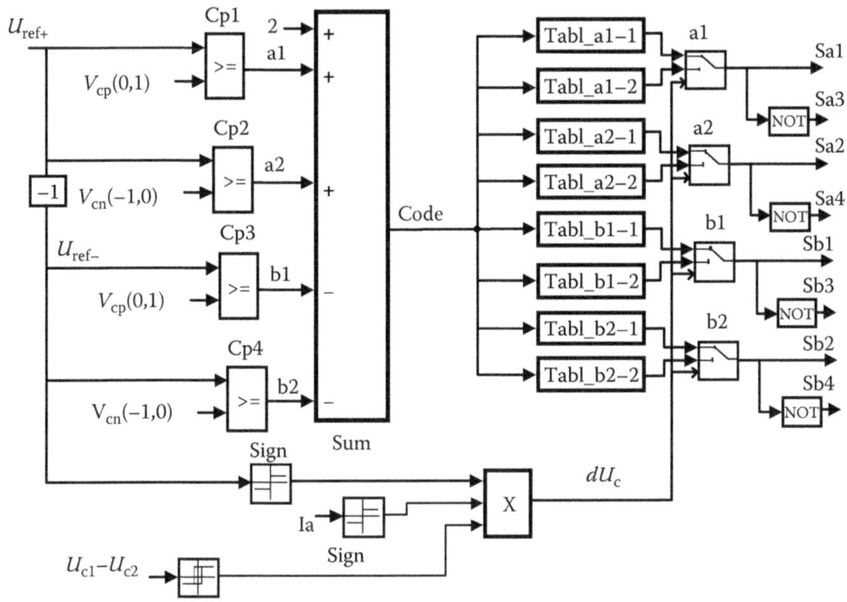

FIGURE 3.39 Diagram of **5L-HNPC** inverter pulse fabrication.

source with the constant voltage. Otherwise, the control system must provide voltage balancing, for instance, with the utilization of the circuits shown in Figure 3.39. The output signals of the comparison elements are summed by the adder Sum; its output is a five-level signal *Code* (Figure 3.38); the levels of this signal correspond to the demanded levels of the phase voltage (in Figure 3.39, the signal *Code* is shifted by two units up, so that it changes from 0 to 4). The tables are used to convert the signal *Code* in the gate pulses; one of two tables can be utilized for each device depending on the capacitor voltage difference dU_c. For example, from Table 3.2, it follows that for the control of the device $a1$, the output of the tables must be [0 0 0 11] or [0 0 0 0 1] under *Code* changing from 0 to 4.

The signal dU_c is formed depending on the sign $U_{c1}-U_{c2}$ and on the sign of the load power; the latter is equal to the logic product of the signs of the voltage and the current. Instead of the output voltage, the modulation signal can be used. If, for example, $U_{c1} > U_{c2}$ and the power sign is plus, the pulse sequence has to be utilized that loads the capacitor C_1, but if the power sign is minus, the gate sequence must be used, under which the capacitor C_2 is active, because the voltage across it will increase (the instant power transfers in the network), and the voltage across C_1 will decrease.

It is worth noting that the states that correspond to the voltages $\pm V_d/2$ determine the switching frequency of the devices, because the output voltage fabrication occurs by switching-over either between levels 0 and $\pm V_d/2$ or between $\pm V_d/2$ and $\pm V_d$. In the PWM considered earlier, the intervals when each of the four possible states that correspond to $\pm V_d/2$ is realized and are nearly equal to one another that provides the same switching frequency of each device. When capacitor voltage balancing is used, the difference of the switching frequencies is possible. Because for the first variant, the capacitor voltage difference rises rather slowly, it is possible to suggest the method when the best part of the time the first PWM variant is utilized, and when the capacitor voltage difference becomes rather large, the second variant is used for the short time.

The described system is realized in model **HNPC_5L**. The inverter converts DC voltage of 8 kV to AC voltage and sends the electric power in the network. The fabrication of the DC voltage, which is common for all phases, is not modeled. The inverter power is 22 MW. The three-phase transformer with the separated secondary windings is used; the transformer power is 30 MVA; the phase voltages are 5.3 kV/35 kV; the high-voltage output windings are connected in the star.

The model of each inverter phase (**5L_PhaseA**, **5L_PhaseB**, **5L_PhaseC**) contains the inverter semiconductor devices and the capacitors according to the diagram in Figure 3.37, and also PWM according to the diagram in Figure 3.39 (the subsystem **Phase_Control** that consists of some subsystems in turn). The subsystem **Compare** contains the generators of two triangle carrier waveforms, the blocks for their comparison with the modulation signals, and also the adder. The subsystem **Tables** converts the signal *Code* in the gate signals for the devices a1, a2, b1, b2, depending on the signal dU_c. The subsystem **Pulses1** takes four signals for device control either from subsystem **Compare** or from the system **Tables** and converts them in the direct and inverse gate pulses. Moreover, the subsystem **Phase_Control** contains the circuits for producing the signal dU_c according to Figure 3.39 and for a choice of PWM variant. The switch **Switch2** and the tumbler **Switch1** are intended for that. If the latter is in the low position and has the constant **Contr_Mode** = 1, the first PWM variant works, and when this constant is zero, the second variant, with balancing capacitor voltages, is active. When the tumbler is in the top position, the combined system is realized: when the capacitor voltage difference, by absolute value, is getting more than the reference in the block **Relay1**, the system that has used the first PWM variant changes for the second variant, until this difference will not become less than the value at which **Relay1** turns off.

The subsystem **Count** (only for phase *A*) counts the switching frequencies of the devices a1, a2, b1, and b2 and the average value and shows them on the displays. The measurement duration (taken as 0.25 s) and the measurement starting time (set as 0.75 s) are defined by the block **Timer**.

Two options for control system are provided in the subsystem **Control**; they are selected with the tumbler **Switch**. In the first option, the harmonic signal (perhaps, with the units for addition of the third harmonic) that has the assigned amplitude and frequency comes to PWM input; in the second option, the output of the control system of the currents I_d and I_q that are aligned in the direction of the network voltage space vector and perpendicular to it is sent to PWM input.

Let us carry out the simulation under following adjustment: **Switch** in the subsystem **Control** is in the top position; the amplitude is 1; the frequency is 50 Hz; the phase is 30° in the block **Three-phase Programmable Source**; **Gain1** = 0 (in order to eliminate the third harmonic); F_c = 1000 Hz (the carrier frequency) in the option *File/Model Properties/Callbacks*; the tumbler **Switch1** is in the low position; and the constant **Contr_Mode** = 1 in the subsystems **Phase_Control**. After the simulation with the duration of 1 s, it is seen on the displays that the device switching frequencies

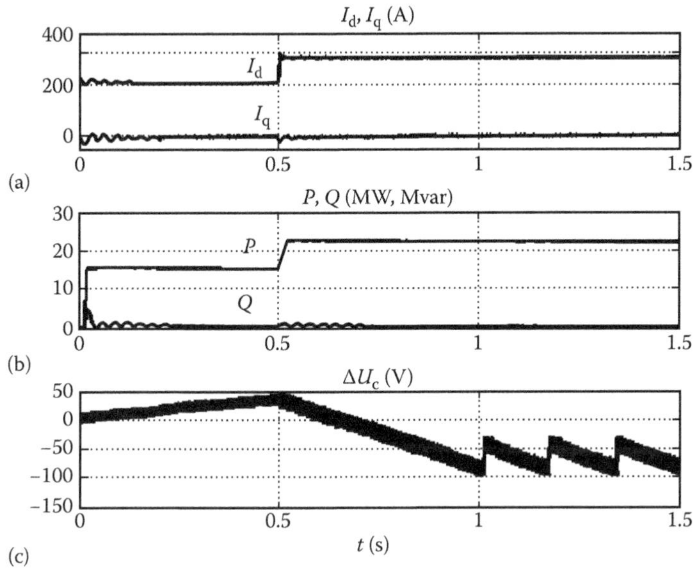

FIGURE 3.40 (a) Inverter currents I_d and I_q, (b) network power, and (c) capacitor voltage difference for the model 5L-HNPC.

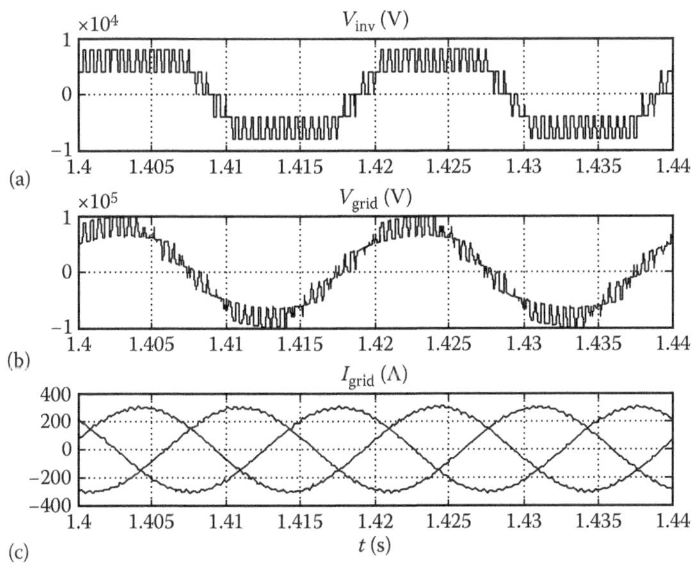

FIGURE 3.41 (a) Inverter phase voltage, (b) network phase-to-phase voltage, and (c) current.

are almost the same (486–514 Hz), and the average frequency is 500 Hz. The capacitor voltage difference is increasing continuously and, for 1 s, reaches 170 V. To repeat the simulation with **Contr_Mode** = 0, one can see that the switching frequencies of the devices noticeably differ (394–618 Hz) under the average frequency of 505 Hz, but the capacitor voltage difference is not more than 100 V. To repeat simulation once more, with the tumbler **Switch1** in the top position, it will be seen that the average switching frequency is 510 Hz and the device switching frequencies do not differ essentially (498–522 Hz); at this, the voltage difference changes in the range of 40–100 V.

The plots of the network power, the inverter currents I_d and I_q, and the capacitor voltage difference are depicted in Figure 3.40 when the tumbler **Switch** in the subsystem **Control** is in the low position, under I_d reference changing from 200 to 300 A at $t = 0.5$ s, and I_q reference is zero; the third harmonic is included.

The inverter phase voltage, the phase-to-phase voltage, and the network current are shown in Figure 3.41. It can be found, with the help of the option *FFT Analysis*, that the higher harmonics are located around the frequency of 2000 Hz and around multiples of it; THD = 2.3%.

3.4.7 Matrix Converter Simulation

The matrix converters convert AC voltage having changeable amplitude and frequency in AC voltage with the other values of amplitude and frequency without the intermediate conversion to DC voltage. Such a converter consists of nine bidirectional switches, which give a possibility to connect each input phase (the grid phase) to any load phase (Figure 3.42). In most cases, such switches are formed with two devices IGBT/Diode connected in opposition, with the common collector or emitter.

When the switch states are selected, two things must be taken into account: a load phase cannot be connected to two different grid phases simultaneously, because these phases will form a short circuit; the load phase cannot be remained unconnected, because owing to its inductance, the overvoltage will have a place that can destroy the semiconductor switches. Thus, 27 matrix converter (MC) states exist. Six of them occur when each load phase is connected to the different input phases; for instance, the switches S11, S22, S33 or S13, S22, S31 are closed. At this, the full not-regulated grid voltage comes to the load. These states are not employed for control. Three states take place when three load phases are connected to the same grid phase (the switches S11, S21, S31 or the switches S12, S22, S32, or the switches S12, S23, S33

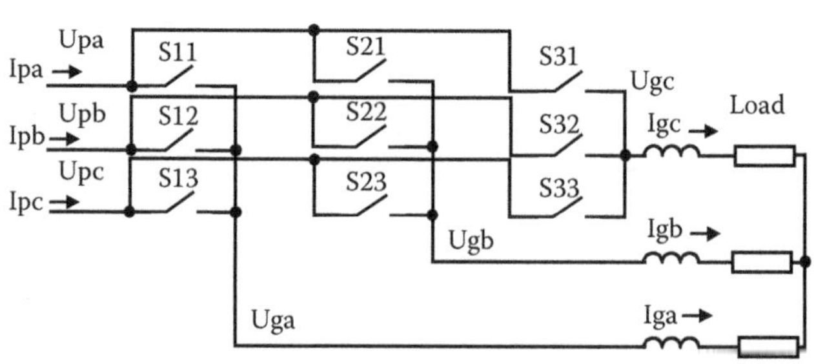

FIGURE 3.42 MC diagram.

TABLE 3.3
MC States

V_k	Turned On	Phase	I	Phase	V_k	Turned On	Phase	I	Phase
				xy = ab					
+1	11, 22, 32	0	I_a	$-\pi/6$	−4	11, 22, 31	$-\pi/3$	$-I_b$	$5\pi/6$
−1	12, 21, 31	0	I_a	$5\pi/6$	+7	12, 22, 31	$4\pi/3$	I_c	$-\pi/6$
+4	12, 21, 32	$2\pi/3$	I_b	$-\pi/6$	−7	11, 21, 32	$\pi/3$	$-I_c$	$5\pi/6$
				xy = bc					
V_k	Turned On	Phase	I	Phase	V_k	Turned On	Phase	I	Phase
+2	12, 23, 33	0	I_a	$\pi/2$	−5	12, 23, 32	$-\pi/3$	$-I_b$	$-\pi/2$
−2	13, 22, 32	0	$-I_a$	$-\pi/2$	+8	13, 23, 32	$4\pi/3$	I_c	$\pi/2$
+5	13, 22, 33	$2\pi/3$	I_b	$\pi/2$	−8	12, 22, 33	$\pi/3$	$-I_c$	$-\pi/2$
				xy = ca					
V_k	Turned On	Phase	I	Phase	V_k	Turned On	Phase	I	Phase
+3	13, 21, 31	0	I_a	$7\pi/2$	−6	13, 21, 33	$\pi/6$	$-I_b$	$-\pi/2$
−3	11, 23, 33	0	$-I_a$	$\pi/6$	+9	11, 21, 33	$4\pi/3$	I_c	$7\pi/6$
+6	11, 23, 31	$2\pi/3$	I_b	$7\pi/6$	−9	13, 23, 31	$\pi/3$	$-I_c$	$\pi/6$

are closed). At this, the load voltage is zero. In the rest of the 18 cases, one load phase is connected to some grid phase, and the two other load phases are connected to another grid phase. The voltages and the currents, and also the switch positions that provide them, are given in Table 3.3; see Figure 3.43 as well. At this, the amplitude of the MC voltage space vector $\mathbf{U}_g$ is proportional to the running phase-to-phase grid voltage: $U_g = 2/3U_{xy}$, $xy = ab$, bc, or ca, depending on the combination of the switches, as follows from Table 3.3. The phase of $\mathbf{U}_g$ depends on the combination of the switches as well.

Let, for example, the switches S13, S23, and S32 be on. Then the windings of phases A and B are connected in parallel and in series with the winding C. The voltage $-U_{bc}$ is applied between phases A (and B) and C. Then the phase voltages

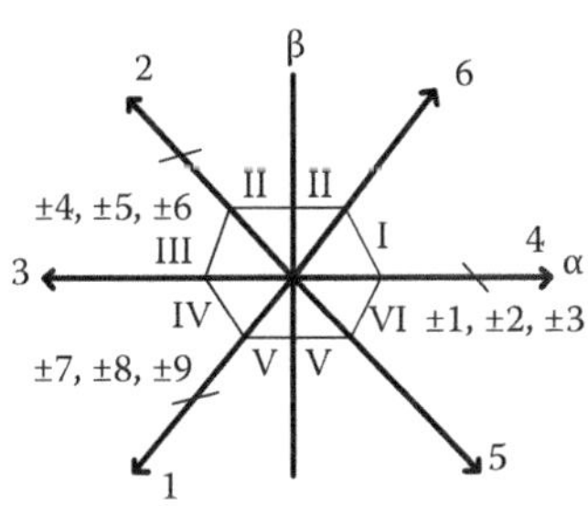

FIGURE 3.43 MC voltage space vectors.

are $U_a = U_b = -U_{bc}/3$, $U_c = 2U_{bc}/3$, $U_\alpha = -U_{bc}/3$, $U_\beta = (-U_{bc}/3 - 2U_{bc}/3)/\sqrt{3} = -U_{bc}/\sqrt{3}$, $U_g = 2/3U_{bc}$, and $\angle U_g = 4\pi/3$. Let the switches S12, S22, and S33 be on. Then, as before, the windings of phases *A* and *B* are connected in parallel and in series with the winding C, and the voltage U_{bc} is applied between phases *A* (and *B*) and *C*. Then the phase voltages are $U_a = U_b = U_{bc}/3$, $U_c = -2U_{bc}/3$, $U_\alpha = U_{bc}/3$, $U_\beta = (U_{bc}/3 + 2U_{bc}/3)/\sqrt{3} = U_{bc}/\sqrt{3}$, $U_g = 2/3U_{bc}$, and $\angle U_g = \pi/3$.

The amplitude of the space vector input current **I*** is proportional to the running value of the load phase current: $I^* = 2/\sqrt{3}\ I_x$, $x = a, b$, or c depending on the switching combination, and its phase depends on the switching combination too. For example, for the first switch combination mentioned earlier, $I_b^* = I_c$, $I_c^* = -I_c$, $I_a^* = 0$, so that $I_\alpha^* = 0$, $I_\beta^* = 2/\sqrt{3}I_c$, $\angle U_g = \pi/2$. For the second combination, $I_b^* = -I_c$, $I_c^* = I_c$, $I_a^* = 0$, so that $I_\alpha^* = 0$, $I_\beta^* = -2/\sqrt{3}I_c$, $\angle U_g = -\pi/2$.

The MC voltage space vectors resemble the inverter voltage space vectors, but the same vector direction can be received with different switch combinations, and the vector amplitude and even its direction can change for the opposite one.

The similarity of the voltage space vectors of a VSI and an MC gives an opportunity to use, for the MC state choice, the same methods, and as for the VSI, with necessary alterations. The current state of the grid voltage space vector must be taken into account. The three-phase grid voltage can be divided into six sectors as depicted in Figure 3.44. The MC states, under which the input phase-to-phase voltages have the constant sign in the sector and are maximal by the amplitude, have to be chosen in every sector.

For example, let vector 6 be realized (Figure 3.43), and the grid voltage is in the second sector (Figure 3.44). The states ±7, ±8, and ±9 give the vectors that are directed in the direction of vector 6. The state ±7 corresponds to the voltage U_{pab} that changes the sign in the second sector; the states +8 and −9 give the negative value (taking into account the input voltage polarity and the space vector phase; it means that the corresponding vectors are directed along vector 1), and only the feasible states −8 and +9 remain. At this, the current phase is $-\pi/2$ for the state −8 and $7\pi/6$ for the state +9. When these states are chosen by turns, the average current phase is $4\pi/3$; therefore, the current vector is aligned with the voltage vector.

Furthermore, for control MC voltage, the SVM method is used [2,11].

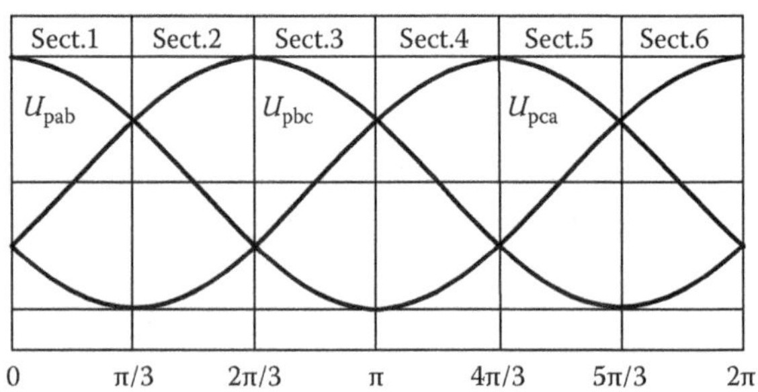

FIGURE 3.44 Grid voltage sectors.

Unlike VSI, not two but four active MC states are applied for suitable time intervals within each period T_c; the zero state is used to complete T_c. These states are listed in Table 3.4 for different sectors of the input U_p and output U_g voltages.

Let us designate the angle that determines the vector $\mathbf{U_g}$ position in the sector as $\Delta\theta$; it is counted off from the sector center, $-\pi/6 < \Delta\theta < \pi/6$. To designate the angle that is counted off from the center of the voltage U_p as $\Delta\alpha$ (Figure 3.44), for the implementation of the relative intervals of the states given in the Table 3.4, the following equations are valid [2,11]:

$$\tau 1 = -1.155 \text{ s } m \cos[(\pi/3) + \Delta\theta] \cos[\Delta\alpha + (\pi/3)]$$

$$\tau_2 = 1.155 \text{ s } m \cos[(\pi/3) + \Delta\theta] \cos[\Delta\alpha - (\pi/3)]$$

TABLE 3.4
Feasible MC States

States	$Z = 0$	$Z = 1$
$\begin{pmatrix} +1 & -3 \\ -7 & +9 \end{pmatrix} \times (-1)^z$	$U_g = 1$ and $U_p = 1$ $U_g = 4$ and $U_p = 4$	$U_g = 1$ and $U_p = 4$ $U_g = 4$ and $U_p = 1$
$\begin{pmatrix} +2 & -3 \\ -8 & +9 \end{pmatrix} \times (-1)^z$	$U_g = 1$ and $U_p = 2$ $U_g = 4$ and $U_p = 5$	$U_g = 4$ and $U_p = 2$ $U_g = 1$ and $U_p = 5$
$\begin{pmatrix} -1 & +2 \\ +7 & -8 \end{pmatrix} \times (-1)^z$	$U_g = 1$ and $U_p = 3$ $U_g = 4$ and $U_p = 6$	$U_g = 4$ and $U_p = 3$ $U_g = 1$ and $U_p = 6$
$\begin{pmatrix} -7 & +9 \\ +4 & -6 \end{pmatrix} \times (-1)^z$	$U_g = 2$ and $U_p = 1$ $U_g = 5$ and $U_p = 4$	$U_g = 5$ and $U_p = 1$ $U_g = 2$ and $U_p = 4$
$\begin{pmatrix} -8 & +9 \\ +5 & -6 \end{pmatrix} \times (-1)^z$	$U_g = 2$ and $U_p = 2$ $U_g = 5$ and $U_p = 5$	$U_g = 5$ and $U_p = 2$ $U_g = 2$ and $U_p = 5$
$\begin{pmatrix} +7 & -8 \\ -4 & +5 \end{pmatrix} \times (-1)^z$	$U_g = 2$ and $U_p = 3$ $U_g = 5$ and $U_p = 6$	$U_g = 5$ and $U_p = 3$ $U_g = 2$ and $U_p = 6$
$\begin{pmatrix} +4 & -6 \\ -1 & +3 \end{pmatrix} \times (-1)^z$	$U_g = 3$ and $U_p = 1$ $U_g = 6$ and $U_p = 4$	$U_g = 6$ and $U_p = 1$ $U_g = 3$ and $U_p = 4$
$\begin{pmatrix} +5 & -6 \\ -2 & +3 \end{pmatrix} \times (-1)^z$	$U_g = 3$ and $U_p = 2$ $U_g = 6$ and $U_p = 5$	$U_g = 6$ and $U_p = 2$ $U_g = 3$ and $U_p = 5$
$\begin{pmatrix} -4 & +5 \\ +1 & -2 \end{pmatrix} \times (-1)^z$	$U_g = 3$ and $U_p = 3$ $U_g = 6$ and $U_p = 6$	$U_g = 6$ and $U_p = 3$ $U_g = 3$ and $U_p = 6$

$$\tau_3 = 1.155\ s\ m \cos[(\pi/3) - \Delta\theta] \cos[\Delta\alpha + (\pi/3)]$$

$$\tau_4 = -1.155\ s\ m \cos[(\pi/3) - \Delta\theta] \cos[\Delta\alpha - (\pi/3)], \tag{3.21}$$

where m is the voltage transfer ratio that is equal to the ratio of the prescribed amplitude of the output voltage to the input voltage amplitude, $m \leq 0.866$, $s = (-1)^{p+g}$, and p and g are the numbers of the U_p and U_g sectors, respectively.

Two of quantities computed by Equation 3.21 are >0, and two of them are <0. In this case, the negative states are selected from Table 3.4.

When MC operates, the load voltage consists of a number of the input voltage segments that alternate with the big frequency; the high-frequency component of this voltage is usually smoothed by the load inductance. The grid current also consists of the load current segments alternating with the big frequency. The high-frequency component of this current actuates the large voltage oscillations across the grid inductance and essentially deteriorates the MC operation and other loads that are connected to the grid; therefore, the L–C filter is usually set at the MC input.

The MC model is realized in the model **Matr_Conv1**. Its main part is the subsystem **Matrix** containing nine blocks, each of them has two blocks **IGBT/Diode** connected in opposite; the blocks are connected according to the diagram in Figure 3.42. MC is supplied from the source of 380 V through the filter. As the load, the four-pole IG 110 kVA and 400 V is used.

Three subsystems are placed in the subsystem **Control_&_Measurement**. The subsystem **Comp_time** calculates the times τ_1–τ_4 by Equation 3.21; this subsystem is triggered (with external start). The calculations in it are carried out once for the sampling period T_c—at its beginning. The subsystem **Time/Code** fabricates the code that determines which MC switches must be turned on at the present time. The following order of implementation of the intervals τ_1–τ_4 is accepted: τ_1, τ_3, τ_4, τ_2. The subsystem **Code/Pulses** converts the fabricated codes in the logic signals that control the MC switch states and consists of a number of the tables. The detailed description of these subsystems is given in Ref. [2].

Induction generator (IG) is modeled in the mode when its rotation speed is determined by the external signal (see Chapter 4). The subsystem of the controlled three-phase generator defines the IG supplying frequency; the amplitude and frequency of the generator output signal can change. Under simulation, the amplitude is taken as 0.75, whereas the frequency is controlled by the power controller. Its reference that is equal to zero initially, at $t = 0.7$ s, increases to −1 (the base value is 60 kW) and decreases to −0.7 at $t = 2$ s. At this, the rotation frequency that was equal to 118 rad/s, at $t = 2.5$ s increases to 120 rad/s. The variations of the IG speed and its torque and the grid power are shown in Figure 3.45. It is seen that after the minimization of the initial big error, the reference power value is maintained under both reference changing and perturbation (rotation speed changing).

The grid current distortion is an important characteristic of MC. The plots of the grid current under both with and without the harmonic filter are shown in Figure 3.46; the filter is tuned for the fifth harmonic. The THD depends on the power of the filter. The readers who take a special interest in this problem can experiment with different parameters of inductance, capacitance, resistance of the parallel resistors, and the harmonic filter for investigation of their influence on the grid current form.

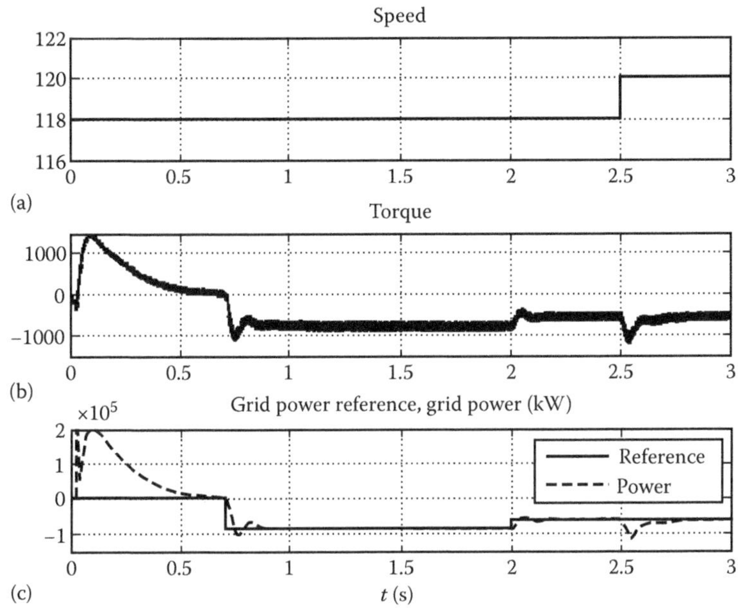

FIGURE 3.45 IG (a) speed and (b) torque and (c) grid power for the model **Matr_Conv1**.

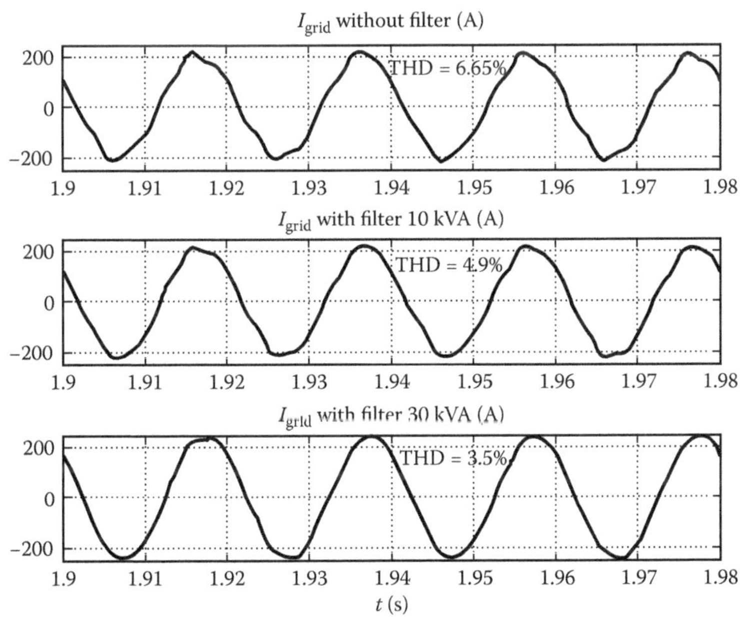

FIGURE 3.46 Grid current with and without the harmonic filter for the model **Matr_Conv1**.

In order to obtain the maximum wind energy harvesting in WG systems, the generator rotating speed change is used. The IG speed regulator can be utilized for that. Such a system is realized in the model **Matr_Conv2**. The main circuits are the same; the difference is in the subsystem **IM_Control** that carries out the so-called indirect vector control; its ground is given, for example, in Ref. [3]; the block diagram is depicted in Figure 3.47.

The control of the components of the stator currents I_d and I_q that are aligned with the rotor flux linkage vector Ψ_r and perpendicular to it is carried out in this system. The position of Ψ_r is defined by the angle θ:

$$\theta = \int \omega_s \, dt = \int (\omega_m + \Delta\omega_s)\,dt, \quad (3.22)$$

where ω_m is the IG angular rotation speed (electrical) and $\Delta\omega_s$ is the absolute slip.

The first value is measured or estimated; the second one is calculated as

$$\Delta\omega_s = \frac{R_r L_m I_q}{L_r \Psi_r}. \quad (3.23)$$

Here R_r, L_m, and L_r are the rotor winding resistance, the magnetizing inductance, and the full rotor inductance, respectively; the meaning of these quantities is considered in the next chapter. The rotor flux linkage Ψ_r is computed as

$$\Psi_r = \frac{L_m I_d}{1 + sT_r}, \quad (3.24)$$

where $T_r = L_r/R_r$.

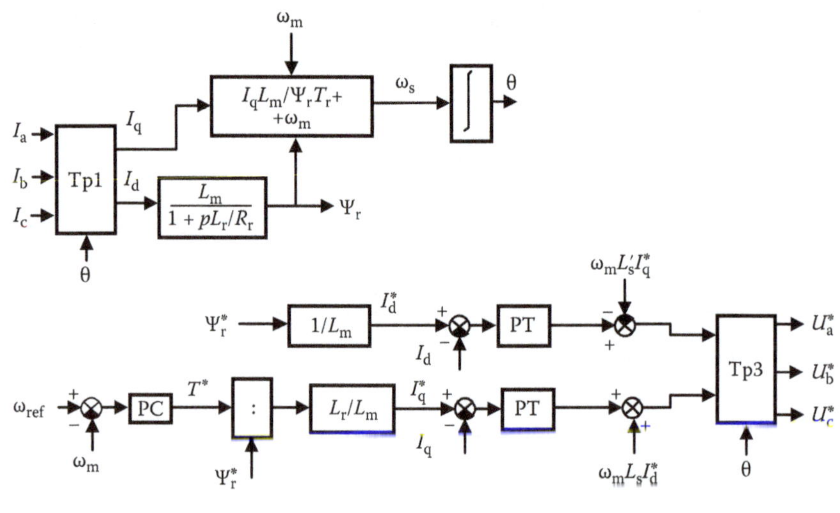

FIGURE 3.47 Block diagram of the indirect vector control system.

The references for the current regulators of I_d and I_q are

$$I_d^* = \frac{\Psi_r^*}{L_m} \tag{3.25}$$

$$I_q^* = \frac{T_e^* L_r}{L_m \Psi_r}, \tag{3.26}$$

where T_e^* is the IG reference electromagnetic torque, in our case, the output of the speed controller, and Ψ_r^* is the reference rotor flux linkage.

Because the control systems of the currents I_d and I_q turn out to be coupled, for their decoupling, the following signals are added to the current controller outputs:

$$I_{dcomp} = -\omega_m L_s' I_q^*, \quad I_{qcomp} = \omega_m L_s I_d^*, \tag{3.27}$$

where L_s is the full stator inductance and L_s' is the transient inductance.

Running ahead (see Chapter 4), it can be noted that $L_s = L_m + L_{1s}$, $L_r = L_m + L_{1r}$, $L_s' \approx L_{1s} + L_{1r}$, L_{1r}, L_{1s} are the leakage inductances of the corresponding windings. Simulation in pu is carried out in the model under consideration.

The following process is simulated: the initial IG excitation under zero speed reference; speeding up to the speed of 0.75 with the load torque of 0.1 at $t = 1$ s; the formation, appearance, origination, and occurence of the active generator torque that is equal to −1 at $t = 2$ s. It can be imagined that firstly, WG speeds up to the given speed with the blades in the passive position; afterward, the blades are set in the active position when the wind energy converts to the electric one. The IG rotation speed and its torque in pu, the power that is sent to the grid, and the grid current are shown in Figure 3.48. The latter is seen to be nearly sinusoidal: THD = 2.9%.

The described MC is characterized as the direct MC. Not long ago, the so-called indirect (or dual-bridge) MC (IMC) was developed, whose diagram is depicted in Figure 3.49, [12–14]. It consists of six bidirectional switches 1–6, such as in MC, and of six IGBT/Diode, as in the usual inverter, so that the total IGBT number is 18 for both MC units. The IMC structure is analogous to the structure of the usual two-level inverter with an active input rectifier; the difference is that, for the former, there are no capacitors and other reactive elements in the DC link. In comparison with MC, IMC has a number of advantages: an intricate procedure of switching-over is necessary in MC, which is not demanded in IMC, because the rectifier switches can be turned over under a zero current, a simpler control system.

The rectifier operation is being considered. Its switches switch over in such a way that the output voltage is maximal at every time, without a shift between input voltages and currents. In each 60° sector of the input voltages, a certain polarity has

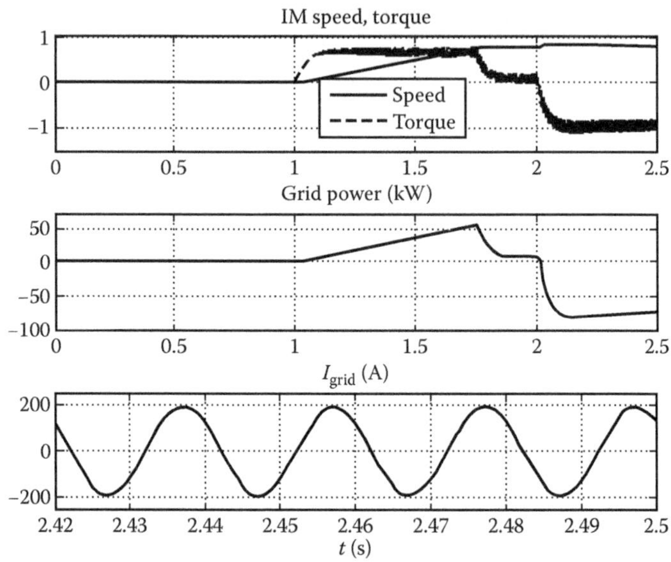

FIGURE 3.48 Processes in the model **Matr_Conv2**.

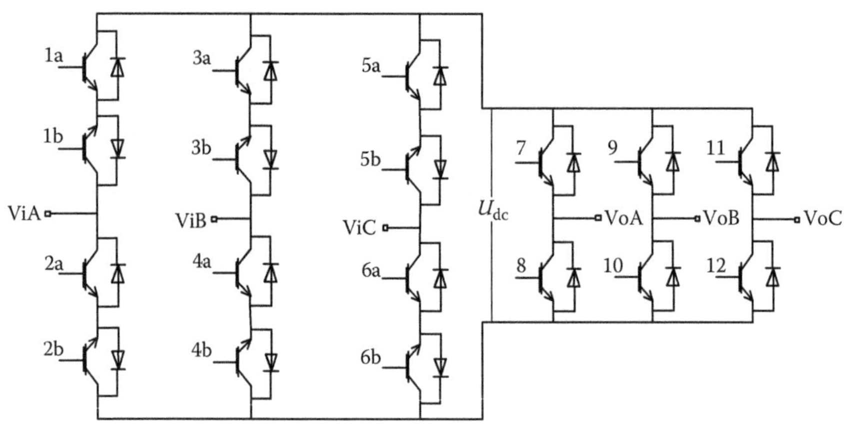

FIGURE 3.49 IMC diagram.

one of three voltages, for example, V_a; the rest have another polarity (Figure 3.50). The sector center coincides with the maximum of the first voltage. One switch that is connected to the output terminal plus and one switch that is connected to the terminal minus are turned on; the rest of the switches are turned off. Then, in this sector, the switch that belongs to V_a remains turned on, but the switches of other phases switch over in such a manner that the position of the current space vector is aligned with the voltage space vector. For that, the relationships for the switching relative

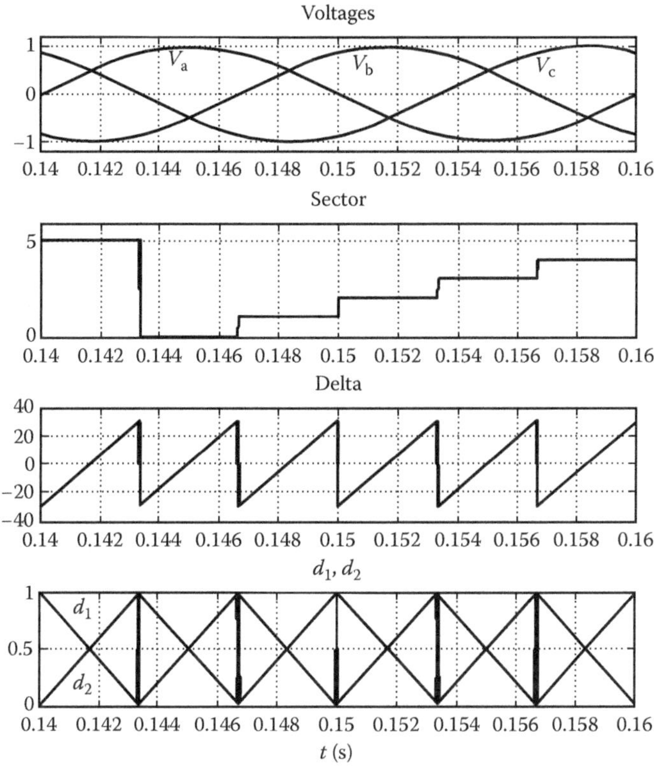

FIGURE 3.50 Voltages in IMC.

intervals d_1 and d_2 have to be fulfilled that are analogous to that given in Equation 3.8 (with substitution $\gamma = \delta + (\pi/6)$ because γ changes in the range $[0, \pi/3]$ and δ in $[-\pi/6, \pi/6]$). Because $d_1 + d_2 = 1$, then

$$d_1 = \sin(\pi/6 - \delta)/\cos\delta, \quad d_2 = \sin(\pi/6 + \delta)/\cos\delta \tag{3.28}$$

The dependence of the switches states on the sector number is given in Table 3.5.

TABLE 3.5
Dependence of the Switch States on the Sector Number

Sector	1	2	3	4	5	6
d_1	1 −1 0	1 0 −1	0 1 −1	−1 1 0	−1 0 1	0 −1 1
d_2	1 0 −1	0 1 −1	−1 1 0	−1 0 1	0 −1 1	1 −1 0

Note: 1: the top switch is closed; −1: the low switch is closed; 0: both switches are open.

The mean value of the output (in the DC link) voltage for a modulation period (in one sector):

$$U_{dc}^{*} = d_1(V_a - V_b) + d_2(V_a - V_c) = V_m\{d_1(\cos\delta - \cos[\delta - (2\pi/3)] + d_2(\cos\delta -$$
$$\cos[\delta + (2\pi/3)]\} = V_m\sqrt{3}[d_1\cos\{[\delta + (\pi/6)] + d_2\cos[(\pi/6) - \delta)]\} = 1.5V_m/\cos\delta. \tag{3.29}$$

Because of

$$\frac{3}{\pi}\int_{-\pi/6}^{\pi/6}\frac{1}{\cos\delta}d\delta = 1.049,$$

the minimal, the average minimal for the modulation period, the total average, and the maximal DC voltages are equal to $\sqrt{3}/2\,V_m$, 1.5 V_m, 1.57 V_m, and $\sqrt{3}\,V_m$, respectively; for instance, under V_{rms} = 380 V, we receive 268, 465, 487, and 536 V, respectively.

As for the inverter, SV PWM can be used. According to Section 3.2, the maximal value of the amplitude of the output phase voltage is

$$U_{out\,max} = \frac{U_{dc\,mean}^{*}}{\sqrt{3}}. \tag{3.30}$$

Under V_{rms} = 380 V, $U_{out\,max} = 487/\sqrt{3}$ = 282 V.

The usual PWM does not give the possibility to switch over the rectifier switches in times when the current does not flow through them. For that, the durations of the active inverter states must be changed: instead of the sequence $V_0(t_0/4)$, $V_i(t_1/2)$, $V_{i+1}(t_2/2)$, $V_7(t_0/2)$, $V_{i+1}(t_2/2)$, $V_i(t_1/2)$, $V_0(t_0/4)$, the sequence $V_0(d_1t_0/2)$, $V_i(d_1t_1)$, $V_{i+1}(d_1t_2)$, $V_7(t_0/2)$, $V_{i+1}(d_2t_2)$, $V_i(d_2t_1)$, $V_0(d_2t_0/2)$ is used; the inverter state V_7 is realized at the switch-over moment when the current in the DC link is zero.

The described system is realized in the model **IMatr_Conv1**. The subsystem for rectifier **Control_R** consists, in turns, of two subsystems and the one-dimensional (1D) tables that convert the wanted code of the rectifier state in the sequence of the gate pulses. The first subsystem calculates the sector number and the d_1 and d_2 durations. This subsystem is the triggered one: it carries out computation once for the modulation period, at its beginning. The second subsystem fabricates the demanded state code according to Table 3.6.

The contents of 1D tables can be determined from comparison of Tables 3.5 and 3.6.

Two variants for the inverter control are employed in the model. When the tumbler **M_Switch1** is in the right position, the gate pulses are fabricated with the standard block SV PWM. To carry out the simulation with the reference amplitude of 1 and frequency of 30 Hz (the block **Reference**), it will be obtained after data processing with the help of **Powergui**: V_m = 333 V (the first axis of the scope **Rect**), $U_{dc\,mean}^{*}$ = 524 V (the third axis of the scope **Control**), $U_{out\,max}$ = 302 V (the first axis

TABLE 3.6
Rectifier State Encoding

Sector	1	2	3	4	5	6
d_1	6	1	2	3	4	5
d_2	1	2	3	4	5	6

of the scope **Load**), THD = 54%. Some deviation of the received results from the theoretical relationships given earlier is explained by the fact that the voltage that supplies the rectifier bridge is very much distorted with the voltage drop across the reactance of the input filter caused by the current harmonics. To connect the ideal source direct to the rectifier input and to repeat simulation, one can be convinced that these relationships are true. According to the scope **Input**, the phase shift between the grid voltage and current is 18°, and at IMC input, only 4°.

It follows from Figure 3.51 that switches 1 and 5 can switch over when the DC current is not zero.

The subsystem **Control_I** has two subsystems; the first one, **Control1**, calculates the intervals of the active states V_i and V_{i+1} by Equation 3.8 and the sector number where these vectors is located. At this, as in the standard block, the sequence of utilization of these states changes in each sector (V_i and V_{i+1} or V_{i+1} and V_i). The subsystem **Control2** computes the intervals, at which the states V_0, V_i, V_{i+1}, and V_7 are implemented, according to the rule given earlier. The computed values are compared with a linear-increasing signal that has the amplitude of 1 and the frequency of F_c

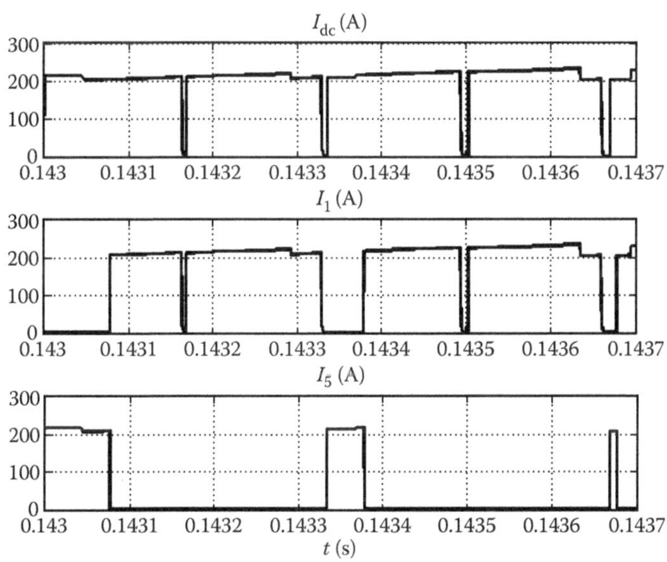

FIGURE 3.51 Current commutation with the usual PWM.

(accepted F_c = 3000 Hz); when this signal reaches the next of the computed values, the adder output increases by 1, so that the number *NN* is equal to the number of running intervals (seven intervals). The table with two inputs (2–D table) converts the number *NN* and the sector number into the demanded inverter states, and 1D tables TabA, TabB, and TabC convert the inverter state code in gate pulses.

Because of the obligatory presence of zero intervals, the modulation factor $m < 1$, its maximal value depends on system switching-over speed and on modulation frequency and is defined by the following relationship, where $t_0 > 0$:

$$t_0 = 1 - m \cos [\gamma - (\pi/6)] \tag{3.31}$$

To carry out simulation with **M_Switch1** in the left position and m = 0.95, it can be found the $U^*_{\text{dc mean}} = 526\,\text{V}$; $U_{\text{out max}}$ = 287 V (it must be theoretically 0.95 × 0.577 × 526 = 288 V); THD = 60%; for load current, THD = 3%; for the input current THD = 4.5%. The load voltage and current, and also the DC voltage, are shown in Figure 3.52; the grid voltage and current are shown in Figure 3.53; the switching-over of the rectifier switches is shown in Figure 3.54. It is seen that these switchings take place when the DC current is zero.

The model **IMatr_Conv2** demonstrates the IMC operation with the IG with squirrel cage rotor. IG control system and operation modes are the same as in the model **Matr_Conv2**. The plots of the IG speed and torque, grid power, and current are displayed in Figure 3.55. The grid current is sinusoidal in fact: THD = 1.9%. Thus, IMC, in comparison with MC, ensures less grid current distortion, with lighter units for filtration.

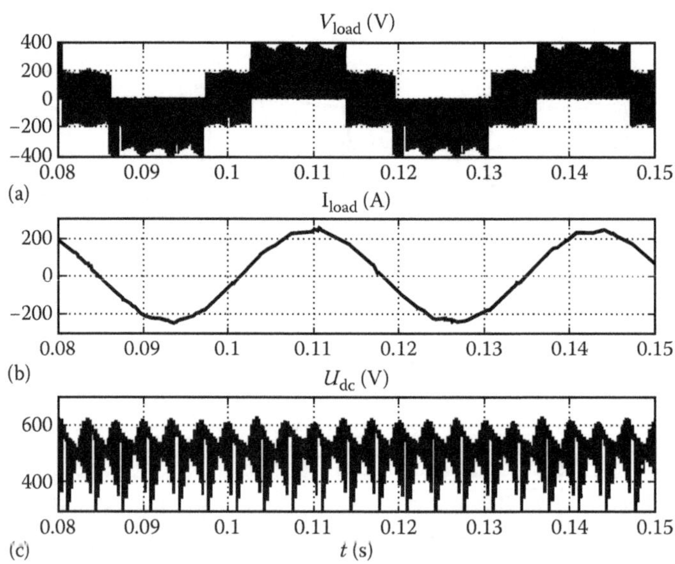

FIGURE 3.52 Load (a) voltage and (b) current, and (c) DC voltage for the model **IMatr_Conv1**.

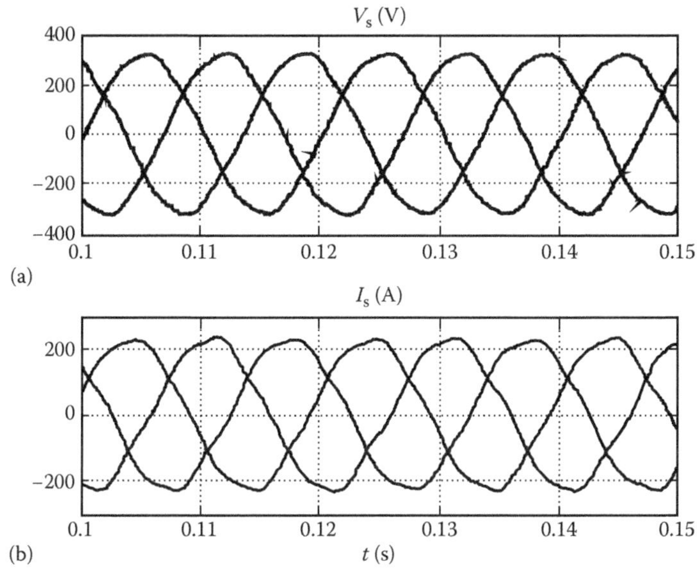

FIGURE 3.53 Grid (a) voltage and (b) current for the model **IMatr_Conv1**.

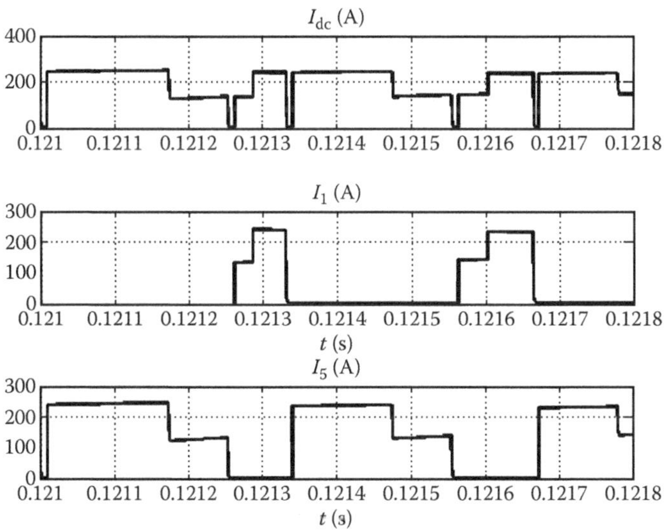

FIGURE 3.54 Current commutation with the new PWM.

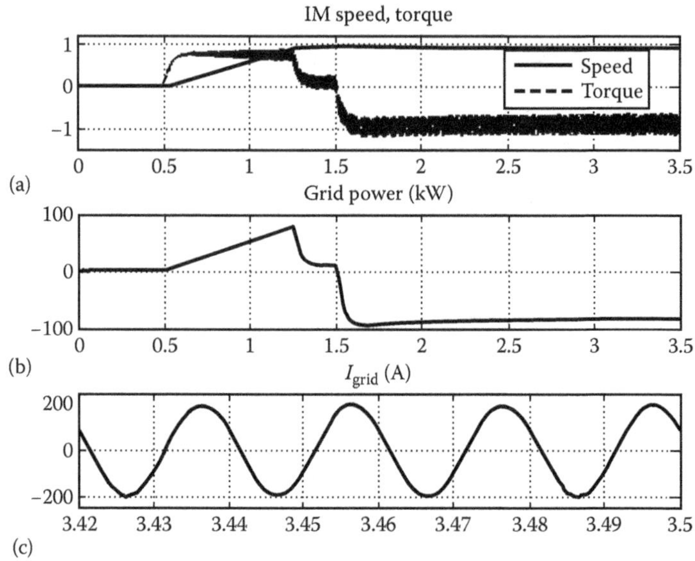

FIGURE 3.55 (a) IG speed and torque, (b) grid power, and (c) current **IMatr_Conv2**.

3.5 FACT SIMULATION

Improving the quality of the transmitted electrical energy gives the possibility to cut costs demanded for its production. The main indicators of the quality are voltage deviation from the nominal value at the receiving end or/and along the line, frequency deviation from the nominal value, contents of the higher harmonic in the line voltage and current, and relative value of the transmitted reactive power. Most of the measures have to decrease the values of these indicators.

New power electronic components have been utilized lately for electric power quality improvement. The power systems with such devices are called FACT: flexible AC transmission systems. The SVC of the reactive power, the STATCOM of the reactive power, the active filters, the static synchronous series compensator, and the unified power flow controller belong to such units. The simulation of a number of such units is considered in the demonstration models in SimPowerSystems, and in Ref. [2] as well. In RES, including WG farms, the SVC and the STATCOM are used for the voltage control in the common point of the connection (PCC) of individual RES (for instance, WGs) that form a cluster (farm, park) [15].

3.5.1 Static Var Compensator Simulation

The SVC is a unit that controls the amount of the reactive power that is taken off from the grid or is delivered to it. The control is carried out by connection and disconnection, with the help of the thyristor switches, the capacitors, and the reactors in the secondary winding of the matching transformer, whose primary winding is connected to the three-phase network, or by the change of the equivalent value of inductance, by the thyristors firing angle control, which are set in series to the reactor (Figure 3.56).

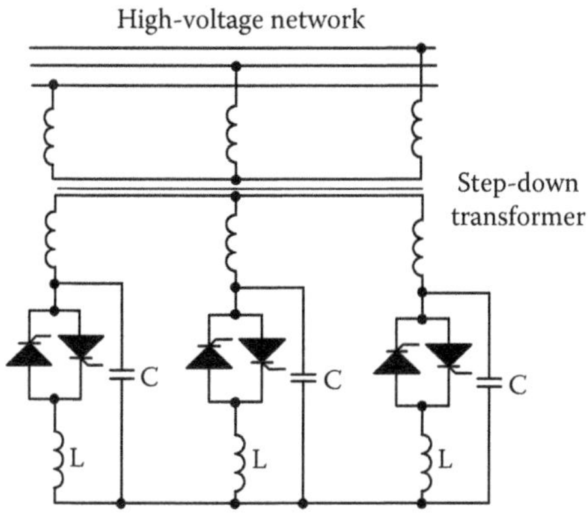

FIGURE 3.56 SVC diagram.

There is an SVC model in SimPowerSystems that implements the SVC steady characteristics for the complex vectors, and it can be used for simulation only in Phasor mode. The model details are not included. The current space vector that is fabricated by SVC is determined as $\mathbf{I} = jB\mathbf{U}$, where $\mathbf{U}$ is the voltage vector in PCC and B is the resulting SVC susceptance. The model can operate in two modes: constant susceptance (var control) and voltage regulation [1,2].

The demonstration model **power_svc_1tcr3tsc.mdl** (SVC-detailed model) is included in SimPowerSystems, in which the SVC contains three capacitor banks that can be connected or disconnected independently. SVC intends for the voltage control in the power system. When it is necessary to lower the voltage, these capacitors are disconnected in its turn that gives the possibility to dispense with less inductance. The thyristors with pulse control are used for switching over the capacitor circuits. A model with the simplified version of such a unit is given in Ref. [2].

In this book, SVC with the analogous circuits is utilized for the reactive power compensation that consumes from the grid by IG with the squirrel cage rotor. Such a generator, producing the active power, demands considerable reactive power that impacts negatively on the total efficiency of the electrical power unit.

In the model **SVC1**, a WG farm with the total power of 150 MVA is modeled with the generalized IG with the voltage of 35 kV. Saturation is taken into account.

The series R–L circuit models the total WG farm reactance from the IG terminals to the step-up transformer that is connected with the 500 kV network via the transmission line that is 25 km in length. The existence of a small resistor in the source 500 kV is caused by the fact that without this resistor, the source is connected in parallel to the capacitors of the transmitted line model; it is impermissible in SimPowerSystems (under simulation start, the error report will be produced). The capacitor bank $\mathbf{C_f}$ with power of 60 Mvar partly compensates the reactive power.

To carry out simulation when the breaker **Br** is turned off and IG torque $T_m = -1$, the following values of the active/reactive powers appear at the low-voltage and high-voltage windings and in the network respectively: −1.44/1.15; −1.4/1.46; −1.4/1.2) × 100 MW/Mvar.

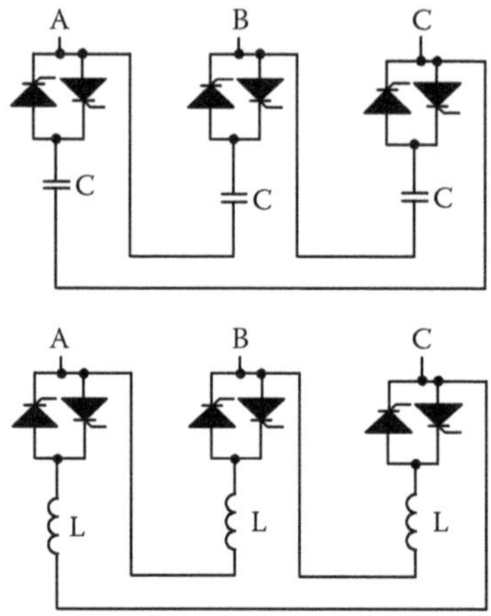

FIGURE 3.57 SVC with controlled L and C.

In order to compensate for the reactive power completely, SVC is set that consists of the reactor block (thyristor-controlled reactor [TCR]) and the three capacitor blocks (TSC1, TSC2, TSC3); each block has the power of 50 Mvar. Both the reactor and capacitor blocks are connected in the delta that excludes the third harmonic currents (Figure 3.57). So the SVC reactive power can change from –50 Mvar (inductive) under α_{TSR} = 0° and with the turned-off thyristor-switched capacitor (TSC) to +150 Mvar (capacitive) under α_{TSR} = 90° and with the turned-on TSC. The small inductors in the capacitor circuits (not shown in the figure) intend to limit through current under possible malfunctions. Taking as the base power P_b = 50 Mvar and bearing in mind that the transformer leakage inductance for this base power is 0.15/4 ≈ 0.04 pu, the equivalent susceptance B in pu, referring to the transformer primary winding, is

$$B = \frac{Q^*}{1 - 0.04Q^*}, \tag{3.32}$$

where Q^* is the reactive power in pu.

It may be found that B changes from −1/(1 + 0.04) = −0.96 to 3/(1 − 3 × 0.04) = 3.4.

The capacitance of capacitor C under the star connection is computed from the relationship

$$Q_c = U_s^2 \omega_s C \tag{3.33}$$

from where $C = 5 \times 10^7/(314 \times 35^2 \times 10^6) = 130$ μF. Under the delta connection C = 43.3 μF. It is accepted as C = 42.5 μF.

The inductance of the inductor L under the star connection is calculated from the relationship

$$Q_l = \frac{U_s^2}{\omega_s L} \tag{3.34}$$

from where $L = 35^2 \times 10^6/(5 \times 10^7 \times 314) = 0.078$ H. Under the delta connection $L =$ 0.234 H. It is taken as $L = 240$ mH.

The SVC control system consists of the reactive power controller, the devices that decide which SVC power blocks have to be active in the present time, and the subsystem with the firing pulse generators.

The output of the power controller defines the value B referred to the SVC transformer primary side. The little deadband is set in the error path in order to avoid frequent switching-over ("jitter"). The control system calculates the firing angle α_{TCR} and forms the signals for the activation of the capacitor blocks. This part of the scheme is depicted in Figure 3.58. Firstly, the value B is reduced to the transformer secondary winding by the inversion formula in Equation 3.32. If $B < 0$, this value, through the summer and the limiter, comes to the amplifier G with a gain = $1/B_{max}$ (in the case $B_{max} = 1$), so that B^* changes from −1 to 0. The relationship between the SVC inductance in pu and the firing angle α is given in Ref. [16]:

$$B^* = \frac{\pi - 2\alpha - \sin(2\alpha)}{\pi} \tag{3.35}$$

In the table, the solution of the equation that is inverse to Equation 3.35 is programmed, so that the angle α is fabricated at the table output. It is worth noting that the other relationship is realized in the subsystem table in fact, because, under its creation, it was supposed that $90° < \alpha_1 < 180°$; it means that the firing angle is computed off from the zero value of the phase voltage; at this, $\alpha = \alpha_1 - 90°$. Therefore, to obtain the demanded value α, the table output decreases by 90°.

When the signal B reaches zero, **Relay 1** is switched on and sends the signal to turn on the first capacitor block. Simultaneously, the summer output decreases by the value that is equal to the relative value of this block reactive power, this output becomes negative, and the TCR gets in the operation active range. If the capacitance of one capacitor block is not sufficient, signal B will continue to rise; when it reaches the value that is equal to the relative reactive power of the capacitor block, the second

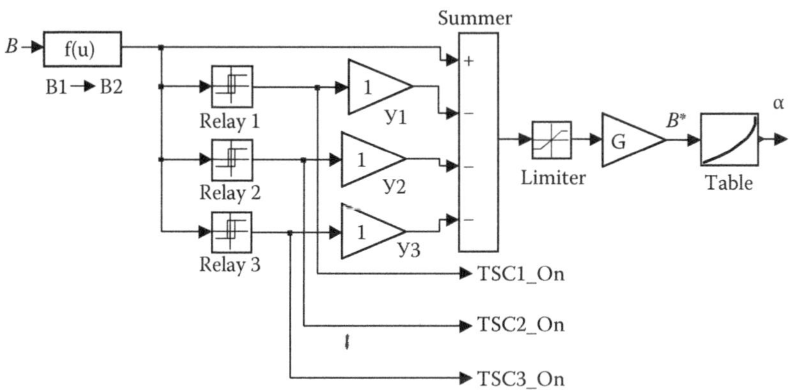

FIGURE 3.58 SVC control.

block is turned on, and the summer output decreases by the same value, and so on. If turning on all the capacitor blocks is not sufficient, the value $B > 0$, and the TCR angle $\alpha = 90°$, it is turned off completely.

The outputs of TSC come to the subsystem **Pulse Generators** that contains four firing pulse generators. One of them controls TCR and works with the firing angle α. For this system, the natural firing point coincides with the moment when the phase voltage is maximal. The other three generators are synchronized in antiphase to the first one and operate with $\alpha \approx 0$ [2].

The following process is simulated: the system starts with IG torque $T_e = -1$; **Br** is turned on at $t = 0.6$ s; T_e decreases to -0.5 at $t = 1.5$ s; the subsequent decrease to -0.1 at $t = 2.5$ s. The torque, the active and reactive IG powers, the reactive powers on the transformer high-voltage side and in the network, the angle α change, and the number of the active TSC (multiplied by 30 for the scale correspondence) are shown in Figure 3.59. The value of the controlled reactive power is seen to be about zero, in the bound of the deadband, but the network reactive power essentially differs from zero. If it is desirable to decrease this quantity in addition, it is possible to set the model of the transmission

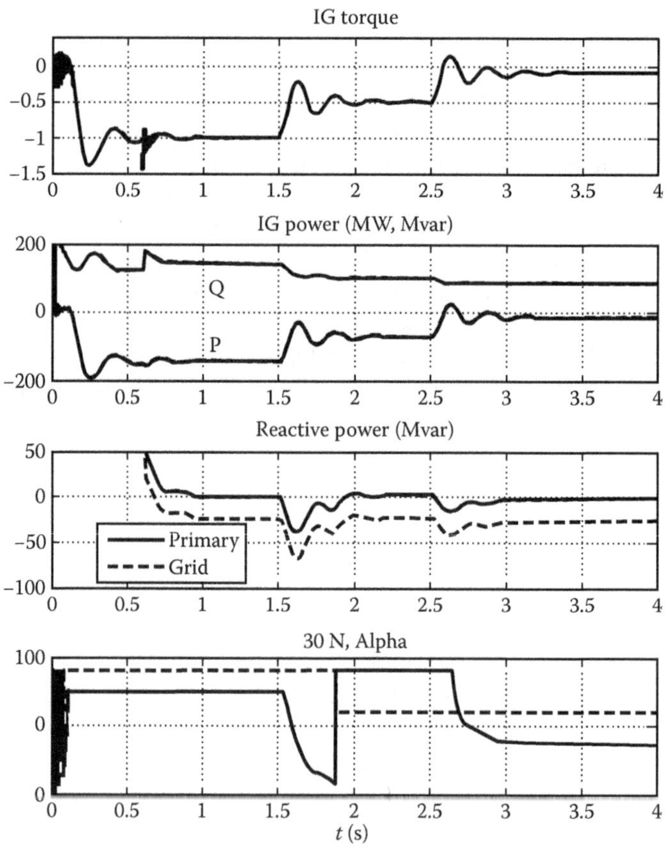

FIGURE 3.59 Processes in SVC with switching capacitors.

line in the controller feedback circuit, in order to take into account the alternation of the reactive power in the line. It is also seen that the number of the active TSC is 3 or 2.

3.5.2 STATCOM Simulation

3.5.2.1 STATCOM with Modular Inverter

The STATCOM controls the voltage at the point where it is connected to the power system (PCC), consuming or generating the reactive power, and is a VSI. In comparison with SVC, STATCOM reacts faster and can be used for harmonic filtration. Two STATCOM types are considered in SimPowerSystems: with a constant voltage V_{dc} in the DC link and with the changeable V_{dc}. In the first case, STATCOM consists of VSI with IGBT devices, with PWM. It is connected to the power system with the help of the transformer or the reactor; the capacitor is set in the DC link. The STATCOM control system controls the I_d and I_q components of its current that are aligned along and perpendicular to the space vector of the network voltage. For that, the I_d control loop is inner for the V_{dc} controller, and the I_q control loop is inner for the network voltage controller. There are alternatives to work with I_q constant or with the constant reactive power.

When STATCOM operates with the changeable V_{dc}, PWM is not used; its output voltage has a stepped form. The amplitude of this voltage is proportional to the voltage across the capacitor. In order to decrease the content of the harmonics in the output voltage and current, the multiphase schemas with several inverters are employed. For instance, four three-level inverters with four transformers are used; the transformers have the phase shifts 7.5° one about another; therefore, a 48-step voltage is fabricated having very little content of the higher harmonics. In order to control the reactive power, the voltage V_{dc} must be changed; for that, the voltage phase angle α that is usually nearly zero is shifted, producing, by such a way, a flow of the active power, which changes V_{dc}.

The new class of the STATCOM is developed that is intended for the operation in the distributed grids (but not for main transmission lines), the so-called DSTATCOM. They have less power (not more than some Mvar) and are connected to the line with relatively low voltage (3–35 kV). PWM is used in DSTATCOM.

The block **Static Synchronous Compensator** is included in SimPowerSystems that models a three-phase STATCOM of the first type in the Phasor mode. The inverter components and the harmonic fabrication are not modeled. The demonstration model **power_statcom.mdl** that illustrates the utilization of this block is included in SimPowerSystems as well. The modeled power system consists of two equivalent sources with the power of 3 and 2.5 GVA and with the voltage of 500 kV; these sources are connected with the line 600 km in length. The source voltages are selected in such a way that the transmitted power from left to right is 930 MW. The STATCOM with power of ±100 Mvar is set in the line center; the DC voltage is 40 kV and the capacitor has a capacitance of 375 μF. STATCOM is connected to the line with the help of the transformer having an equivalent impedance of 0.22 pu under power 100 MVA.

The demonstrational model **power_statcom_gto48p.mdl** uses the second principle of control: the VSI with the stepped output voltage (without PWM), with the reactive power control by V_{dc} changing; this model takes into account the details of the unit operation. Its power circuits utilize four three-level VSIs with the common capacitors in the DC link. The three-phase **Zigzag Phase-Shifting Transformers** are set at the inverter outputs that have the phase shift +7.5° or −7.5°. The additional phase shift by 30° resulted from the delta connection of the secondary windings of two transformers; so the primary windings have the phase shifts as 7.5°, −22.5°, −7.5°, −37.5°. The transformer primary windings are connected in series, and the modulation signals for VSI are shifted by 0°, −30°, −15°, −45°, respectively, so that all fundamental harmonics are in phase, and most of the higher harmonics are compensated mutually: only harmonics of the orders 23, 25, 47, 49, or more remain.

The phase-to-phase voltages have a form of rectangular waves. To use a stepped voltage, instead of rectangular, that is obtained by decreasing the intervals from 180° to $\sigma = 172.5°$, when the voltage $\pm V_{dc}/2$ is applied, the 23rd and 25th harmonics reduce essentially, and most considerable, 47th and 49th remain. The detailed description of this and some other STATCOM types are given in Refs. [1,2].

The model **power_dstatcom_pwm.mdl** is included in SimPowerSystems that uses the first principle. STATCOM is intended for the load voltage control when both the load parameters and the supplying voltage change. STATCOM consists of two two-level VSIs with the common capacitor; the primary windings of the three single-phase transformers are connected to the VSI terminals of the same name; the secondary windings are connected in the star, forming the output phase-to-phase voltage of 25 kV. The VSIs are controlled in antiphase that results in the decrease in the higher harmonics in the output voltage. The control system regulates the current components I_d and I_q, depending on the DC and load voltages, as it was said earlier. The detailed description of this model is given in Ref. [1].

In the developed model, STATCOM with the modular multilevel inverter is used. The employment of such an inverter for STATCOM is very worthwhile because, with the increase in the module number, it is possible to dispense with the transformer for the connection to the high-voltage network. In the model **Statcom_modinv**, such STATCOM is utilized for voltage control at the point where WGs are connected to the grid and the load and for reduction of the reactive power in the grid.

The WG farm with IG is modeled as one equivalent IG with the power of 50 MVA and 575 V. The capacitor bank with the reactive power of 15 Mvar is set for the partial compensation of the reactive power. The IG voltage is transformed to the level of 25 kV and via the line that is 1 km in length is connected to the load of 40 MW. At this point (designated as B25), the WG farm is connected to the power system; this point is connected to the 120 kV network with the help of the 25 km line and the transformer of 25 kV/120 kV. The network power is 5 GVA. The IG torque that was initially equal to −0.2 increases to −1 at $t = 2$ s and afterward decreases to −0.5 at $t =$ 20 s. Without STATCOM, the load voltage differs from the nominal one essentially (see Figure 3.60 further); the network reactive power is 15 Mvar at $t = 20$ s.

STATCOM is built as shown in Figure 3.28, but the DC source is absent. The number of the modules in a semiphase is four; the module nominal voltage is 2200 V. For module control, the same principle is used, as for the model **M2CDR1**. The inverter is connected to the PCC with the help of the 4.6 kV/25 kV step-up transformer.

The STATCOM control system is depicted in Figure 3.61. Block PLL fabricates the signals sin x and cos x, where x is a position of the voltage V_{abc} space vector in the

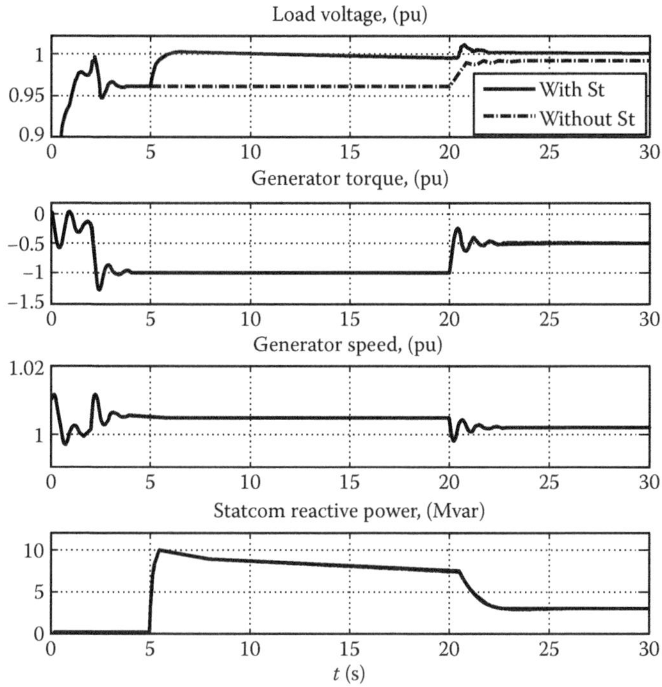

FIGURE 3.60 Processes in STATCOM with the modular inverter.

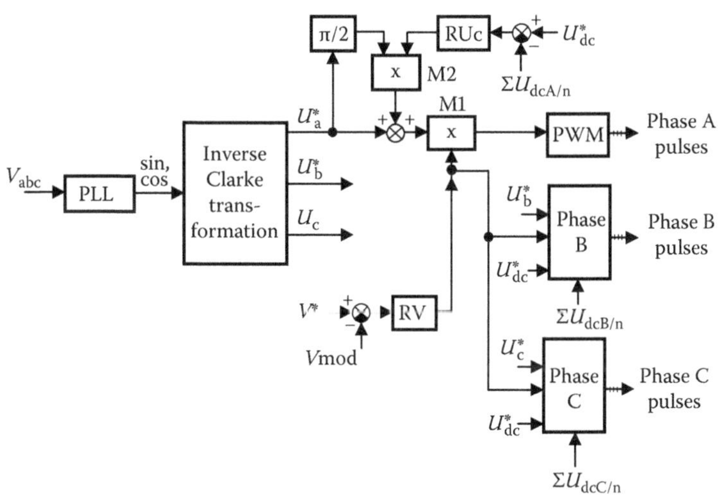

FIGURE 3.61 STATCOM with modular inverter control system.

point B25. With the help of the inverse Clarke transformation, the signals U_a^*, U_b^*, and U_c^* are formed that are in-phase with the corresponding phase voltages but have the amplitude equal to 1. In the units M1, these signals are multiplied by the modulation factor that is determined by the voltage controller output.

The capacitor voltage control is realized in the following way. The reference capacitor voltage and the average value of the capacitor voltages of one phase are compared at the inputs of the capacitor voltage controller. The outputs of the controllers modulate in amplitude the signals that are shifted by 90°, relatively signals U_a^*, U_b^*, and U_c^*, which afterward are added to the formers. In this way, the control signals are formed that relatively shifted the phase voltages and carry out an exchange of an energy between the capacitor and the network with the aim to keep the capacitor reference voltage.

The processes in the system with STATCOM, which is connected to the power system at $t = 5$ s, are shown in Figure 3.60. The load voltage is equal to the reference value. The STATCOM steady reactive power is 8 Mvar, and the network reactive power drops down 5 Mvar. The voltage across the capacitor of one of the modules, the inverter voltage, the currents of the network, and the STATCOM are shown in Figure 3.62a through d. The capacitor voltage deviates not much from the reference

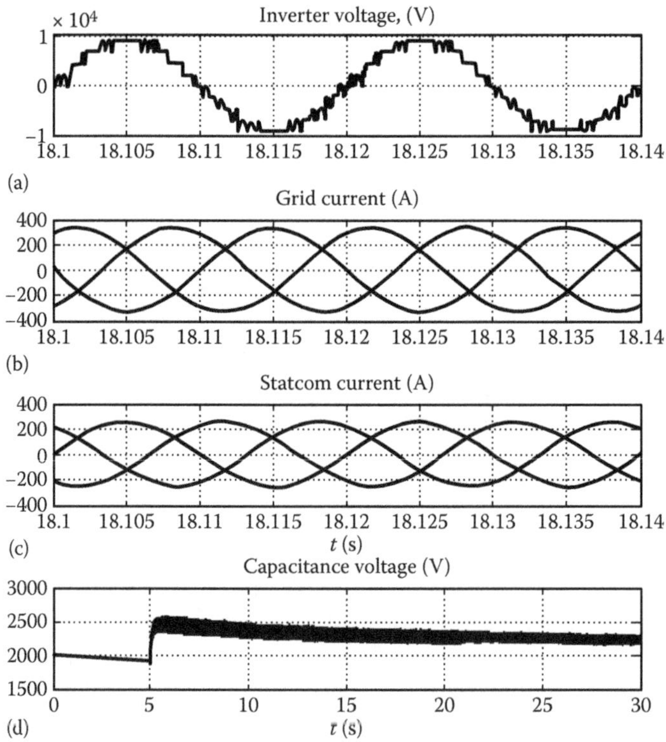

FIGURE 3.62 Capacitor and inverter voltages (a), (d) and currents (b), (c) of the network and STATCOM with the modular inverter.

one; the scope in the top semiphase *A* shows that the voltages of all modules change almost equally. Therefore, the developed circuits ensure balancing of the capacitor voltages.

As has been said earlier, the advantage of the modular multilevel inverter is the possibility to connect it to the network without the matching transformer. Such a system is simulated in the model **STATCOM_modinv1**. The model is simplified in comparison with the previous one, in order to make the simulation faster. STATCOM is employed for keeping the load voltage equal to the reference one. The load has a power of 50 MW and is supplied from the weak grid with nominal voltage of 20 kV having the short-circuit power of 160 MVA. STATCOM has 10 modules in each semiphase, as how it is in the model **M2CDCR1**; the control principle is taken from that model as well. The voltage of every module is 5 kV. STATCOM is connected to the PCC with the help of the reactor with the inductance of 32 mH. The following process is simulated: at t = 5 s, STATCOM is turned on; at t = 18 s, the supply voltage drops down by 10%; at t = 28 s, the supply voltage surges to 1.05 of the nominal value. The load voltage with and without STATCOM, the STATCOM current, and the voltage of one module are shown in Figure 3.63. In the steady state, the voltages across both the load and the module capacitors are equal to the reference values. The scope in the semiphase A1 shows that the voltages across all modules are the same. The phase-to-phase inverter voltage and its current are depicted in Figure 3.64 at different times. The current form is nearly sinusoidal. The plots show that the load voltage reaches its steady value after disturbance rather quickly, but the current reaches its final form during sufficiently long transients.

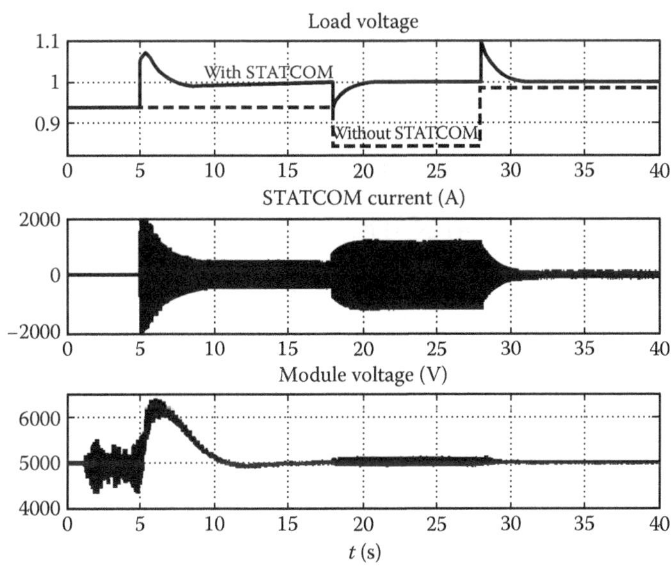

FIGURE 3.63 Processes in STATCOM with the modular inverter with 10 modules in semiphase.

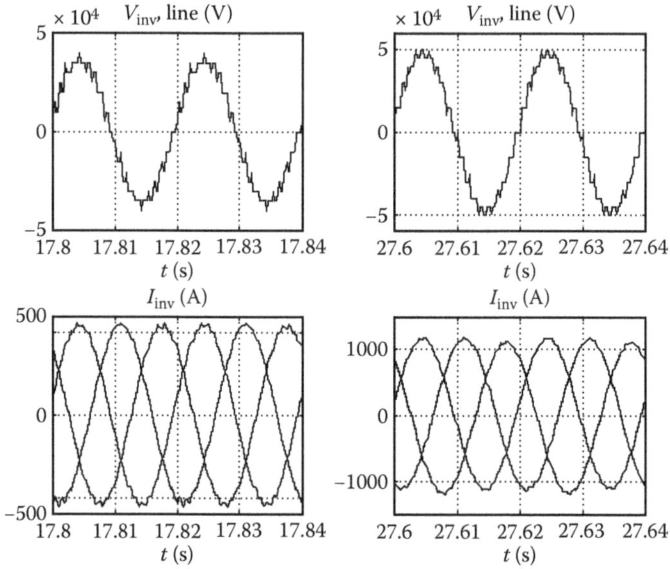

FIGURE 3.64 Phase-to-phase voltages and currents in STATCOM with 10 modules in semiphase.

3.5.2.2 STATCOM with Cascaded H-Bridge Multilevel Inverter Simulation

There are a number of developments of the cascaded inverters for STATCOM where the supply source for each inverter bridge is a capacitor (in the diagram in Figure 3.17, the DC source is replaced with the capacitor). Because STATCOM fabricates the reactive power, the average value of the voltage across the charged capacitor must not change in the ideal case. However, because of inevitable power losses, the control system has to have the circuits for capacitor voltage control. In the model **STATCOM_Hinv1**, the same system is investigated as in the model **STATCOM_modinv**, but the modular multilevel inverter is replaced with the cascaded multilevel inverter that was described in Section 3.4.3. STATCOM has in each phase three single-phase modules; the capacitor voltage is 2.3 kV; the capacitance is 15 mF (specified in the option *File/Model Properties/Callbacks*). STATCOM is connected to the network with help of the 6.3 kV/25 kV step-up transformer. The control system structure is the same as in the model with the modular multilevel inverter (Figure 3.61), and the circuits for the module selection are absent. Together with the voltage control, the control of the reactive power delivered by STATCOM can be realized (the tumbler Sw is in the right position).

Processes in the system under the same conditions as in Figure 3.63 are shown in Figure 3.65. The load voltage (in PCC) and the voltages of the module capacitors are keeping equal to the reference values. The scopes **UdA**, **B**, and **C** confirm that the voltages across all the capacitors are equal.

The process of the reactive power control is shown in Figure 3.66. Its reference increases from 0 to 20 Mvar at $t = 5$ s. The IG torque is equal to −1. The new power steady level is reached sufficiently quickly.

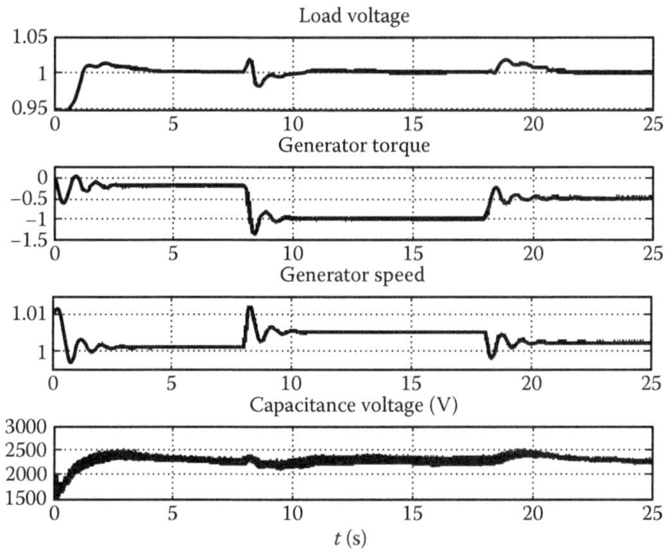

FIGURE 3.65 Processes in STATCOM with cascaded H-bridge multilevel inverter.

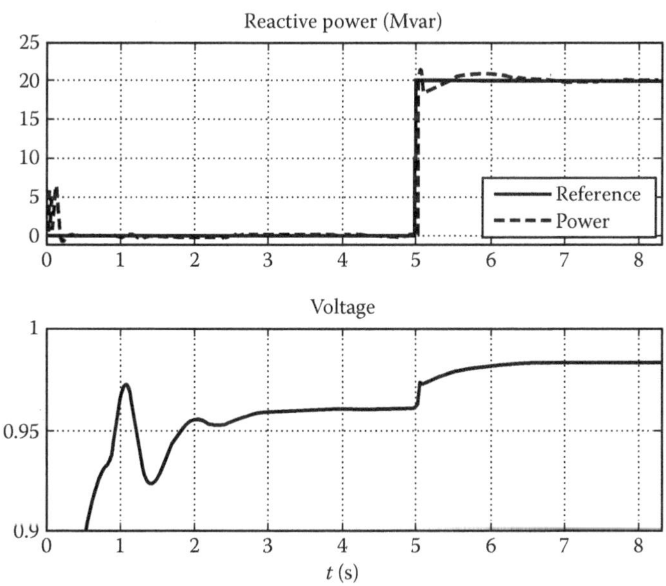

FIGURE 3.66 Reactive power control by STATCOM with cascaded H-bridge multilevel inverter.

STATCOM in **STATCOM_Hinv1** ensures the little content of the higher harmonic in the output voltage, but it is reached at the expense of the high frequency of switching-over inverter devices that makes STATCOM employment difficult for the units of large power. The utilization of the selected harmonic elimination (SHE) method gives an opportunity to have the only device switching-over (switch on/switch off) for the period of the fundamental frequency. It is obtained in the following way.

Each block of the multilevel inverter, which is made as it is depicted in Figure 3.17, switches on once for the half period of the wanted output voltage with the angle θ_i and switches off with the angle $\pi - \theta_i$, producing, through this way, the stepped output voltage as displayed in Figure 3.67. The Fourier series for the inverter phase voltage is

$$V = \frac{4}{\pi} U_{dc} \sum_{i=1}^{n} \left(\cos\theta_i \sin\omega t + \frac{1}{3}\cos 3\theta_i \sin 3\omega t + \frac{1}{5}\cos 5\theta_i \sin 5\omega t + \ldots \right), \quad (3.36)$$

where U_{dc} is the DC voltage of one block and n is the number of the phase blocks.

It is also supposed that $0 < \theta_i < \pi/2$. Then the amplitude of the first harmonic is

$$V_1 = \frac{4}{\pi} U_{dc} \sum_{i=1}^{n} \cos\theta_i, \quad (3.37)$$

and the amplitude of the harmonic with the index k is

$$V_k = \frac{4}{\pi k} U_{dc} \sum_{i=1}^{n} \cos k\theta_i. \quad (3.38)$$

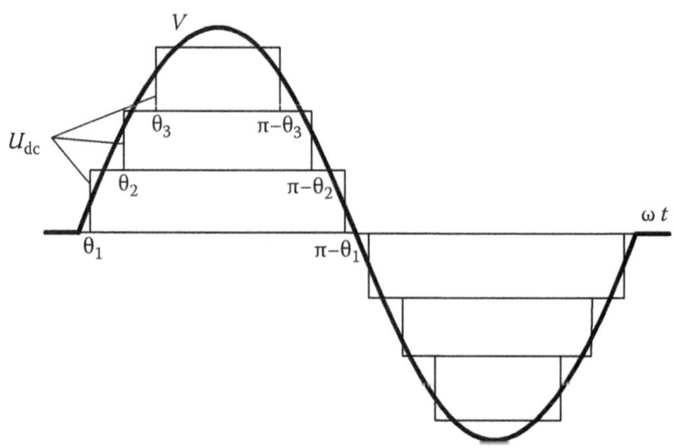

FIGURE 3.67 Voltage forms in STATCOM with SHE.

If the amplitude of the first harmonic has to be V_1, and the amplitude of the kth harmonic has to be zero, the switching on angles must satisfy the following set of equations:

$$\sum_{i=1}^{n} \cos\theta_i = \frac{\pi V_1}{4U_{dc}} = mn \tag{3.39}$$

$$\sum_{i=1}^{n} \cos k\theta_i = 0, \tag{3.40}$$

where m is the modulation factor that is equal to $\pi V_1/4U_d$ and U_d is the total DC voltage of a phase.

In all, $n - 1$ equations as Equation 3.40 can be set up, therefore, under given modulation factor, and $n - 1$ harmonics can be excluded.

The necessary switching angles are computed by the joint solution of Equation 3.39 and $n - 1$ equations of Equation 3.40 for $\theta_1, \ldots, \theta_n$. This system of equations can be solved only by numerical methods, so it cannot be solved online but offline, and the solution results have to be saved as tables. The solution of such sets of equations proves that it exists only in the limiting range of m changing; this fact excludes the utilization of the SHE method in the cases when the m value changes over the extensive range, for example, for the electrical drives, but it is possible in STATCOM. It can be noted that several solutions often exist, in such a case that there is the opportunity to choose one of them, which is preferable by some figure of merit.

It is usually reasonable to exclude the low-order harmonics because the remaining ones are easier to filter. Because the triple harmonics do not manifest in the phase-to-phase voltage of the three-phase system, the harmonics of the orders 5, 7, 11, 13, and so on are excluded.

At first, STATCOM with $n = 3$ is considered, as in the model **STATCOM_Hinv1**. The 5th and 7th harmonics can be excluded; the switching angles are found from the set of the following equations:

$$\cos\theta_1 + \cos\theta_2 + \cos\theta_3 = 3m$$

$$\cos 5\theta_1 + \cos 5\theta_2 + \cos 5\theta_3 = 0$$

$$\cos 7\theta_1 + \cos 7\theta_2 + \cos 7\theta_3 = 0. \tag{3.41}$$

The MATLAB function *fsolve* is used for the solution of this set:

Program Sel_tab_3

```
x0 = [0.71;1.14;1.55];
options = optimset('Display','iter');% Option to display output
global m
for i = 1:48;
```

```
  m = 0.36+i*0.01
[x,fval] = fsolve(@myfun2,x0,options)% Call solver
y(i,:) = x
x0 = x;
end;
plot(y(:,1),'-k')
hold on
plot(y(:,2),'—k')
hold on
plot(y(:,3),'-.k')
title('Teta1, Teta2, Teta3,rad')
xlabel('i = 100(m-0.36)')
legend('Teta1','Teta2','Teta3')
grid
```

In this program, beginning from the given vector of the initial values x_0, the set of equations is solved that is written in the program myfun2, with m step of 0.01. The initial vector for every step is the solution that was received on the previous step. After the computation, the plots of the calculated angles are built (Figure 3.68), and in MATLAB Workspace the 48 × 3 matrixes of these quantities are formed. The solution that satisfies the condition $0 < \theta_1 < \theta_2 < \theta_3 < \pi/2$ exists under $0.38 \le m \le 0.84$.

Program myfun2

```
function F = myfun2(x)
global m
F = [cos(x(1)) + cos(x(2)) + cos(x(3))-3*m;
cos(5*x(1)) + cos(5*x(2)) + cos(5*x(3));
cos(7*x(1)) + cos(7*x(2)) + cos(7*x(3))]
```

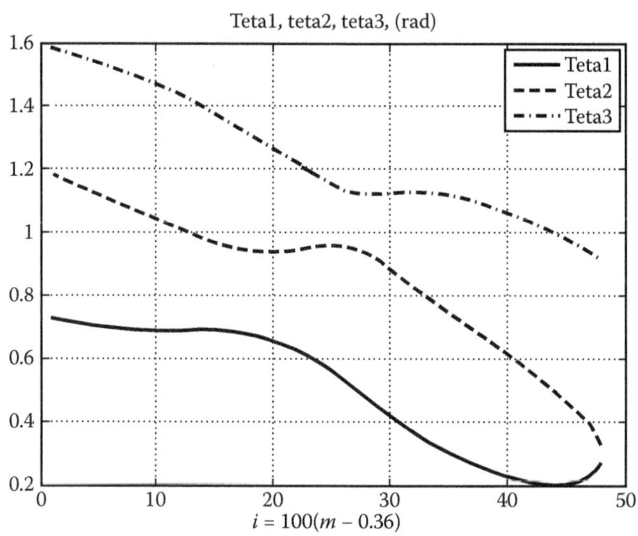

FIGURE 3.68 SHE switching angles for $n = 3$.

The model **STATCOM_3_SHEM_DC** is intended for the demonstration of the main features of the considered method. STATCOM has to keep the load voltage. The model is analogous to be utilized in the model **STATCOM_Hinv1**, but the capacitors are replaced with DC sources of 2000 V. The modulation factor is determined as a sum of some constant value and the controller (of the voltage or the power) output. This value is converted in the number i for input in the tables. In the control system of each phase, the given sinusoidal signal having the amplitude 1 is transformed into in-phase triangular signal with the amplitude of $\pi/2$, with the help of the function arcsin. In each phase, there are three tables for the angles θ_1, θ_2, and θ_3 corresponding to input i. The triangular signal is compared with the table outputs, fabricating the gate pulses in the needed moments of time.

In order to examine the operation of the isolated inverter, the breakers **Br1** and **Br2** are open, and the tumbler **SW_Contr** is in the right position. In carrying out simulation with different m values, one can observe the changes of the phase, the phase-to-phase voltages, and the harmonics.

The phase and the phase-to-phase voltages are shown in Figures 3.69 and 3.70 for three m values. Note that under minimal value $m = 0.38$, the third voltage step turns out to be very narrow and disappears, under accepted discreteness. The processing of these plots with the help of the option *Powergui/FFT* gives the following values of the fundamental harmonic amplitudes of the phase-to-phase voltages: (numerator—measured value, denominator—design one) and the relative amplitudes of the 5th and 7th harmonics: $m = 0.84$, $V_1 = 11{,}110/11{,}107$ V, 0.03%, 0.01%; $m = 0.64$, $V_1 = 8476/8463$ V, 0.03%, 0.03%; and $m = 0.4$, $V_1 = 5293/5289$ V, 0.03%, 0.04%.

For the closed system simulation, the breaker **Br2** closes, and both tumblers in the control system are set in the left position. The breaker **Br1** closes at $t = 2.1$ s. As a

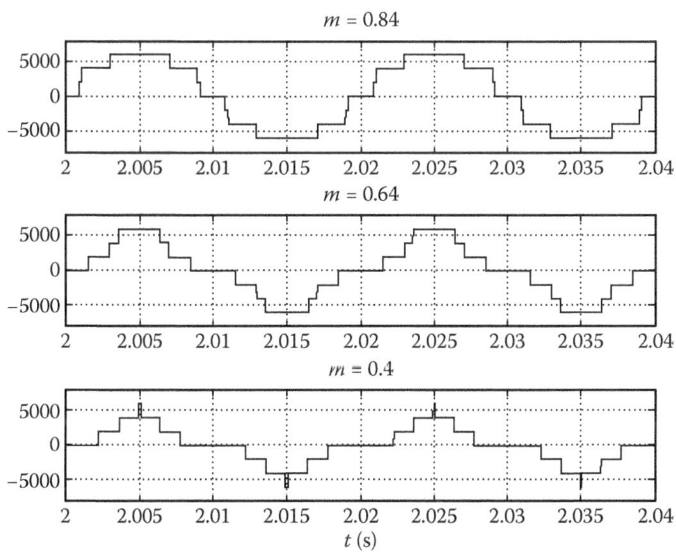

FIGURE 3.69 Phase voltage in STATCOM with SHE for $n = 3$.

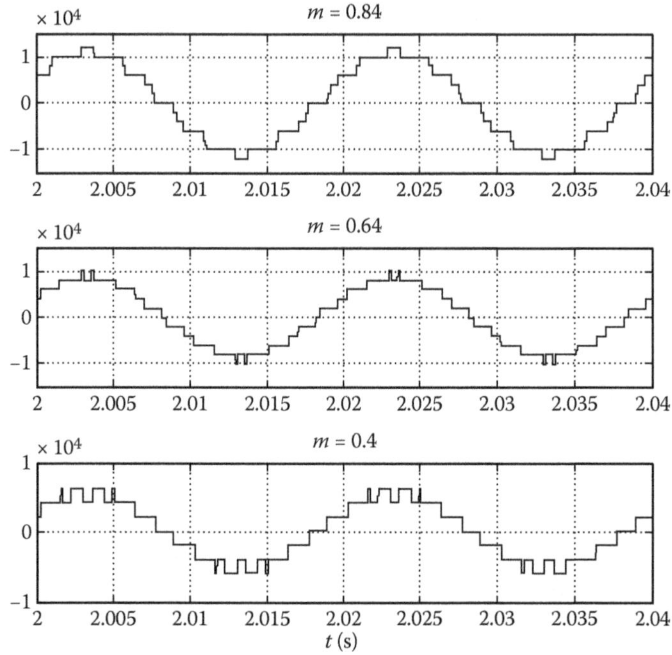

FIGURE 3.70 Phase-to-phase voltage in STATCOM with SHE for $n = 3$.

result of simulation, the curve of the PCC voltage change and also the change of the table input *i* are obtained, which are depicted in Figure 3.71a and b.

In reality, the capacitors are used as the sources for module supply. At this, the problems arise both to keep the total capacitor voltage (6 kV in the case) and to keep individual capacitor voltages. The first problem is decided as for the model **STATCOM_Hinv1** by the addition of the voltage component, which is perpendicular to the voltage space vector in PCC and changes, with the help of the controller of the total capacitor voltage. The second problem is more complicated and can be decided by the rearrangement of the modules that operate with the given angles. Suppose that the phase current direction is such that the capacitor voltages of the modules, which are active in the moment, increase. Then the module, whose capacitor has the maximal voltage, must be active for the shortest time and work with the switching angle θ_3, the module with the next voltage value—with the switching angle θ_2, and, at last, the module with the least voltage—with switching angle θ_1. If the phase current direction (relative voltage) is such that the capacitor voltages decrease, the inverse condition has a place.

The described principles are realized in the model **STATCOM_3_SHEM**. From the previous one, it distinguishes with the subsystem **Distrib** that is set between the subsystem **Teta_form** that forms the gate pulses for the inverter modules and the proper modules; this subsystem assigns the modules that operate with the switching angles θ_1, θ_2, and θ_3 depending on the module capacitor voltages. It is made in the following way.

Now it is reasonable to look at the subsystem circuits (they are too complicated for the figures in the book). Suppose for clarity that the capacitor voltages are 1950, 2100,

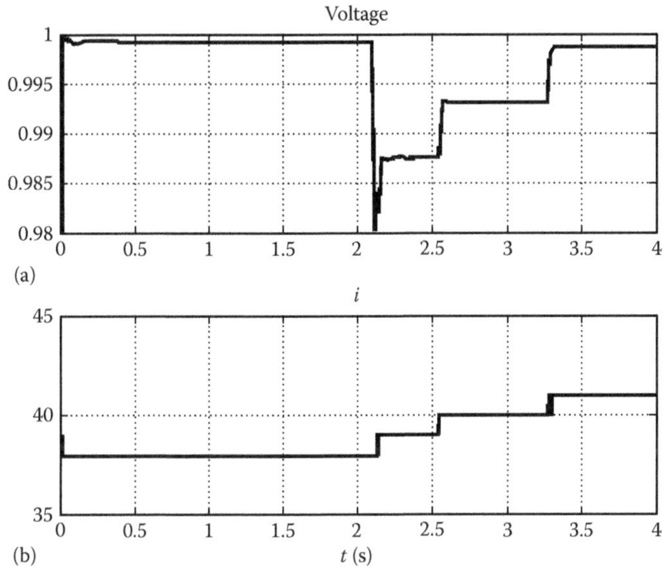

FIGURE 3.71 Voltage control in STATCOM with SHE for $n = 3$. (a) PCC voltage and (b) inputs for table.

and 1900V. The vector [1950, 2100, 1900] comes to the input U_c. The MATLAB function `fcn` is defined as `[B, IX] = sort (u)`. According to this function, the vector of the indexes appears at the output `IX`; at this, the indexes are arranged in the order in which the voltages increase; in the case under consideration, it will be the vector [3, 1, 2]. In the selector, the index order inverts. Depending on the sign **Signiz**, a direct or reverse index vector comes to the input of the triggered subsystem, which is intended for saving the index vector for the time between the consecutive arrival of the pulses **Change** that come when the demanded inverter output voltage crosses zero level; it is in the moment when the count-off of the angles for fabrication of the next half wave of the output voltage begins.

The index vector comes to three blocks for the index selection; the number of the selected index is equal to the number of the block control input IC. Let the index sequence be direct. At the first block IC = 3, the block output = 2. The block for comparison with the number 2 produces the logic signal 1 that sets the **Multiport Switch1** in the first position, so that the switch output (at the gates of the module 2) is the signal that corresponds to θ_3, when the capacitor voltage increase is minimal. At the second block IC = 2, the block output = 1. The corresponding block for comparison, whose output is multiplied by 2, produces the logic signal 1 that sets the **Multiport Switch** in the second position, so that the switch output (at the gates of the module 1) is the signal that corresponds to θ_2. At last, the output of the third block for the index selection = 3. At the output of **Multiport switch2** (at the gates of the nodule 3), whose voltage is minimal, the signal corresponding θ_1 appears, under which the rise of the capacitor voltage is maximal.

Under simulation, it is suggested that the capacitors have been charged preliminary to the voltage of 2 kV. The breaker **Br1** closes at $t = 12$ s. The amplitude of the

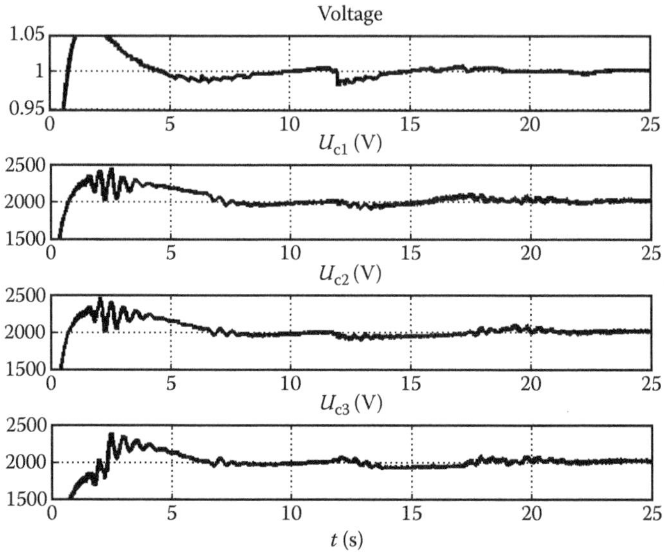

FIGURE 3.72 Capacitor voltage balancing in STATCOM with SHE for $n = 3$.

phase-to-phase voltage in pu in PCC and the voltages across the capacitors of the phase A modules are shown in Figure 3.72. It is seen that after initial transients, the capacitor voltages keep sufficiently accurate. The modulation factor index on the process end (the scope i) is 41; it means that $m = 0.36 + 0.41 = 0.77$. Then the inverter phase-to-phase amplitude must be $4 \times 6000 \times 0.77 \times (1.73/\pi) = 10{,}188$ V; it can be found that, with the help of the option *Powergui/FFT*, $V_{ph\text{-}ph} = 10{,}200$ V. The amplitudes of the 5th and 7th harmonics are nearly zero.

The processes under the reactive power step changing are displayed in Figure 3.73.

With the power rising, the number of the modules in a phase increases too. So STATCOM with $n = 5$ found industrial application [17]. At this, there is an opportunity to exclude not only the 5th and 7th, but also the 11th and 13th harmonics. The function `myfun` that is used under the solution of the system of equation acquires the following form:

```
function F = myfun(x)
global m
F = [cos(x(1)) + cos(x(2)) + cos(x(3)) + cos(x(4)) +
cos(x(5))-5*m;
cos(5*x(1)) + cos(5*x(2)) + cos(5*x(3)) + cos(5*x(4)) +
cos(5*x(5));
cos(7*x(1)) + cos(7*x(2)) + cos(7*x(3)) + cos(7*x(4)) +
cos(7*x(5));
cos(11*x(1)) + cos(11*x(2)) + cos(11*x(3)) + cos(11*x(4)) +
cos(11*x(5));
cos(13*x(1)) + cos(13*x(2)) + cos(13*x(3)) + cos(13*x(4)) +
cos(13*x(5))]
```

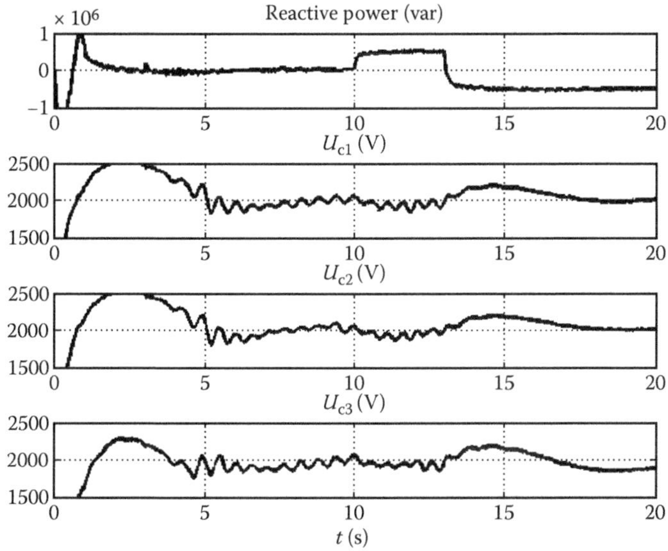

FIGURE 3.73 Reactive power control with STATCOM with SHE for $n = 3$.

The solution of these equations, with the help of the function `fsolve`, proves that the solution exists under $0.44 \le m \le 0.7$ and $0.75 \le m \le 0.84$. There are two opportunities under system development: (1) To change the capacitor voltages in such a manner that the factor m was always in the interval where a solution exists. It can be made rather slowly, and the inverter devices can be utilized not always efficiently. (2) To redefine the switching angles in the intervals where the solution does not exist in some other way. Specifically, it is possible to take $\theta_4 = \theta_5$ in this interval and to define the function `myfun1` in it as follows:

```
function F = myfun1(x)
global m
F = [cos(x(1)) + cos(x(2)) + cos(x(3)) + 2*cos(x(4))-5*m;
cos(5*x(1)) + cos(5*x(2)) + cos(5*x(3)) + 2*cos(5*x(4));
cos(7*x(1)) + cos(7*x(2)) + cos(7*x(3)) + 2*cos(7*x(4));
cos(11*x(1)) + cos(11*x(2)) + cos(11*x(3)) + 2*cos(11*x(4))]
```

It means not to exclude the 13th harmonic. The program `Sel_tab_5_1` is used at that:

```
Sel_tab_5_1
y = zeros(41,5)
x0 = [0.29; 0.46; 0.8; 1.06;1.09];
global m
for i = 1:27;
    m = 0.43+i*0.01
[x,fval] = fsolve(@myfun,x0)
y(i,:) = x
```

```
x0 = x;
end;
x0 = [0.111;0.263;0.41;0.95];
for i = 28:31;
    m = 0.71-0.28+i*0.01
[x,fval] = fsolve(@myfun1,x0)
y(i,1:4) = x
y(i,5) = y(i,4)
x0 = x;
end;
a = y(28:31,1)
b = y(28:31,2)
y(28:31,2) = a
y(28:31,1) = b
x0 = [0.188;0.362;0.592;0.923;1.1];
for i = 32:41;
    m = 0.75-0.32+i*0.01
[x,fval] = fsolve(@myfun,x0)%,options)% Call solver
y(i,:) = x
x0 = x;
end;
plot(y(:,1),'-k')
hold on
plot(y(:,2),'—k')
hold on
plot(y(:,3),'-.k')
hold on
plot(y(:,4),'-ok')
hold on
plot(y(:,5),'-xk')
title('Teta1, Teta2, Teta3,Teta4,Teta5,rad')
xlabel('i = 100(m-0.43)')
legend('Teta1','Teta2','Teta3','Teta4','Teta5')
grid
```

In the interval $i = 28–31$, it turns out that $\theta_2 < \theta_1$; in order to have the former order of the angle change, the columns for these values i are permutated. The switching angles for this method are shown in Figure 3.74. It is also possible to make linear interpolation of the angle values in the interval [0.7–0.74] with the help of the program `Sel_tab_5_2`:

```
y = zeros (41, 5)
x0 = [0.29; 0.46; 0.8; 1.06; 1.09];
global m
for i = 1:27;
    m = 0.43+i*0.01
[x,fval] = fsolve(@myfun,x0)
y(i,:) = x
x0 = x;
end;
x0 = [0.188;0.362;0.592;0.923;1.1];
```

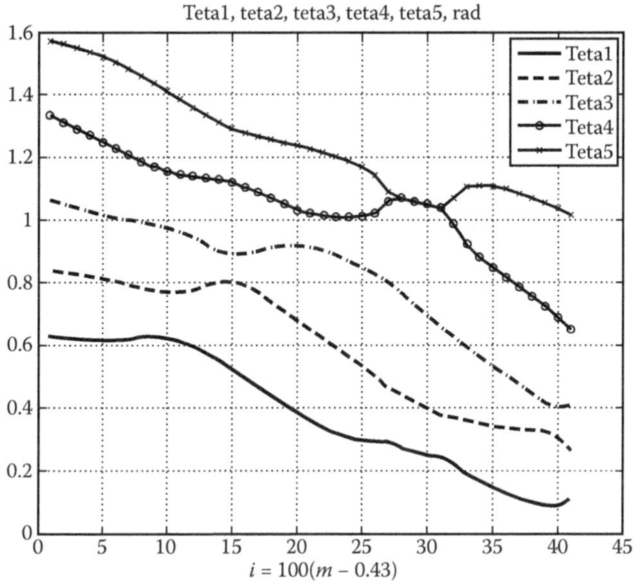

FIGURE 3.74 SHE switching angles for $n = 5$, the first option.

```
for i = 32:41;
    m = 0.75-0.32+i*0.01
[x,fval] = fsolve(@myfun,x0)
y(i,:) = x
x0 = x;
end;
y(28,:) = y(27,:)+(y(32,:)-y(27,:))*0.2
y(29,:) = y(27,:)+(y(32,:)-y(27,:))*0.4
y(30,:) = y(27,:)+(y(32,:)-y(27,:))*0.6
y(31,:) = y(27,:)+(y(32,:)-y(27,:))*0.8
plot(y(:,1),'-k')
hold on
plot(y(:,2),'—k')
hold on
plot(y(:,3),'-.k')
hold on
plot(y(:,4),'-ok')
hold on
plot(y(:,5),' xk')
title('Teta1, Teta2, Teta3,Teta4,Teta5,rad')
xlabel('i = 100(m-0.43)')
legend('Teta1','Teta2','Teta3','Teta4','Teta5')
grid
```

The switching angles that are calculated this way are shown in Figure 3.75.

The simple model **Selection_invest** is used for the primary investigation of the described methods. The sinusoidal signal of 50 Hz is transformed in the triangular

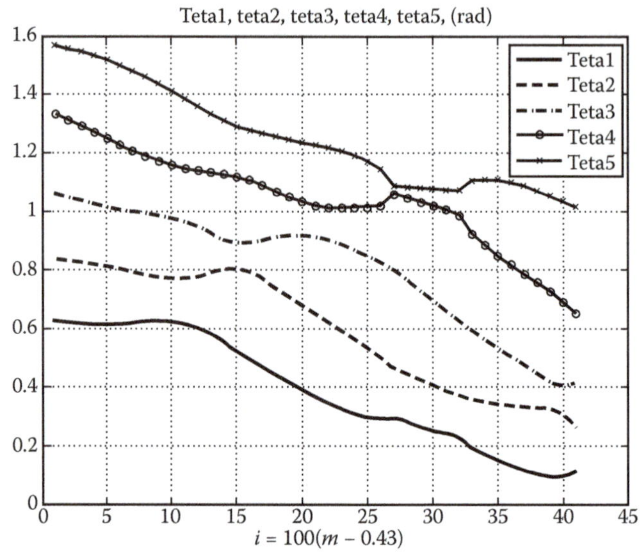

FIGURE 3.75 SHE switching angles for $n = 5$, the second option.

waveform that compares, with the help of the blocks **Relay**, with the angles that are defined by the matrix **y**, having a dimension of 41 × 5, obtained as a result of the execution of the programs given earlier. At the summer outputs, the corresponding stepped signals V_1 and $V2$ are fabricated; at this, the latter signal is shifted about the former by 120° that gives the possibility to form, at the output of the subtracter Sum2, an imitation of the phase-to-phase voltage. The factor m is transformed in the number of the row of the matrix **y**, whose elements determine the switching angles.

Let us write down `run Sel_tab_5_1` in the option *File/Model Properties/Callbacks/Init Fcn* and carry out simulation for m = 0.45, 0.84, 0.73. For that, in the same option, it is necessary to define successively i = 2, 41, 30. The following results are obtained. For m = 0.45, the amplitude of the fundamental harmonic of the phase-to-phase voltage A_1 = 4.964/4.962 (the theoretical value $\sqrt{3} \times (4/\pi) \times 5m$ is indicated in a dominator), the amplitudes of the 5th, 7th, 11th, and 13th harmonics relatively fundamental are respectively 0.02%, 0.07%, 0.02%, and 0.07%, THD = 9.17%. Respectively for m = 0.84, 9.263/9.266, 0%, 0.01%, 0.02%, and 0.03%, THD = 5.3%; for m = 0.73 (there is no solution), 8.051/8.049, 0.01%, 0.05%, 0.06%, and 1.2%, THD = 7.1%.

To replace `run Sel_tab_5_1` for `run Sel_tab_5_2` in the option *File/Model Properties/Callbacks/Init Fcn* and to repeat simulation for m = 0.73, it will obtain 8.062/8.049, 0.49%, 0.05%, 0.1%, and 0.94%, THD = 6.1%. Therefore, the 5th harmonic increases but the total harmonic content slightly decreases.

STATCOM_5_SHEM models the described system. Unlike the previous system, each phase has five modules with the same voltage of 2 kV. The AC voltage is 10.6 kV. The load power is doubled. The same principle is employed for the capacitor voltage balancing, but with complicated circuits because the module number increases.

The process voltage control in PCC is shown in Figure 3.76. The breaker Br1 closes at $t = 15$ s. The PCC phase-to-phase voltage and the voltages across the capacitors of the phase A are shown in Figure 3.76. After the initial transients, the capacitor voltages are kept with sufficient accuracy. The index of the modulation factor at $t = 24.5$ s is 30; it means that $m = 0.30 + 0.43 = 0.73$. The amplitude of the phase-to phase voltage has to be $4 \times 10{,}000 \times 0.73 \times \sqrt{3}/\pi = 16{,}099$ V; the option *Powergui/FFT* gives a result $V_{ph\text{-}ph} =$ 16,160 V. The amplitudes of the 5th, 7th, 11th, and 13th harmonics are equal to 0.17%, 0.15%, 0.1%, and 1.34%, respectively. The large value of the 13th harmonic is explained by the fact that m is exactly in the interval where the solution is absent. For instance, at $t = 17.5$ s, when $m = 0.75$, the relative harmonic amplitudes are 0.05%, 0.08%, 0.01%, and 0.04%. The phase and the phase-to-phase voltages are depicted in Figure 3.77.

The processes under the change of the reference of the reactive power are shown in Figure 3.78.

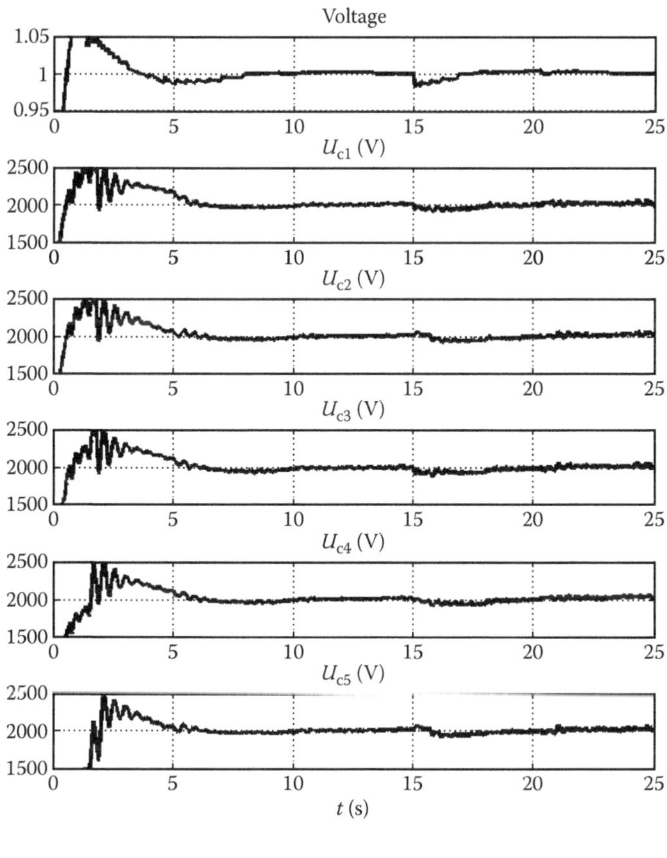

FIGURE 3.76 Phase-to-phase voltage and the voltages across the capacitors in STATCOM with SHE, $n = 5$.

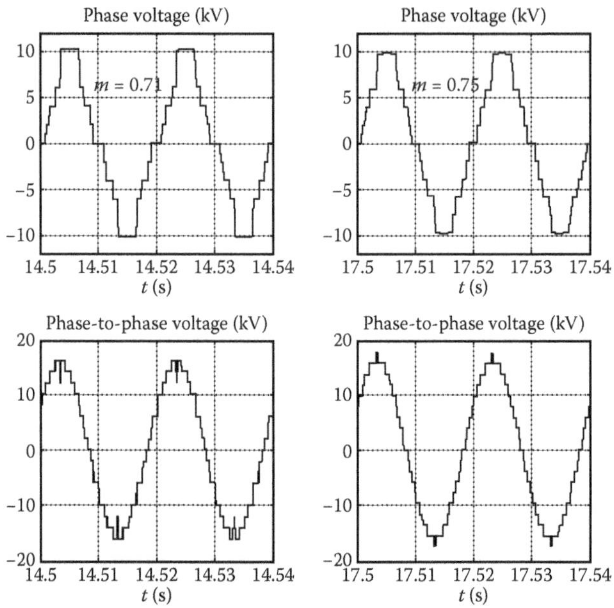

FIGURE 3.77 Phase and phase-to-phase voltages in STATCOM with SHE, $n = 5$.

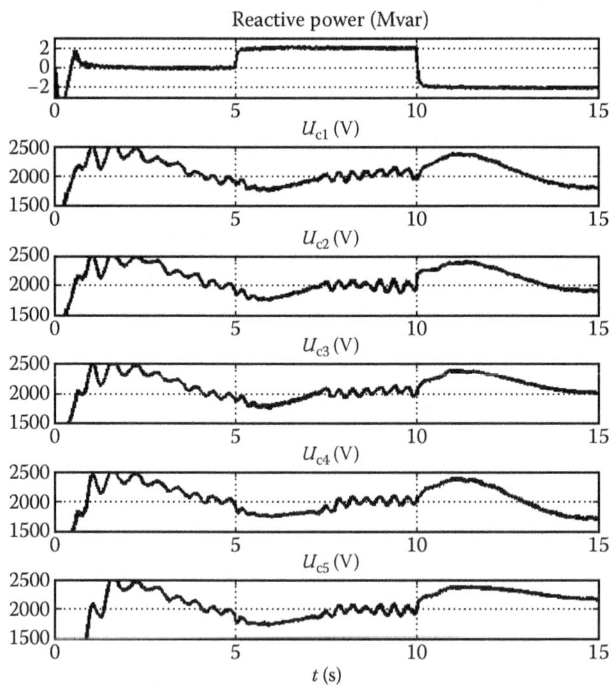

FIGURE 3.78 Reactive power changing with STATCOM with SHE, $n = 5$.

REFERENCES

1. MathWorks, *SimPowerSystems™: User's Guide.* MathWorks, Natick, MA, 2011–2016.
2. Perelmuter, V. *Electrotechnical Systems, Simulation with Simulink and SimPowerSystems*, CRC Press, Boca Raton, FL, 2012.
3. Bose, B. K. *Modern Power Electronics and AC Drives.* Prentice Hall PTR, Upper Saddle River, NJ, 2002.
4. Xu, L. Grid integration of offshore wind farms. In *Advances in Wind Energy Conversion Technology.* Springer-Verlag, Berlin, Heidelberg, 2011.
5. Zhang, R., Boroyevich, D., Prasad, H., Mao, H., Lee, F. C., and Dubovsky, S. A three-phase inverter with a neutral leg with space vector modulation. *Conference Record of IEEE APEC*, 1997, vol. 2 pp. 857–863.
6. Kim, J.-H., and Sul, S.-K. A carrier-based PWM method for three-phase four-leg voltage source converters. *IEEE Transactions on Power Electronics*, 19(1), 2004, 66–75.
7. Peng, F. Z. Z-source inverter. *IEEE Transaction on Industry Application*, 39(2), 2003, 504–510.
8. Ding, X., Qian, Z., Yang, S., Cuil, B., and Peng, F. Z. A direct peak DC-link boost voltage control strategy in Z-source inverter. *Proceedings of IEEE Applied Power Electronics Conference, Sydney, New South Wales*, 2007, pp. 648–653.
9. Ding, X., Qian, Z., Yang, S., Cuil, B., and Peng, F. Z. A PID control strategy for DC-link boost voltage in Z-source inverter. *Proceedings of IEEE Applied Power Electronics Conference, Sydney, New South Wales*, 2007, pp. 1145–1148.
10. Blaabjerg, F., Liserre, M., and Ma, K., Power electronics converters for wind turbine systems. *IEEE Transaction on Industry Applications*, 48(2), 2012, 712–719.
11. Casadei, D., Serra, G., Tani, A., and Zarri, L., Matrix converter modulation strategies: A new general approach based on space-vector representation of the switch state. *IEEE Transaction on Industrial Electronics*, 49(2), 2002, 370–381.
12. Wei, L., and Lipo, T. A. A novel matrix converter topology with simple commutation. *Conference Records of 36th IEEE IAS Annual Meeting*, 3, 2001, 1749–1754.
13. Hamouda, M., Fnaiech, F., and Al-Haddad, K. Space vector modulation scheme for dual-bridge matrix converters using safe-commutation strategy. *31st Annual Conference of IEEE Industrial Electronics Society*, 2005, pp. 1060–1065.
14. Kolar, J. W., Friedli, T., Rodriguez, J., and Wheeler, P. W. Review of three-phase PWM AC-AC converter topologies. *IEEE Transactions on Industrial Electronics*, 58(11), 2011, 4988–5006.
15. Chen, Z., Guerrero, J. M., and Blaabjerg, F. A review of the state of the art of power electronics for wind turbines. *IEEE Transactions on Power Electronics*, 24(8), 2009, pp. 1859–1875.
16. Hingorani, N. G., and Gyugyi, L. *Understanding FACTS: Concepts and Technology of Flexible AC Transmission Systems.* IEEE Press, New York, 2000.
17. Gultekin, B., Gerçek, C. O., and Atalik, T. et al. Design and implementation of a 154-kV ±50-Mvar transmission STATCOM based on 21-level cascaded multilevel converter. *IEEE Transactions on Industry Applications*, 48(3), 2012, May/June, 1030–1045.

4 Electric Generators

4.1 INDUCTION GENERATORS

4.1.1 Model Description

Induction generators (IGs) are today the most widespread type of electrical machines in renewable electric sources. The equations for the IG voltages are

$$U_{sa} = R_{sa}i_{sa} + \frac{d\Psi_{sa}}{dt}$$
$$U_{sb} = R_{sb}i_{sb} + \frac{d\Psi_{sb}}{dt} \quad (4.1)$$
$$U_{sc} = R_{sc}i_{sc} + \frac{d\Psi_{sc}}{dt}$$

and

$$U_{ra} = R_{ra}i_{ra} + \frac{d\Psi_{ra}}{dt}$$
$$U_{rb} = R_{rb}i_{rb} + \frac{d\Psi_{rb}}{dt} \quad (4.2)$$
$$U_{rc} = R_{rc}i_{rc} + \frac{d\Psi_{rc}}{dt}.$$

Here U_s and U_r are the stator and rotor phase voltages, respectively, R_s and R_r are the active resistances of the stator and rotor windings, respectively; at that, the resistances of the different phases can be different; Ψ_s and Ψ_r are the flux linkages of the stator and rotor phases, respectively; and i_s and i_r are the phase currents of the stator and the rotor, respectively. All the quantities refer to the stator.

IG modeling is carried out, as a rule, with the transformation of the three-phase IG to two phase, but sometimes the equations of the three-phase machine must be used, for instance, under different resistances of the phase windings.

The transformation to the equivalent two-phase IG is carried out with the replacement of the three-phase quantities, for example, voltages U_a, U_b, and U_c, with two quantities U_q and U_d in the reference frame that rotates with a speed ω by the formulas

$$U_q = 2/3\{U_a \cos\theta + U_b \cos[\theta - (2\pi/3)] + U_c \cos[\theta + (2\pi/3)]\}$$
$$U_d = 2/3\{U_a \sin\theta + U_b \sin[\theta - (2\pi/3)] + U_c \sin[\theta + (2\pi/3)]\}. \quad (4.3)$$

Here θ is an angle between the phase a axis and the axis q of the rotating frame. It is supposed that the IG windings have a star connection with an isolated neutral, so that zero component is absent.

Since the phase-to-phase voltages are applied to the IG, after substitution

$$\begin{aligned}U_{ca} &= -U_{ab} - U_{bc},\\ U_a &= \frac{U_{ab} - U_{ca}}{3},\\ U_b &= \frac{U_{bc} - U_{ab}}{3},\\ U_c &= \frac{U_{ca} - U_{bc}}{3},\end{aligned} \tag{4.4}$$

it comes out for the stator

$$\begin{aligned}U_{qs} &= \frac{1}{3}\left[2\cos\theta U_{abs} + (\cos\theta + \sqrt{3}\sin\theta)U_{bcs}\right],\\ U_{ds} &= \frac{1}{3}\left[2\sin\theta U_{abs} + (\sin\theta - \sqrt{3}\cos\theta)U_{bcs}\right].\end{aligned} \tag{4.5}$$

Analogous for the rotor

$$\begin{aligned}U_{qr} &= \frac{1}{3}\left[2\cos\beta U_{abr} + (\cos\beta + \sqrt{3}\sin\beta)U_{bcr}\right],\\ U_{dr} &= \frac{1}{3}\left[2\sin\beta U_{abr} + (\sin\beta - \sqrt{3}\cos\beta)U_{bcr}\right],\end{aligned} \tag{4.6}$$

where $\beta = \theta - \theta_r$, and θ_r is the rotor angle position.

For the squirrel cage IG, $U_{dr} = U_{qr} = 0$.

Three reference frames can be used: stationary, $\theta = \omega = 0$; synchronous,

$$\theta = \int \omega_s \, dt, \tag{4.7}$$

where ω_s is a synchronous speed (that is the frequency of the supplying stator voltage), $\omega = \omega_s$; connected with the rotor, $\omega = \omega_r = Z_p\omega_m$,

$$\theta = \int \omega_r \, dt, \tag{4.8}$$

where Z_p is the number of the pole pairs.

The transformation of the I_q and I_d currents in the phase currents is carried out by the formulas

$$\begin{aligned} i_{as} &= \cos\theta i_{qs} + \sin\theta i_{ds} \\ i_{bs} &= 0.5\left[\left(-\cos\theta + \sqrt{3}\sin\theta\right) i_{qs} - \left(\sin\theta + \sqrt{3}\cos\theta\right) i_{ds}\right] \\ i_{sc} &= -i_{as} - i_{bs} \end{aligned} \tag{4.9}$$

and analogous for the rotor, by changing θ for β. All the rotor quantities refer to the stator.

In the q-d reference frame, the IG equations are as follows:

$$\begin{aligned} U_{qs} &= R_s i_{qs} + \frac{d\Psi_{qs}}{dt} + \omega\Psi_{ds} \\ U_{ds} &= R_s i_{ds} + \frac{d\Psi_{ds}}{dt} - \omega\Psi_{qs} \end{aligned} \tag{4.10}$$

$$\begin{aligned} U_{qr} &= R_r i_{qr} + \frac{d\Psi_{qr}}{dt} + (\omega - \omega_r)\Psi_{dr} \\ U_{dr} &= R_r i_{dr} + \frac{d\Psi_{dr}}{dt} - (\omega - \omega_r)\Psi_{qr}, \end{aligned} \tag{4.11}$$

$$T_e = 1.5Z_p(\Psi_{ds} i_{qs} - \Psi_{qs} i_{ds}), \tag{4.12}$$

$$T_e = 1.5Z_p[L_m/(L_m + L_{1r})](\Psi_{dr} i_{qs} - \Psi_{qr} i_{ds}), \tag{4.13}$$

$$\begin{aligned} \Psi_{qs} &= (L_m + L_{1s})i_{qs} + L_m i_{qr}, \\ \Psi_{ds} &= (L_m + L_{1s})i_{ds} + L_m i_{dr}, \end{aligned} \tag{4.14}$$

$$\begin{aligned} \Psi_{qr} &= (L_m + L_{1r})i_{qr} + L_m i_{qs} \\ \Psi_{dr} &= (L_m + L_{1r})i_{dr} + L_m i_{ds} \end{aligned} \tag{4.15}$$

Here T_e is the IG electromagnetic torque, $L_s = L_{1s} + L_m$, L_{1s} is the stator leakage inductance, L_m is the magnetizing inductance, and $L_r = L_{1r} + L_m$, L_{1r} is the rotor leakage inductance.

The electromechanical equations are

$$J\frac{d\omega_m}{dt} = T_e - f\omega_m - T_m, \tag{4.16}$$

$$\frac{d\theta_r}{dt} = Z_p\omega_m, \tag{4.17}$$

where J is a total moment of inertia of IG and connected parts, f is a viscous friction factor, T_m is a load torque, and $T_m < 0$ in generator mode.

IG equations may be written as follows:

When introducing the vector $\mathbf{\Psi} = [\Psi_{qs}, \Psi_{ds}, \Psi_{qr}, \Psi_{dr}]$, the vectors **U** and **I** that are arranged by the same way, the diagonal 4 × 4 matrix of the resistances **R**, the matrix of the inductances **L**

$$\mathbf{R} = \begin{bmatrix} R_s & 0 & 0 & 0 \\ 0 & R_s & 0 & 0 \\ 0 & 0 & R_r & 0 \\ 0 & 0 & 0 & R_r \end{bmatrix} \mathbf{L} = \begin{bmatrix} L_m + L_{ls} & 0 & L_m & 0 \\ 0 & L_m + L_{ls} & 0 & L_m \\ L_m & 0 & L_m + L_{lr} & 0 \\ 0 & L_m & 0 & L_m + L_{lr} \end{bmatrix} \tag{4.18}$$

and also the matrix

$$\mathbf{W} = \begin{bmatrix} 0 & \omega & 0 & 0 \\ -\omega & 0 & 0 & 0 \\ 0 & 0 & 0 & \omega - \omega_r \\ 0 & 0 & \omega_r - \omega & 0 \end{bmatrix}, \tag{4.19}$$

then Equations 4.10 and 4.11 may be written as

$$\frac{d\mathbf{\Psi}}{dt} = \mathbf{U} - \mathbf{RI} - \mathbf{W\Psi}, \tag{4.20}$$

and Equations 4.14 and 4.15 as

$$\mathbf{\Psi} = \mathbf{LI}. \tag{4.21}$$

After substitution of Equation 4.21 in Equation 4.20, we obtain

$$\frac{d\mathbf{\Psi}}{dt} = (\mathbf{U} - \mathbf{RL}^{-1}\mathbf{\Psi} - \mathbf{W\Psi})\omega_b. \tag{4.22}$$

In Equation 4.22 and further, all given and calculated values are in pu; specifically, the speed is defined in the ratio of the base speed ω_b that is usually the stator nominal frequency ω_s. IG torque in pu is

$$T_e = \Psi_{ds} i_{qs} - \Psi_{qs} i_{ds}; \tag{4.23}$$

Equation 4.16 changes as

$$2H \frac{d\omega_r}{dt} = T_e - f\omega_r - T_m, \tag{4.24}$$

where T_m is given in pu as well; the inertia constant

$$H = \frac{J\omega_b^2}{2Z_p^2 P_b}, \tag{4.25}$$

where P_b is the nominal IG power (VA).

The base values for IG are defined as follows. The main quantities are the base power P_b, the base voltage V_b, and the base frequency ω_b. P_b is the rated IG power (VA); V_b is the rms of the rated phase voltage. The base value of the current is $I_b = 1/3P_b/V_b$, while that of the torque is $T_b = Z_p P_b/\omega_b$, and that of the resistance is $R_b = V_b/I_b$. The inductances in pu are the ratio of their reactances under the base frequency to R_b. When the instantaneous values of the voltage, the current, and the flux linkage are considered, their rms values multiplied by $\sqrt{2}$ are accepted for the base values.

The IG model using either International System of Units (SI units) or pu units can be chosen in SimPowerSystems, but the simulation is carried out always in pu units; only the units for the torque T_m and for the output values are different. The dialog box has four pages. On the first page, one of the manufactured induction machines with power from 3.7 to 160 kW can be selected for modeling.

The models of the electrical machines can be used in two modes: with the given torque T_m or with the given rotational speed; in the latter case, the mechanical part of the model is disregarded, and the input T_m is replaced by the input *w* on the block picture. This choice and also two following choices are fulfilled on the first page as well. The IG with wound rotor or with squirrel cage rotor can be chosen for the simulation. There is the opportunity to model the induction machine with a double squirrel cage.

One of three reference frames can be selected in the field *Reference frame* for modeling. The general recommendations are as follows [1]: The *Stationary* reference frame is used if the stator voltages are either unbalanced or discontinuous and the rotor voltages are balanced (or 0); the *Rotor* reference frame is used if the rotor voltages are either unbalanced or discontinuous and the stator voltages are balanced; and the *Stationary* or the *Synchronous* reference frame are used if all the voltages are balanced and continuous.

The parameters of IG are put into the corresponding fields on the second page; their meanings are clear from relationships given earlier. If a saturation is modeled, the 2 × n matrix is put down in the next field where the first row defines the stator currents and the second one, the stator voltages under these currents in the no-load condition; n is the number of the points of the magnetization curve. The first point corresponds to the point where the effect of saturation begins. At this, information about L_m disappears.

The sampling time can be specified on the third page under simulation in the discrete mode: when *Sample time* = −1, the sampling time is defined by **Powergui**, or it can be greater by whole times that makes the simulation faster. The integration method under model discretization is specified on this page too.

On the last page, *Load Flow*, the value of the IG power is determined that is utilized by the **Powergui** option *Load Flow Tool.*

This parameter is used for model initialization only and has no impact on the block model or on the simulation performance.

The IG model consists of four subsystems: the power block for connection with the other SimPowerSystems models, the measurement, the electrical part, and the mechanical part. The model of the electrical part solves Equation 4.22, computes the torque by Equation 4.23 and currents by the equation that is reverse to Equation 4.21, and fulfills the calculations necessary for saturation. The relationships 4.24 and 4.17 are carried out in the model of the mechanical part.

Under saturation consideration, the magnetization curve is modeled for the total magnetic flux in the no-load machine. It is carried out in the following way (Figure 4.1).

It may be written as [2]

$$\Psi_{mq} = L_a\left(\frac{\Psi_{qs}}{L_{ls}} + \frac{\Psi_{qr}}{L_{lr}}\right) \tag{4.26}$$

$$\Psi_{md} = L_a\left(\frac{\Psi_{ds}}{L_{ls}} + \frac{\Psi_{dr}}{L_{lr}}\right) \tag{4.27}$$

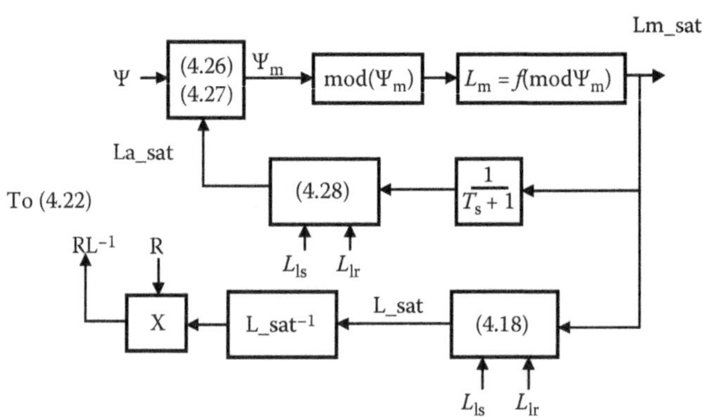

FIGURE 4.1 Block diagram of saturation calculations.

$$L_a = \left(\frac{1}{L_m} + \frac{1}{L_{ls}} + \frac{1}{L_{lr}} \right)^{-1}. \tag{4.28}$$

At first, the components Ψ_{mq} and Ψ_{md} are calculated and afterward the flux module Ψ_m that determines the main inductance taking saturation into account with the employment of the table, which is fabricated by data in the dialog box. The value of this inductance is utilized under the calculations by Equations 4.18, 4.21, and 4.22.

The IG states can be measured by Simulink blocks with the aid of the blocks **Bus Selector** from the Simulink library. Their inputs are connected to the output *m*. At this, the window (windows) is formed, in which the IG quantities available for measurement are indicated; the necessary variables are selected in each particular case.

The modeling of the diesel generator set with the IG is further considered.

The units with the diesel engine (D) and the SG are the most widespread independent supply source. At this, the output voltage frequency is determined by the diesel rotation speed, and the output voltage control is carried out with the action on SG excitation current. The shortcoming of such an arrangement is SG complexity and insufficient reliability. In recent years, owing to the development of power electronics, alternative variants appeared: with SG with permanent magnets (PMSG) and with squirrel cage IG. The VSIs are usually used in such sets. When IG is employed, the following problems arise: the initial IG excitation; the frequency control with action on the engine rotation speed, taking in mind that for IG, the rotation speed and the stator voltage frequency are different owing to IG slip; and the output voltage control. These problems can be decided by using the circuits shown in Figure 4.2 [3].

The squirrel cage IG is driven with the diesel engine D that is equipped, as usual, with the sensor of the rotation speed ω_D. The diesel has the speed control system; in order to

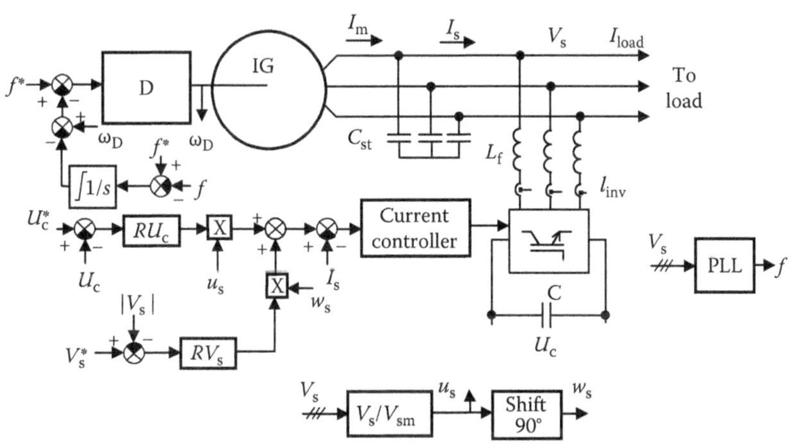

FIGURE 4.2 Block diagram of the system diesel IG.

have the wanted frequency at the IG output, the feedback signal is adjusted by the integral of the difference of the given and actual frequencies. The latter signal is fabricated by PLL. The capacitors C_{st} form the self-excitation circuit for the initial excitation.

The library IG model does not provide a self-excitation function. There is the possibility to define the initial stator currents that under IG start leads to self-excitation, but the obtaining process does not correspond to the actual process, when, during IG rotation, the residual flux induces electromotive force (EMF) in the rotor windings that causes the current in them. That, in its turn, creates the magnetic flux that induces the voltage in the stator windings and the current flow in its windings with the shift of 90°. That causes the flux rise, which leads to voltage increase and so on.

The IG model, which takes into account the self-excitation process, is developed for the model **Diesel_IG**. It operates in the stationary reference frame in pu. The induction machine model structure that was utilized in the previous versions of SimPowerSystems (for instance, in version R2010a, partly in 2012b) is used in this model that consists of six subsystems. The model equations are given in Refs. [2,4]. The subsystem **Uabc—Fabc** computes the stator flux linkages by Equation 4.1; the subsystem **Stator** calculates Ψ_{ds} and Ψ_{qs} by the formulas that are analogous to Equation 4.5, and the currents I_{ds} and I_{qs} by the following:

$$I_{qs} = \frac{\Psi_{qs} - \Psi_{mq}}{L_{ls}}$$
$$I_{ds} = \frac{\Psi_{ds} - \Psi_{md}}{L_{ls}} \tag{4.29}$$

In comparison with the library model, the unit for the rotor supply is added in the subsystem **Rotor**; this unit produces in the rotor windings the voltage with the amplitude of F_0 ($F_0 = 0.02$ is accepted) during T_0 ($T_0 = 1$ s is accepted); relationships 4.30–4.33 are realized in the subsystem **Flux** [2]:

$$\Psi_{mq} = L_a\left(\frac{\Psi_{qs}}{L_{ls}} + \frac{\Psi_{qr}}{L_{lr}}\right) - \frac{L_a}{L_m} f(\Psi_{mq}) \tag{4.30}$$

$$\Psi_{md} = L_a\left(\frac{\Psi_{ds}}{L_{ls}} + \frac{\Psi_{dr}}{L_{lr}}\right) - \frac{L_a}{L_m} f(\Psi_{md}), \tag{4.31}$$

where

$$f(\Psi_{mq}) = \Psi_{mq}\frac{f(\Psi_m)}{\Psi_m} \tag{4.32}$$

$$f(\Psi_{md}) = \Psi_{md}\frac{f(\Psi_m)}{\Psi_m}. \tag{4.33}$$

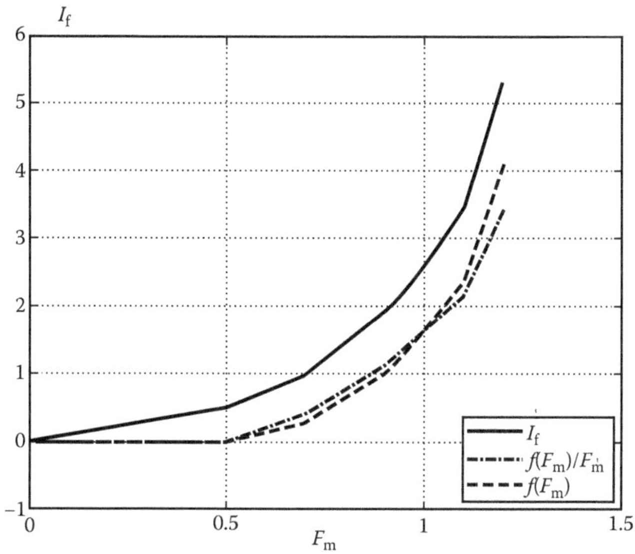

FIGURE 4.3 Computation of the magnetizing curve.

The function $f(\Psi_m)$ is the difference between the coordinates of the straight line having a slope of 45° and the magnetizing curve in pu (Figure 4.3) (the magnetizing current is equal to the stator current because the no-load machine is considered). It is better to set in the table the function $\dfrac{f(\Psi_m)}{\Psi_m}$.

The actual stator currents are calculated in the subsystem **Iqd—Iabc** by Equation 4.9; the subsystem **Mex** is the same as in the library model. Before simulation start, the program `IM_MODEL_0.m` has to be executed that is fixed in the option *File/Model Properties/Callbacks/InitFcn*. The IG parameters, including the magnetizing curve, are specified in this program; the base values are calculated; and the data that are necessary for saturation consideration are computed as well.

The voltage control is fulfilled with VSI that is connected in parallel to an IG output. The active and reactive components of the I_s current are controlled in the VSI control system with the aid of the high-speed regulator of the three-phase current. At this, the control of the active component depends on the voltage U_c across the inverter capacitor, and the reactive component is controlled, depending on the IG output voltage.

The aforesaid is carried out in the following way. The three-phase quantities u_s with the amplitude equal to 1 that is in-phase with the voltage V_s are computed in the block **Vs/Vsm**:

$$u_{sa,b,c} = \frac{V_{sabc}}{V_m} \tag{4.34}$$

$$V_m = \sqrt{\frac{2}{3}\left(V_{sa}^2 + V_{sb}^2 + V_{sc}^2\right)} \tag{4.35}$$

The three-phase system w_s leads the system u_s by $\pi/2$:

$$w_a = \frac{u_{sc} - u_{sb}}{\sqrt{3}},\ w_b = -\frac{w_a}{2} + \frac{\sqrt{3}}{2}u_{sa},\ w_c = -\frac{w_a}{2} - \frac{\sqrt{3}}{2}u_{sa} \tag{4.36}$$

The quantities u_s are multiplied by the output of the U_c voltage controller, producing the reference for the active component of I_s. Analogously, the quantities w_s are multiplied by the output of the V_s voltage controller, producing the reference for the reactive component of I_s. The results of the multiplications are summed up, forming the current reference signal for current regulator mentioned earlier that acts to the inverter gate pulses.

In the model under consideration, the IG has a power of 160 kW under the voltage of 400 V, and it is driven with the diesel, whose model is shown in Figure 4.4; it is the revised model from the demonstration model **power_machines**. It is necessary to say that the diesel control systems can be very different, and if one tries to utilize this model for a certain system, the diesel model structure and parameters have to be closely analyzed. The three-phase capacitor bank is set for IG self-excitation; the bank has the delta-connection and the phase capacitance of 1.3 mF. The inverter is connected to the IG by means of the reactor with the inductance of 2 mH. The capacitance of the inverter capacitor is 10 mF.

The inverter control system is made as depicted in Figure 4.2; the reference for the DC inverter voltage is set as 800 V; the rms of output voltage is 380 V. The hysteresis current regulator is used for the current I_s. With the help of the breakers **Br**, **Br1**, and **Br2**, the diverse load types can be connected.

At first, the balanced three-phase load is simulated that is applied as a step. The breakers on-times are set as follows: **Br** is 7 s, **Br1** is 4 s, **Br2** is 40 s, and the single-phase loads **Load1** are equal to 2Ω. The IG speed, the voltage frequency, the rms of output voltage, and the inverter capacitor voltage are shown in Figure 4.5. One can see that the system variables recover during a rather short time after load application. The scopes **Scope** and **Scope1** display that the load voltage and the IG stator current are very close to a sinusoid.

To repeat simulation with different resistances of the **Load1** phases (for instance, 2Ω, 4Ω, 1.33Ω), it can be seen that at t = 9 s, the load currents are essentially

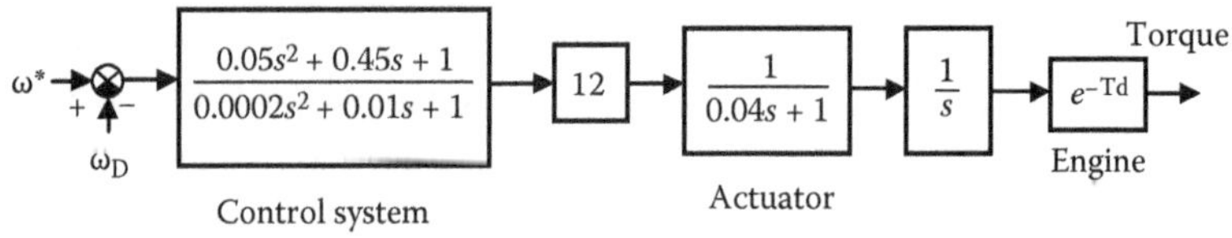

FIGURE 4.4 Model of the diesel.

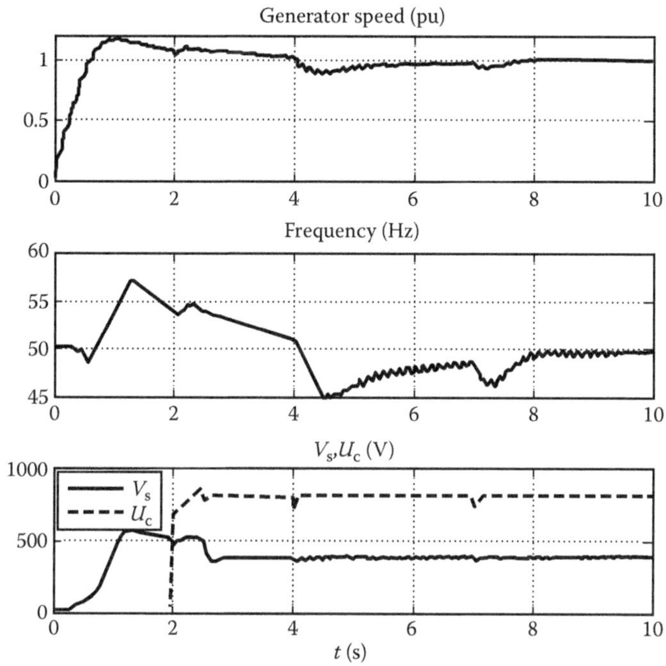

FIGURE 4.5 Processes in the set diesel-IG.

asymmetrical (the amplitudes are 254, 200, 274 A), but the load voltages and the IG currents are practically symmetrical (the phase voltage amplitudes are 314, 310, 309 V; the amplitudes of IG currents in pu are 1.21, 1.17, 1.22), that is, the system carries out the symmetrization under unbalanced load.

Under the simulation of the nonlinear balanced load, the breakers on-times are set, for instance, as **Br1** = 40 s, **Br2** = 4 s; all the phases are closed in **Br2**. When a capacitance of the load capacitors is 1 μF, the THD is 17.4% for the load current but only 1.3% for the load voltage and 2.8% for the IG current; therefore, the harmonics are effectively filtered. The most severe case is when the load is nonlinear and unbalanced. For this, phase *C* in **Br2** is kept open. Under the load capacitance of 500 μF, the THD for load phase currents are 17.2%, 14.1%, 1%; for the load phase voltages, 2.1%, 1%, 1.4%; for IG phase currents, 11.9%, 4.4%, 9.2%; at that, the main contribution gives the third harmonic.

The library model of IG, but with wounded rotor, is used in the model **Diesel_IG1**. Its parameters are the same as for the investigated IG. For self-excitation modeling, two voltage-controlled sources, placed in the subsystem **SelfExcit**, are connected to the rotor windings. The voltage amplitude is 6 V (about 2%); the frequency is proportional to the IG speed. When the self-excitation process comes to the end, the rotor windings are shorted, and subsequently the IG operates as squirrel cage one.

To fulfill with this model the same actions as with the previous one, the obtained results will be almost similar.

4.1.2 IG with Pole Number Switching

It will be shown in Chapter 6 that for every wind speed, the optimal WG rotating speed exists when the produced power is maximal. But the rotary speed of the squirrel cage IG is almost constant that does not give the possibility to utilize the wind energy to the best advantage. Different steps can be taken, in order to eliminate this disadvantage if only partly, among them, the employment of IG with pole number changing.

In principle, it can be obtained in two ways: with the stator with two windings or by the connection change of the single stator winding; the second possibility is considered at present as the main one. Usually, switching from $2Z_p = 4$ to $2Z_p = 6(4/6)$ is used; IG (4/8) has better merit, but the WG power is proportional to the wind speed in the third power, so that WG with $2Z_p = 8$ produces the optimal power under the wind speed decreasing by eight times when the power level is too low.

IG EMF may be written as [5]

$$E = 2\sqrt{2}fwk_w l_\delta \tau B_\delta, \tag{4.37}$$

where f is the voltage frequency, w is the number of the phase turns, and k_w is the winding coefficient that is equal to the ratio of the vector sum of the EMF of the phase wires connected in series to their algebraic sum; k_w is always <1 and depends on the winding arrangement, l_δ is the stack length, τ is the pole pitch, and B_δ is the air gap fundamental flux density.

Because it is preferable, under pole number changing, at the same voltage to have the same induction, the relationship has to be kept:

$$w_1 k_{w1} Z_{p2} = w_2 k_{w2} Z_{p1}; \tag{4.38}$$

indexes 1 and 2 are related to the windings with different pole number. If $Z_{p1} = 2$ and $Z_{p2} = 3$, then it must be

$$w_1 k_{w1} = (2/3) w_2 k_{w2}. \tag{4.39}$$

The winding circuits with minimal switching over are preferable, for example, under $Z_p = 3$; all the phase turns connect in series, and under $Z_p = 2$ in parallel, that is, $w_1 = w_2/2$. Then $k_{w1} = 4/3k_{w2}$. The IG is described in Ref. [6] with $k_{w1} = 0.831$, $k_{w2} = 0.644$, that is, relationship 4.39 is kept rather accurately.

Let us see how the parameters of the equivalent circuit change. From the relationships given in Ref. [5], it follows that the magnetizing inductance is

$$X_m = \frac{K_g w^2 k_w^2}{Z_p^2}, \tag{4.40}$$

where K_g is a constant value for the certain machine; therefore, for the case under consideration,

$$\frac{X_{m1}}{X_{m2}} = (9/16)\frac{k_{w1}^2}{k_{w2}^2}. \tag{4.41}$$

Under switching over 4 → 6 (from $Z_p = 2$ to $Z_p = 3$), the stator resistance increases four times; the rotor referred resistance increases proportional to

$$\frac{w_2^2 k_{w2}^2}{w_1^2 k_{w1}^2} = 4\frac{k_{w2}^2}{k_{w1}^2}. \tag{4.42}$$

The rotor flux leakage is referred to the stator with the aid of the same relationship. The formulas for the computation of the stator leakage reactances are rather complicated, but, from the relationships given in Ref. [5], one can make a conclusion that they are proportional to w^2/Z_p; therefore, under switching-over 4 → 6, they increase in the ratio of 8/3. The values of the leakage reactance changing that are obtained in this way give fairly good conformity with the results of measurements given in Ref. [6].

The IG model with pole number changing is given in the model **IG_switchpole**. It is based on the IG model with self-excitation from the model **Diesel_IG**. The signal P actuates the switching over from $Z_p = 2$ (under $P = 1$) to $Z_p = 3$ ($P = 0$). The measured IG quantities are outputted in pu in ratio to the base values at $Z_p = 2$. The program `IM_MODEL_1.m` is executed under the model start, in which, by the IG parameters given in pu under $Z_p = 2$, the parameters in pu under $Z_p = 3$ are calculated. The main parameter that influences the computation results is the IG power under $Z_p = 3$ because the power under $Z_p = 2$ is taken as the nominal one. It is taken in the program P_n ($Z_p = 3$) = $0.5P_n(Z_p = 2)$.

The WG operation with the network of a great power is simulated in simplified in the model **IG_switchpole**; more details of the model of such a system and the necessary explanations will be given in Chapter 6. In the WG turbine model, the nominal IG power of 160 kW is reached at the wind speed of 10 m/s. In the model, the wind speed that was initially equal to 10 m/s, at $t = 1.5$ s, decreases to 7 m/s; the pole number changing 2 → 3 has place at $t = 3$ s. The power that WG sends to the network is displayed in Figure 4.6. Firstly, this value is set as 155 kW; afterward, at $t > 1.5$ s, it decreases to 27 kW; and at $t > 3$ s, it increases to 52 kW. Because the WG power decreases in proportion to the wind speed drop in the third power, the maximal power under wind speed of 7 m/s is equal to $155 \times 7^3/10^3 = 53.1$ кВт that is achieved by the IG with pole number switching.

4.1.3 Model of Six-Phase IG

At present, the trend occurs to increase a WG single-unit power. The up-to-date WGs are provided with semiconductor units, in order to increase their efficiency;

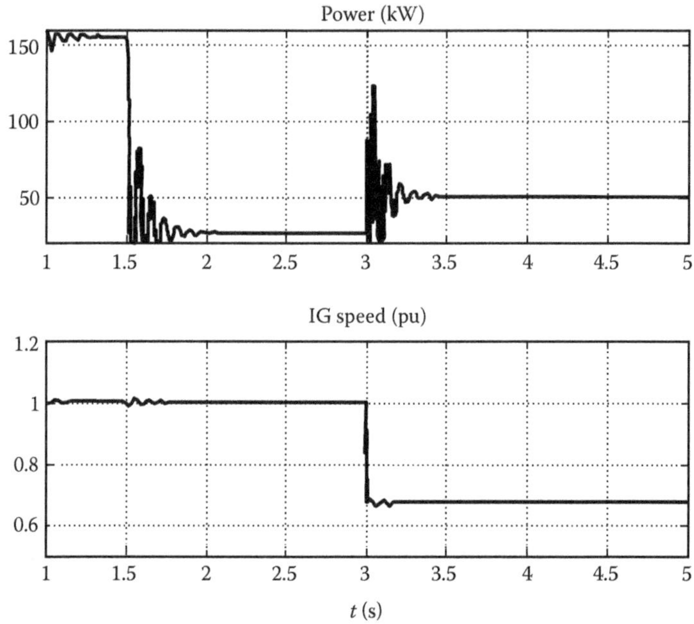

FIGURE 4.6 Power changing in WG with pole number switching.

these units are placed between the IG stator and the network. At that, the power of the individual semiconductor devices is not sufficient, and they must be connected in series or (and) in parallel that leads to some problems under the practical realization of such systems. More preferable is the converter parallel connection that gives an opportunity, with proper control, to decrease the content of the higher harmonics caused by the set. However, the stray currents appear in this case, which circulate between converters; these currents are caused by the difference of voltage instantaneous values and by the parameter spread.

Special steps are taken to exclude these currents; for instance, the multiwinding transformer can be set between IG stator and converters, but such a way makes the set much more expensive. The other approach is to use the multiphase, for example, a six-phase IG. Such IG provides more reliability, because the IG can be kept in operation under opening one or some IG phase windings.

The six-phase IG has two systems of the three-phase stator windings; the shift between them is 30° (electrical). The arrangement of IG windings is shown in Figure 4.7; each three-phase system has its isolated neutral. The IG model equations are given in Refs. [7,8] (see Ref. [4] too). With the aim to make the investigation and the simulation simpler, a six-dimension system of equations (for phases A, B, C, X, Y, Z) that describes the considered IG in the stationary reference frame can be changed by three two-dimension subsystems (α, β), (μ_1, μ_2), and (z_1, z_2) with the use of the transformation matrix

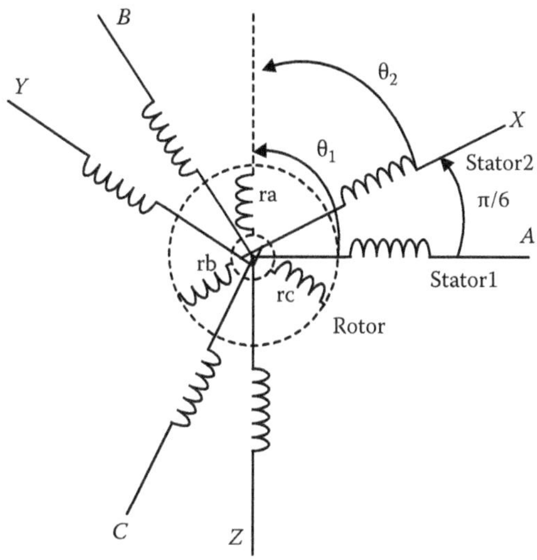

FIGURE 4.7 Winding arrangement in the six-phase IG.

$$\mathbf{T} = \frac{1}{3} \begin{vmatrix} 1 & -0.5 & -0.5 & \sqrt{3}/2 & -\sqrt{3}/2 & 0 \\ 0 & \sqrt{3}/2 & -\sqrt{3}/2 & 0.5 & 0.5 & -1 \\ 1 & -0.5 & -0.5 & -\sqrt{3}/2 & \sqrt{3}/2 & 0 \\ 0 & -\sqrt{3}/2 & \sqrt{3}/2 & 0.5 & 0.5 & -1 \\ 1 & 1 & 1 & 0 & 0 & 0 \\ 0 & 0 & 0 & 1 & 1 & 1 \end{vmatrix} \quad (4.43)$$

Thereby, three independent systems of equations are obtained for the variables (α, β), (μ_1, μ_2), and (z_1, z_2). At this, only variables (α, β) determine the IG flux and the torque; hence, only they must be taken into consideration for system synthesis; the equations for these variables are the same as the ones applicable for the usual IG:

$$v_{s\alpha} = (R_s + sL_s)i_{s\alpha} + sMi_{r\alpha}$$

$$v_{s\beta} - (R_s + sL_s)i_{s\beta} + sMi_{r\beta}$$

$$0 = sMi_{s\alpha} + \omega_r Mi_{s\beta} + (R_r + sL_r)i_{r\alpha} + \omega_r L_r i_{r\beta}$$

$$0 = -\omega_r Mi_{s\alpha} + sMi_{s\beta} - \omega_r L_r i_{r\alpha} + (R_r + sL_r)i_{r\beta}, \quad (4.44)$$

where s is the Laplace transformation symbol, ω_r is the rotor rotation speed, and $L_s = L_{ls} + M$, $L_r = L_{lr} + M$, and $M = 3\ L_{ms}$ are the total inductance (usually, IG $M = 1.5\ L_{ms}$).

The equations for the flux linkage are

$$\begin{aligned}\Psi_{s\alpha} &= (M + L_{1s})i_{s\alpha} + Mi_{r\alpha} \\ \Psi_{s\beta} &= (M + L_{1s})i_{s\beta} + Mi_{r\beta}\end{aligned} \tag{4.45}$$

$$\begin{aligned}\Psi_{r\alpha} &= (M + L_{1r})i_{r\alpha} + Mi_{s\alpha} \\ \Psi_{r\beta} &= (M + L_{1r})i_{r\beta} + Mi_{s\beta}.\end{aligned} \tag{4.46}$$

For the IG, the torque is

$$T_e = 3Z_p(\Psi_{s\alpha}i_{s\beta} - \Psi_{s\beta}i_{s\alpha}). \tag{4.47}$$

The variables of this system have the harmonics of the order of $k = 12n + 1$, $n = 0, 1, \ldots$ More details about this system is described in Ref. [4].

The model of the six-phase IG is used in the model **IG_6Ph**. This model is an extension of the IG with self-excitation from the model **Diesel_IG**. The driving torque comes to input T_m (with the sign minus) from the Simulink system. The phase-to-phase voltages U_{ac}, U_{bc}, U_{xz}, and U_{yz} are measured and, with the aid of the matrix 4.43, transformed in the voltages $v_{s\alpha}$, $v_{s\beta}$, $v_{\mu1}$, $v_{\mu2}$, v_{z1}, v_{z2}. Using these quantities, the IG fluxes and torque are computed and also the currents $i_{s\alpha}$, $i_{s\beta}$, $i_{\mu1}$, $i_{\mu2}$ (currents i_{z1} and i_{z2} are equal to zero for three-phase windings with isolated neutrals). Using these currents, with the help of the inverse transformation, the IG currents are calculated that IG sends to the external circuits or consumes from them. Five groups of the signals are collected at the output m1: three phase *A*, *B*, *C* currents; three phase *X*, *Y*, *Z* currents; two (in the axes α and β) stator flux linkages; the same for the rotor flux linkages; and the same for the rotor currents. The IG speed, its torque, and the rotor angle are collected at the output m2. All the quantities, except the latter, are given in pu. As the base values, the base values of the equivalent three-phase IG of the same power are selected. The program `IM_MODEL_6Ph` has to be executed before simulation start. A detailed model description is given in Ref. [4].

In the model **IG_6Ph**, IG is driven with the diesel as in the model **Diesel_IG**. On the whole, the set is intended for DC voltage fabrication. In principle, such a system unlikely has the practical sense and is given here in order to check the operation of the developed IG model, whose utilization will be demonstrated in Chapter 6. After simulation execution, one can see by **Scope1** the mutual arrangement of the stator currents, and by the first two axes of **Scope2**, the mutual shift of the stator phase-to-phase voltages. It is seen on the third axis of this scope that the rectified voltages of each diode bridge have the pulsations with the frequency of 300 Hz and with peak-to-peak amplitude of 50/434 = 12%, whereas the total rectified voltage has the pulsations with frequency of 600 Hz and with peak-to-peak amplitude of 24/730 − 3%.

4.1.4 Brushless Doubly Fed IG Simulation

The wound rotor IG is utilized widespread in WG; the use of the semiconductor converters in the rotor circuits that have a power of about 30% of the total WG power gives the possibility to use the wind energy to the limit. The shortcoming of such WG is the slip rings and the brushes that demand regular maintenance. This deficiency is especially important for the remote sets. Therefore, it is reasonable to have IG of such a kind but without contacts on the rotor.

Such IG has two three-phase stator windings but with a different number of the pole pairs. One of the windings is supplied with the grid voltage with the frequency f_s and has p pole pairs; the other is supplied with the voltage having controlled the amplitude and the frequency f_c and has q pole pairs. The first winding is the power one and the second is the control winding. Since the windings have different pole numbers, they do not link directly, but they are coupled through the rotor that has $n = p + q$ pole numbers (not pairs). In order to produce the co-operative torque with both windings in the rotor, the conditions $p\omega_r = \omega_s$ and $q\omega_r = \omega_c$ have to be kept, from which

$$\omega_r = \frac{\omega_s \pm \omega_c}{p+q} \tag{4.48}$$

or in the ratio to ω_s,

$$\omega_r^* = \frac{1 \pm \omega_c^*}{p+q}. \tag{4.49}$$

The sign minus means that the voltage space vector of the control winding rotates in the opposite direction about the voltage space vector of the power winding. Two types of the rotor are usually used: the rotor with the winding as in IG and the rotor with salient poles as in the reluctance motor; only the first type is considered further. Each rotor pole consists of some closed turns (loops) as shown in Figure 4.8 for two neighboring rotor poles, three turns in each pole.

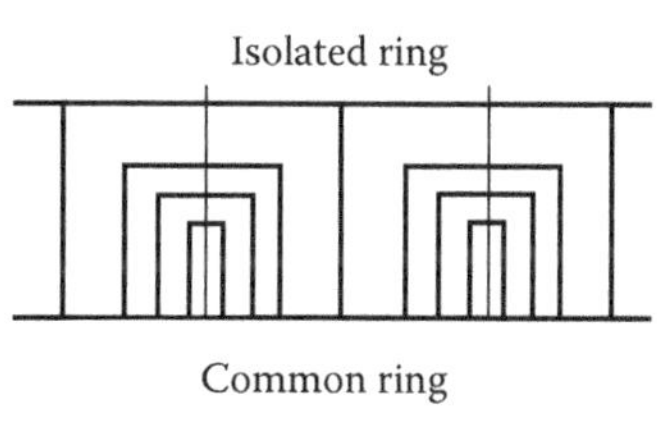

FIGURE 4.8 Arrangement of the rotor winding.

This machine can operate in three modes: (1) asynchronous mode with open control winding; (2) operation with natural speed $\omega_r = \omega_s/(p + q)$ when the control winding has DC supply; and (3) synchronous mode when relationship 4.48 is fulfilled. In order to reach this mode, the IG has to run at the speed close to the synchronous one at first.

There are a lot of theoretical works devoted to the investigation of such machines, but the industrial specimens are absent today. Only experimental devices of low power are available. One can hope that the developed model attracts new researchers of this, theoretically, very interesting kind of electrical machine.

Because the power winding is supplied with the grid voltage, it is reasonable to model the IG in the reference frame aligned with the power voltage space vector. It is often supposed for the simplification of the machine dynamic behavior that the rotor circuits have only one turn for a pole, although there are the works taking into account the rotor complex construction [9].

Then the machine behavior can be described with the system of equations [10]:

$$\begin{aligned} U_{qp} &= R_p i_{qp} + \frac{d\Psi_{qp}}{dt} + \omega\Psi_{dp} \\ U_{dp} &= R_p i_{dp} + \frac{d\Psi_{dp}}{dt} - \omega\Psi_{qp} \end{aligned} \tag{4.50}$$

$$\begin{aligned} U_{qr} &= R_r i_{qr} + \frac{d\Psi_{qr}}{dt} + (\omega - \omega_r)\Psi_{dr} \\ U_{dr} &= R_r i_{dr} + \frac{d\Psi_{dr}}{dt} - (\omega - \omega_r)\Psi_{qr} \end{aligned} \tag{4.51}$$

$$\begin{aligned} U_{qc} &= R_c i_{qc} + \frac{d\Psi_{qc}}{dt} + \omega\Psi_{dc} \\ U_{dc} &= R_c i_{dc} + \frac{d\Psi_{dc}}{dt} + \omega\Psi_{qc} \end{aligned} \tag{4.52}$$

$$\begin{aligned} \Psi_{qp} &= L_p i_{qp} + L_{pr} i_{qr}, \\ \Psi_{dp} &= L_p i_{dp} + L_{pr} i_{dr} \end{aligned} \tag{4.53}$$

$$\begin{aligned} \Psi_{qc} &= L_c i_{qc} + L_{cr} i_{qr}, \\ \Psi_{dc} &= L_c i_{dc} + L_{cr} i_{dr} \end{aligned} \tag{4.54}$$

$$\Psi_{qr} = L_r i_{qr} + L_{pr} i_{qp} + L_{cr} i_{qc}$$
$$\Psi_{dr} = L_r i_{dr} + L_{pr} i_{dp} + L_{cr} i_{dc} \quad (4.55)$$

The winding stray coupling is neglected [10]. After substitution of Equations 4.53 through 4.55 into Equations 4.50 through 4.52 and bearing in mind $U_{dr} = U_{qr} = 0$, we obtain the following:

$$U_{qp} = R_p i_{qp} + L_p \frac{di_{qp}}{dt} + L_{pr} \frac{di_{qr}}{dt} + \omega(L_p i_{dp} + L_{pr} i_{dr})$$
$$U_{dp} = R_p i_{dp} + L_p \frac{di_{dp}}{dt} + L_{pr} \frac{di_{dr}}{dt} - \omega(L_p i_{qp} + L_{pr} i_{qr}) \quad (4.56)$$

$$U_{qc} = R_c i_{qc} + L_c \frac{di_{qc}}{dt} + L_{cr} \frac{di_{qr}}{dt} + \omega(L_c i_{dc} + L_{cr} i_{dr})$$
$$U_{dc} = R_c i_{dc} + L_c \frac{di_{dc}}{dt} + L_{cr} \frac{di_{dr}}{dt} - \omega(L_c i_{qc} + L_{cr} i_{qr}) \quad (4.57)$$

$$0 = R_r i_{qr} + L_r \frac{di_{qr}}{dt} + L_{pr} \frac{di_{qp}}{dt} + L_{cr} \frac{di_{qc}}{dt} + (\omega - \omega_r)(L_r i_{dr} + L_{pr} i_{dp} + L_{cr} i_{dc})$$
$$0 = R_r i_{dr} + L_r \frac{di_{dr}}{dt} + L_{pr} \frac{di_{dp}}{dt} + L_{cr} \frac{di_{dc}}{dt} - (\omega - \omega_r)(L_r i_{qr} + L_{pr} i_{qp} + L_{cr} i_{qc}) \quad (4.58)$$

To introduce the following 6 × 6 matrixes,

$$\mathbf{Q} = [L_p\, 0\, 0\, 0\, L_{pr}\, 0; 0\, L_p\, 0\, 0\, 0\, L_{pr}; 0\, 0\, L_c\, 0\, L_{cr}\, 0; 0\, 0\, 0\, L_c\, 0\, L_{cr};$$
$$L_{pr}\, 0\, L_{cr}\, 0\, L_r\, 0; 0\, L_{pr}\, 0\, L_{cr}\, 0\, L_r] \quad (4.59)$$

$$\mathbf{R_s} = [R_p\, 0\, 0\, 0\, 0\, 0; 0\, R_p\, 0\, 0\, 0\, 0; 0\, 0\, R_c\, 0\, 0\, 0; 0\, 0\, 0\, R_c\, 0\, 0;$$
$$0\, 0\, 0\, 0\, R_r\, 0; 0\, 0\, 0\, 0\, 0\, R_r] \quad (4.60)$$

$$\mathbf{D} = [0\, \omega L_p\, 0\, 0\, 0\, \omega L_{pr}; -\omega L_p\, 0\, 0\, 0 - \omega L_{pr}\, 0; 0\, 0\, 0\, \omega L_c\, 0\, \omega L_{cr}; 0\, 0 - \omega L_c\, 0 - \omega L_{cr}\, 0;$$
$$0 (\omega - \omega_r) L_{pr}\, 0 (\omega - \omega_r) L_{cr}\, 0 (\omega - \omega_r) L_r;$$
$$-(\omega - \omega_r) L_{pr}\, 0 - (\omega - \omega_r) L_{cr}\, 0 - (\omega - \omega_r) L_r\, 0] \quad (4.61)$$

and vectors

$$\mathbf{I} = [i_{dp}\ i_{qp}\ i_{dc}\ i_{qc}\ i_{dr}\ i_{qr}]^T,\ \mathbf{U} = [U_{dp}\ U_{qp}\ U_{dc}\ U_{qc}\ 0\ 0]^T,$$

$$\Psi = [\Psi_{dp}\ \Psi_{qp}\ \Psi_{dc}\ \Psi_{qc}\ \Psi_{dr}\ \Psi_{qr}],$$

the machine equations may be written as

$$\mathbf{U} - \mathbf{R_s I} + \mathbf{DI} = \mathbf{Q}\frac{d\mathbf{I}}{dt} \tag{4.62}$$

$$\Psi = \mathbf{QI} \tag{4.63}$$

For the simulation, Equation 4.62 is written in the standard form in the state space:

$$\frac{d\mathbf{I}}{dt} = \mathbf{Q}^{-1}\mathbf{U} + \mathbf{Q}^{-1}(\mathbf{D} - \mathbf{Rs})\mathbf{I}, \tag{4.64}$$

where

$$\mathbf{D} = \omega \mathbf{D}_1 + \omega_r \mathbf{D}_2 \tag{4.65}$$

$$\begin{aligned} \mathbf{D}_1 = [&0\ L_p\ 0\ 0\ 0\ L_{pr}; -L_p\ 0\ 0\ 0 - L_{pr}\ 0; 0\ 0\ 0\ L_c\ 0\ L_{cr}; 0\ 0 - L_c\ 0 - L_{cr}\ 0;\\ &0\ L_{pr}\ 0\ L_{cr}\ 0\ L_r; -L_{pr}\ 0 - L_{cr}\ 0 - L_r\ 0] \end{aligned} \tag{4.66}$$

$$\begin{aligned} \mathbf{D}_2 = [&0\,0\,0\,0\,0\,0; 0\,0\,0\,0\,0\,0; 0\,0\,0\,0\,0\,0; 0\,0\,0\,0\,0\,0;\\ &0 - L_{pr}\ 0 - L_{cr}\ 0 - L_r; L_{pr}\ 0\ L_{cr}\ 0\ L_r\ 0]. \end{aligned} \tag{4.67}$$

The IG torque is calculated as

$$T_e = 1.5p\,(\Psi_{dp} i_{qp} - \Psi_{qp} i_{dp}) + 1.5q\,(\Psi_{qc} i_{dc} - \Psi_{dc} i_{qc}). \tag{4.68}$$

The three-phase voltages U_p and U_c are converted in the voltages U_{dp}, U_{qp}, U_{dc}, and U_{qc} with the help of the Park transformation with angles $\theta_p = \omega_s t$ and $\theta_c = \theta_p - (p + q)(\theta_r + \delta) + q\gamma$, respectively, where θ_r is the rotor angle—the integral of its rotation speed, δ is the initial rotor position, and γ is the angle between the axes of the stator windings (further, it is taken as zero).

The obtained variables of the rotor are transformed in the real quantities by the reverse Park transformation with the angle $\theta_p - p\theta_r$. So the utilization of the rotor position sensor is supposed. Before simulation start, the program `Double_Im.m` has to be executed that is fixed in the option *Model Properties/Callbacks/InitFcn.*

The diagram of the machine model is depicted in Figure 4.9. It is necessary to bear in mind that firstly, the machine must be accelerated until the natural speed, and only afterward can it be put in the synchronous mode. For that, the rated voltage is given to the power winding, and the DC voltage supplies the control wind; under acceleration, the switch **Switch** is set in the low position, determining the constant angle for transformation; the DC voltage comes to the input U_{cabc}.

The diverse control systems for brushless doubly fed IG are developed; the modification of the scalar control is used in the model described further (Figure 4.10). The cascaded control system is used; the innermost loop controls the current I_{cdq}. The quantities having the amplitudes equal to 1 as in

$$I_{crefd} = \sin[(\omega_s + \omega_c)t - (p + q)\theta_r],$$

$$I_{crefq} = \cos[(\omega_s + \omega_c)t - (p + q)\theta_r] \tag{4.69}$$

are employed to form the reference. These quantities are constant in the steady state and assign the components I_{cdq}; they are modulated in amplitude by the torque controller. The latter can be either the outermost loop or the loop that is inner for the speed controller (SC). The output of the current controller defines the voltage U_{cabc} by means of VSI with PWM. At this, the frequency $\omega_c = (p + q)\,\omega_r - \omega_s$.

In WG with the wound rotor IG, two VSIs connected back to back are usually used as shown in Figure 6.3b; the VSI that is connected to the IG rotor (VSI-Ge) is responsible for the IG speed, and the VSI that is connected to the grid (VSI-Gr) is responsible for the transfer of the produced electric energy to the grid. Two models of the considered IG are given in the following. One of them does not use the blocks of SimPowerSystems and is intended for the development and the parameter choice

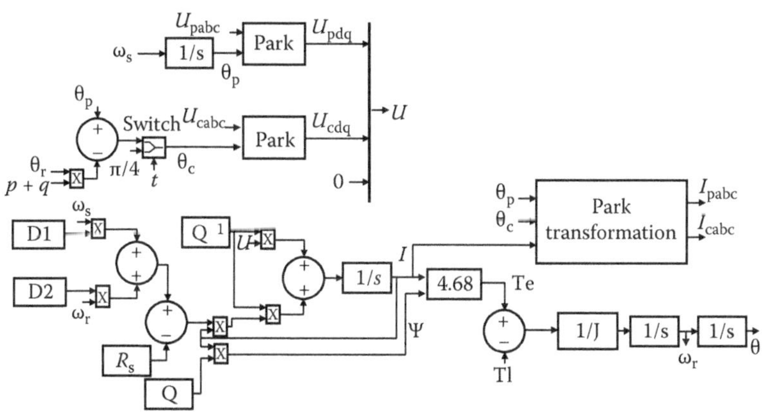

FIGURE 4.9 Model diagram of brushless doubly fed IG.

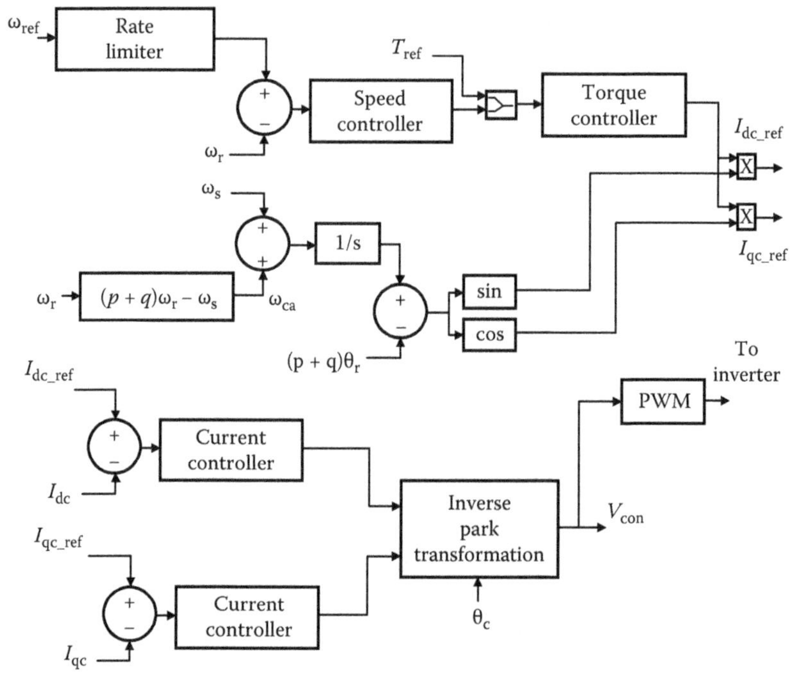

FIGURE 4.10 Block diagram of the control system for brushless doubly fed IG.

of the control system, because the simulation of this system is executed much faster. The second model is the full system model with two VSIs. The IG parameters are given in the program `Double_Im.m`; $p = 2$ and $q = 6$, so that the natural speed is 3000/8 = 375 r/min. The speed change up and down about the natural speed is investigated in these models.

In the model **IM_double_1**, the power winding is supplied from the three-phase controlled source; the current controller, by means of the block of the reverse Park transformation, is connected to the control winding. During the start ($t < 2$ s), the DC voltage is given the control winding with the use of the scheme that imitates the power of this winding from the VSI that has devices 1, 4, and 6 (Figure 3.1b) in the conducting state. The driving torque under start is equal to 50 Nm and increases to 1000 Nm at $t = 1.5$ s. At $t = 2$ s, the reference speed is 275 r/min; at $t = 3$ s, 475 r/min; and at $t = 4$ s, 275 r/min again.

In the model **IM_double_2**, the devices for the connection with the power circuits that supply the machine are added: the voltage sensors and the controlled current sources. The power winding is powered from the voltage source of 600 V; the circuits of the control winding are supplied from the step-down transformer with the secondary voltage of 110 V. The control system of the VSI-Ge is described earlier; the control system of the VSI-Gr is the same as in the cascade Scherbius and will be described in Chapter 6. At the start initial time that lasts 0.5 s, the gate pulses come to the VSI-Ge through the **Switch**; these pulses set

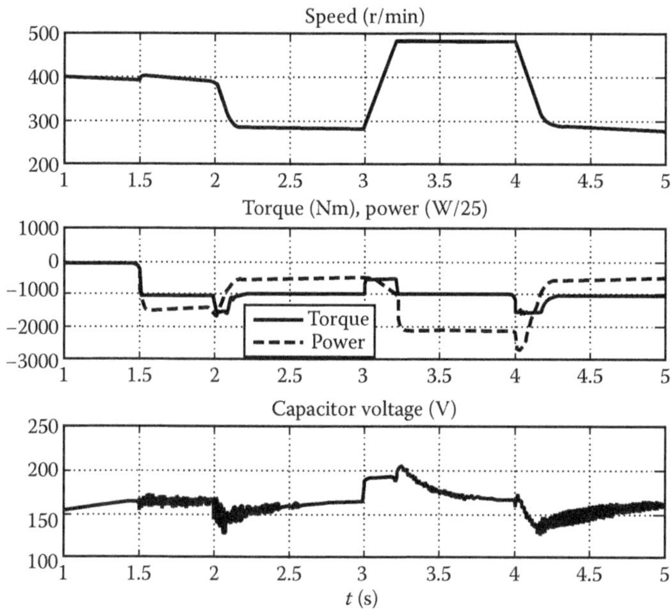

FIGURE 4.11 Processes in brushless doubly fed IG.

the transistors T1, T4, and T6 in the conduction state, fabricating the constant current in the control winding.

The processes of changing speed, torque, power, and capacitor voltage are shown in Figure 4.11.

4.2 SYNCHRONOUS GENERATORS

4.2.1 SG Model

It is accepted in SimPowerSystems that the salient pole synchronous machine (SM) has on the rotor, besides the excitation winding, two damper windings, whose axes are aligned along the direct and quadrature axes. It is supposed that round rotor SM has two damper windings in the axis. Only the equations for salient pole SM are given further.

The SM modeling is carried out in the rotating reference frame aligned with the rotor; the three-phase SM is replaced by the equivalent two-phase SM, whose phase quantities are designated as q and d.

For the transition to the two-phase SM, three-phase quantities are transformed with the aid of Equations 4.3 and 4.5. The equations for the stator voltages are Equation 4.10. There are three single-phase windings on the rotor: the quadrature damper winding with parameters R_{kq} and L_{1kq}, with i_{kq} as its current, and Ψ_{kq} as its flux linkage; the direct damper winding with parameters R_{kd} and L_{1kd}, with i_{kd} as its current, and Ψ_{kd} as its flux linkage; the excitation winding with parameters R_f and L_{1f}, and Ψ_f and i_f as its flux linkage and current. Denoting $\mathbf{R}_r$ as the resistance matrix, $\boldsymbol{V}_{qdr}$, $\boldsymbol{I}_{qdr}$, and $\boldsymbol{\Psi}_{qdr}$ as the rotor voltage, current, and flux linkage matrixes, respectively, arranged as follows: $[\Psi_{kq}, \Psi_f, \Psi_{kd}]$, and so on, where $\mathbf{V}_{qdr} = [0, U_f, 0]$ (U_f is the excitation voltage), then the rotor voltage equation is

$$\mathbf{V}_{qdr} = \mathbf{R}_r\mathbf{I}_{qdr} + \frac{d\mathbf{\Psi}_{qdr}}{dt}. \tag{4.70}$$

Equations 4.10 and 4.70 can be united in one matrix equation of the fifth order that connects the voltages and the flux linkages:

$$\mathbf{V} = \mathbf{RI} + \frac{d\mathbf{\Psi}}{dt} + \mathbf{W\Psi}, \tag{4.71}$$

where $\mathbf{V} = [U_{qs}\ U_{ds}\ 0\ U_f\ 0]$; the vectors $\mathbf{I}$ and Ψ are arranged in the same way; $\mathbf{R}$ is the diagonal matrix of the resistances; and the matrix $\mathbf{W}$ consists of zero elements except $w(1, 2) = \omega_r$ and $w(2, 1) = -\omega_r$.

The equations for the flux linkages are as follows:

$$\begin{aligned}
\Psi_{qs} &= L_q i_{qs} + L_{mq} i_{kq},\ L_q = L_{sl} + L_{mq} \\
\Psi_{ds} &= L_d i_{ds} + L_{md} i_{kd} + L_{md} i_f,\ L_d = L_{sl} + L_{md} \\
\Psi_{kq} &= L_{kq} i_{kq} + L_{mq} i_{qs},\ L_{kq} = L_{lkq} + L_{mq} \\
\Psi_f &= L_{fd} i_f + L_{md} i_{kd} + L_{md} i_{ds},\ L_{fd} = L_{lf} + L_{md} \\
\Psi_{kd} &= L_{md} i_f + L_{kd} i_{kd} + L_{md} i_{ds},\ L_{kd} = L_{lkd} + L_{md}
\end{aligned} \tag{4.72}$$

These equations may be written in the matrix form as Equation 4.21; after substitution in Equation 4.71, it results in

$$\frac{d\mathbf{\Psi}}{dt} = (\mathbf{U} - \mathbf{RL}^{-1}\mathbf{\Psi} - \mathbf{W\Psi}). \tag{4.73}$$

After the solution of Equation 4.73, the currents and the torque can be found as

$$\mathbf{I} = \mathbf{L}^{-1}\Psi \tag{4.74}$$

$$T_e = 1.5Z_p(\Psi_{ds} i_{qs} - \Psi_{qs} i_{ds}). \tag{4.75}$$

Because all calculations in the SM model are carried out in pu, the right side of Equation 4.73 is multiplied by ω_b as in Equation 4.22, and the multiplier $1.5Z_p$ in Equation 4.75 falls out.

Additionally, the mutual fluxes Ψ_{mq} and Ψ_{md} are computed as follows:

$$\Psi_{mq} = X_{aq}\left(\frac{\Psi_{qs}}{L_{sl}} + \frac{\Psi_{kq}}{L_{lkq}}\right) \tag{4.76}$$

$$\Psi_{md} = X_{ad}\left(\frac{\Psi_{ds}}{L_{sl}} + \frac{\Psi_{kd}}{L_{lkd}} + \frac{\Psi_{f}}{L_{lf}}\right) \tag{4.77}$$

$$X_{aq} = \left(\frac{1}{L_{mq}} + \frac{1}{L_{sl}} + \frac{1}{L_{lkq}}\right)^{-1} \tag{4.78}$$

$$X_{ad} = \left(\frac{1}{L_{md}} + \frac{1}{L_{sl}} + \frac{1}{L_{lkd}} + \frac{1}{L_{lf}}\right)^{-1} \tag{4.79}$$

The stator currents are computed after i_{qs} and i_{ds} calculation by the transformation in Equation 4.9.

The rotor rotation speed ω_r and the position θ are calculated as in Equations 4.17 and 4.24.

It is necessary to say that this way to model the SM is realized in the last versions of SimPowerSystems (particularly, in R2014a); in the previous versions (R2012b and earlier), the SM equations were solved by the method that was adopted from Ref. [2] (the relevant equations are given in Ref. [4] as well). Of course, the results of the simulation are the same.

The SM parameters can be specified in SI units, in pu, or as standard values. For that, the proper SM model has to be chosen from SimPowerSystems library, but the simulation is always carried out in pu. Unlike IG, not a torque but a power that is formed in Simulink and equal to $T_m\omega_r$ comes to the input *P*m. The power is positive in the generator operation mode. The SM dialog box looks like the one for IG to a considerable degree. On the first page, one of the existing SM with power of 2–2.5 MVA can be selected and also the rotor type: salient pole or round. The SM parameters are specified on the second page; their meanings are clear from the relationships and explanations given earlier for IG.

If the SM parameters are given in SI units, the additional opportunity appears: on the third page of the dialog box, the option *Display nominal field current and voltage producing 1 pu stator voltage*, and on the second page, in the first field, the fourth parameter can be specified—the nominal field current i_{fn} (A) (be reminded that it is the current that produces the nominal stator voltage across the open stator windings). If i_{fn} is not specified, the displayed voltage v_{fdn}, being sent to the input V_f, produces the nominal field current. But it is not the voltage that must be applied to the rotor winding really, because this voltage is referred to the stator. If the actual nominal field current is known and put down as it was explained earlier, the activation of this option gives the actual voltage that must be applied to the rotor.

Let us explain this in more detail. The SM has the nominal power P_n and voltage V_n of 4620 kVA and 6 kV, respectively; $L_{md} = 0.0283$ H; $R_f = 0.211\ \Omega$; and $i_{fn} = 161$ A. The referring factor is

$$\begin{aligned} K_r = N_s/N_f &= 2/3 \times 314 \times \sqrt{3} \times L_{md} i_{fn}/(\sqrt{2}\,V_n) \\ &= 2/3 \times 314 \times 1.732 \times 0.0283 \times 161/1.41/6000 = 0.195. \end{aligned}$$

Then $R_f' = 1.5 \times 0.195^2 \times 0.211 = 0.012\,\Omega$. This value has to be put down in the proper field on the second page. The activation of the considered option displays the voltage that must be applied to the input V_f under simulation as 6.612 V. To repeat the same when the value of i_{fn} = 161 A is added as the fourth parameter in the first field on page two, the option displays the excitation voltage as 33.95 V. At this, 33.95/0.211 = 161 A, as it has to be; just this excitation voltage must be used under simulation in this condition.

When saturation modeling is selected, the saturation of the mutual inductance for no-load condition is simulated; for the salient pole SM, only the saturation for L_{md} is simulated. The saturation curve is considered as the dependence of the mutual flux Ψ_{md} on the current i_{ds} in pu: $\Psi_{mdsat} = f(i_{ds})$, at the curve initial piece $\Psi_{mdsat} = i_{ds}$. Then

$$L_{mdsat} = L_{md}\frac{\Psi_{mdsat}}{i_{ds}}. \tag{4.80}$$

This value is used in Equations 4.73, 4.77, and 4.79 instead of L_{md}.

The SM manufacturers often give the information about SM parameters in a so-called standard form, when, instead of resistances and inductances, the reactances in pu and the time constants are specified. In this case, it is reasonable to choose the **SM model pu standard**. The SM reactances are defined as the product of the corresponding inductance by the synchronous frequency ω_s, divided by the value of the base resistance. The values of the time constants can be specified either under open or under close stator circuit. The formulas for reactances and time constants are given, for example, in Refs. [2,4].

Like for IG, SM states can be measured by Simulink blocks with the aid of the blocks **Bus Selector** that are connected to the output *m*. For measurement, the following are available: the stator phase currents i_{sa}, i_{sb}, and i_{sc}; the excitation current i_{fd}; the mutual fluxes Ψ_{mq} and Ψ_{md}; the rotor rotation speed ω_r; the rotor angle position θ (grad); the electromagnetic torque T_e; the torque angle δ that is equal to atan(v_{ds}/v_{qs}); the stator currents i_{qs} and i_{ds}; the currents of the damper windings; the stator voltages in the rotating reference frame v_{qs} and v_{ds}; the electrical power $P_e = T_e\omega_m$; the active and reactive powers that are equal to $P_{eo} = v_{qs}i_{qs} + v_{ds}i_{ds}$ and $Q_{eo} = v_{qs}i_{ds} - v_{ds}i_{qs}$, respectively; the deviation of the rotation speed from the synchronous one $\delta\omega$; and the deviation angle of the rotor $\delta\theta$ that is an integral from $\delta\omega$. It is necessary to have in mind that the model presumes that the stator currents flow into the windings, but the measured currents i_{sa}, i_{sb}, $i_{sc}i_{qs}$, and i_{ds} are supposed to flow from the SM.

The SM function in the power system under power flow computation (Chapter 1) can be specified on the fourth page of the dialog box. There are three possibilities. In *PV* mode, the active power and the bounds for the reactive power are defined. In *PQ* mode, the values for *P* and *Q* are given. The third possibility is the swing bus.

4.2.2 Excitation System Simulation

The system for powering and controlling the SG excitation winding must keep the output SG voltage under load change and contribute to stability improvement in an emergency. For this aim, the excitation current has to change sufficiently rapidly. The following excitation types are utilized.

The DC generator, which is placed on the SG shaft and simultaneously rotates with it, was used in the obsolete SG. That generator had the shunt winding for self-excitation with a rheostat in its circuit. Today, such exciters are not in production but are utilized in the old sets.

At the present time, the brushless excitation systems are widespread. These systems consist of the inverted SG (ISG), with rotating stator winding and motionless excitation winding. The ISG stator winding voltage is rectified by the diode rectifier that rotates together with the SG shaft and powers the SG excitation winding. The SG voltage control is carried out by the change of the excitation current of ISG with the help of the controller that receives a power either from the SG (in this case, the machines must have an ability to self-excite) or from the auxiliary source, for instance, from PMSG that is set on the same shaft. System shortcomings are a lag of its excitation circuits and the fact that, when it is necessary to kill the SG field rapidly, its excitation winding is shorted, and the current drops with the large time constant. It is possible to use the thyristors instead of the diodes, so that the rectifier is put in inverter mode under the necessity to kill the excitation current rapidly.

The most high-speed excitation system consists of the thyristor rectifier that supplies SG excitation winding with the slip rings; the rectifier receives the supply from the external source that ensures rapid system activation.

The SG model is developed is such a way that its excitation cannot be carried out from the blocks of SimPowerSystems (rectifiers, machines, and others), but only from Simulink blocks, so the equivalent models of the excitation systems can be developed. Eight models of the excitation systems are included in SimPowerSystems that are developed with use of IEEE Standard 421.5 [11].

The diagram of the model **Excitation System** is depicted in Figure 4.12. It is supposed to be the models of the SG excitation from the DC exciter. V_d and V_q are the components of the output SG voltage where the reference voltage is V_{ref}; this voltage is measured with the sensor with time constant T_r. The initial values of SG voltage V_{t0} and of the exciter V_{f0} can be fixed. When the self-excitation process is simulated, the small value V_{f0} (1–2%) and $K_e < 0$ must be selected. The limiter of the output voltage operates in the following way: If $E_{fd} < E_{fmin}$, $V_e = E_{fmin}$; if ($E_{fd} > E_{fmax}$ and $K_p = 0$), $V_e = E_{fmax}$; if ($E_{fd} > E_{fmax}$ and $K_p > 0$), $V_{emax} = K_pV_t$. V_{stab} is a signal of the power system stabilization.

The other models are designated as **DC1A**, **DC2A**, **AC1A**, **AC4A**, **AC5A**, **ST1A**, and **ST2A**. The first and second models simulate the excitation from DC exciter and close to the previous model. The difference between these two models is that in the first one, the exciter output limitations are constant, and in the second model, the low and top limits are in proportion to the value V_t.

The main difference of these models shown in Figure 4.12 is the consideration of the exciter saturation (Figure 4.13). The nonlinear dependence $S_e(V_f)$ is computed as shown in Figure 4.14, where the plots of the dependences of the exciter EMF V_f on I_{fd} without and with saturation are depicted; by using these curves, the dependence $(B - A)/B$ as a function V_f is calculated. This dependence is approximated by two points designated as E_{fd1} and E_{fd2}. The first point is recommended

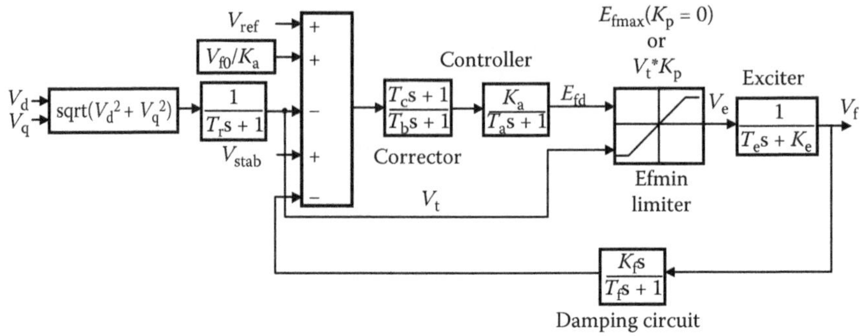

FIGURE 4.12 Diagram of the model Excitation system.

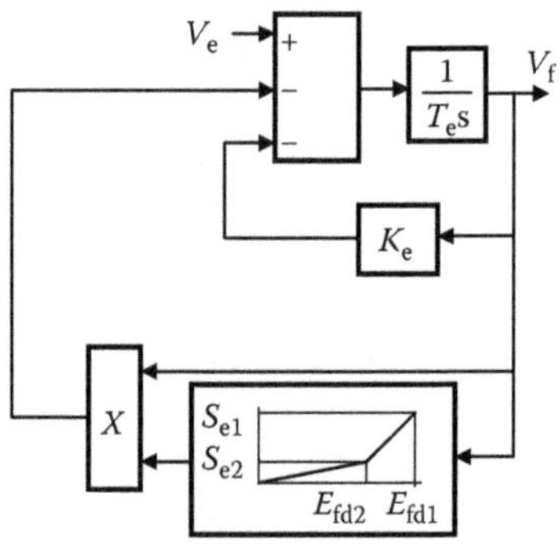

FIGURE 4.13 Diagram of the output part of the models DC1A and DC2A.

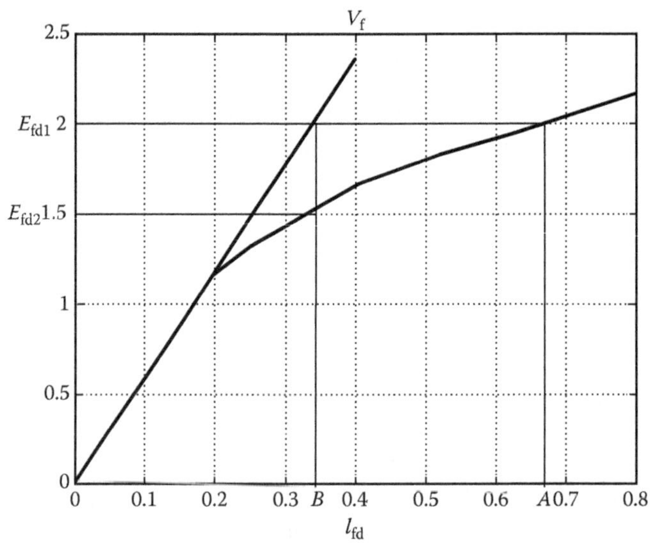

FIGURE 4.14 Saturation curve construction.

to select close to the maximum value of the output voltage, $E_{fd2} \approx 0.75E_{fd1}$. These values are the abscissas of the nonlinear dependence, whose coordinates are S_{e1} and S_{e2}. For the curves in Figure 4.14, $S_{e1} = (0.67 - 0.34)/0.34 = 0.97$, and $S_{e2} = (0.335 - 0.255)/0.255 = 0.31$.

The **AC1A** models excitation from the ISG loaded with the diode rectifier. The regulator receives a power from an external source. The model diagram is not depicted but it looks like (or is similar to) the one depicted in Figure 4.12 to a great degree, but there are two limiters, VA and VR, connected in series. The main distinction lies in the model structure of the exciter proper (Figure 4.15).

The additional effect of the exciter current I_{fd} on the excitation voltage with the factor K_d represents the decrease in the exciter flux linkage with the rise of the exciter load current and depends on its synchronous reactance that is recommended by IEEE [11] to take as 0.38. Moreover, the voltage decrease at the diode rectifier output caused by the diode switching over is taken into account in this model. The value of this decrease is defined by the exciter commutating inductance; as such, the half sum of its direct axis and quadrature axis transient reactances is considered.

When the load current is not too much, the switching over takes place only between two diodes at the same terminal: plus or minus, only one diode conducts at another terminal. Then the output voltage is

$$V_f = E_{d0} - \frac{3}{\pi} x_s I_f, \tag{4.81}$$

where x_s is the commutating reactance, $E_{d0} = \frac{3}{\pi} U_{lm}$, and U_{lm} is the amplitude of the phase-to-phase exciter EMF. Or

$$V_f = E_{d0}\left(1 - \frac{3x_s I_f}{\pi E_{d0}}\right) = E_{d0} K_k. \tag{4.82}$$

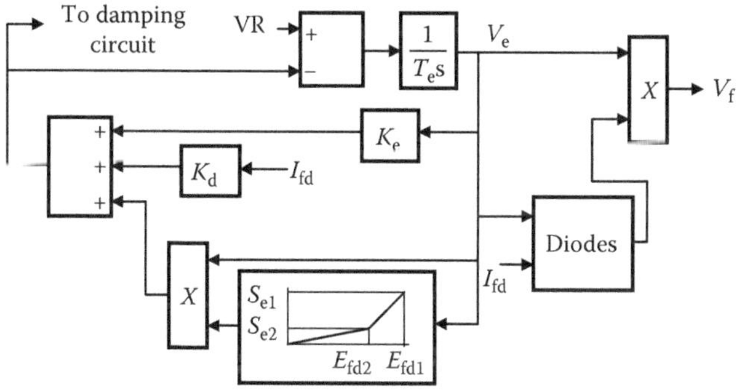

FIGURE 4.15 Diagram of the exciter AC1A.

Because the excitation voltage base value produces the nominal current in the excitation winding, it may be written as

$$K_k = 1 - \frac{3x_s I_f^* I_{fnom}}{\pi E_{d0}^* E_{d0base}} = 1 - \frac{3x_s I_f^*}{\pi E_{d0}^* R_f}. \quad (4.83)$$

If

$$K_c = \frac{3\sqrt{3}x_s}{\pi R_f}, \quad (4.84)$$

then

$$K_k = 1 - \frac{1}{\sqrt{3}} \frac{K_c I_f^*}{E_{d0}^*}. \quad (4.85)$$

These equations are valid until the commutating (overlap) angle is less than 60° that has a place under condition

$$\frac{4x_s I_f}{U_{lm}} < 1 \quad \text{or} \quad \frac{K_c I_f^*}{E_{d0}^*} < \frac{\sqrt{3}}{4} = 0.433. \quad (4.86)$$

Under load current increase, the conducting state of the proper diode at the opposite terminal delays; at this, the commutating angle is 60° as before and

$$V_f = \frac{\sqrt{3}}{2} E_{d0} \cos(\alpha_d + 30°), \quad (4.87)$$

where α_d is the delay angle that satisfies the relationship

$$\sin(\alpha_d + 30°) = \frac{2x_s I_f}{U_{lm}} \quad (4.88)$$

After substitution of Equations 4.87 and 4.88 in Equation 4.82, we receive

$$K_k = \sqrt{0.75 - \left(\frac{K_c I_f^*}{E_{d0}^*}\right)^2}. \quad (4.89)$$

This relationship is valid, until

$$I_f < \frac{\sqrt{3}U_{lm}}{x_s} \tag{4.90}$$

or

$$\frac{K_c I_f^*}{E_{d0}^*} < 0.75. \tag{4.91}$$

When the current increases further, two or three diodes conduct. In that mode,

$$K_k = \sqrt{3}\left(1 - \frac{\sqrt{3}x_s I_f}{U_{lm}}\right) = \sqrt{3}\left(1 - \frac{K_c I_f^*}{E_{d0}^*}\right). \tag{4.92}$$

The relationships given earlier are realized in the exciter model, in the subsystem **Rectifier**. The calculated value $I_{fd}K_c/V_e$ (V_e is the exciter EMF in pu) is compared with the adjoined points of the segments 0.433, 0.75, and 1 and, depending on the results of the comparison, forms the output signal in conformity with the relationship given earlier.

The model **AC4A** models the excitation system with the ISG and the controlled thyristor rectifier. The model diagram is displayed in Figure 4.16. The decrease in the rectified voltage caused by thyristor commutation is modeled with the factor K_c.

The model **AC5A** is a simplified model of the brushless excitation. The regulator is supposed to gain power from a source that is not subjected to external perturbations. The model is often utilized for the systems with other types of excitation, if the detailed information is lacking (Figure 4.17).

ST1A models the excitation system with the thyristor exciter, so that the excitation voltage can be negative (Figure 4.18). The converter is supplied from the independent voltage source or from the SG terminals. In the latter case, the maximum of the excitation voltage is in proportion to the SG voltage V_t. The excitation voltage reduction, because of thyristor commutation, is taken into account with the factor K_c.

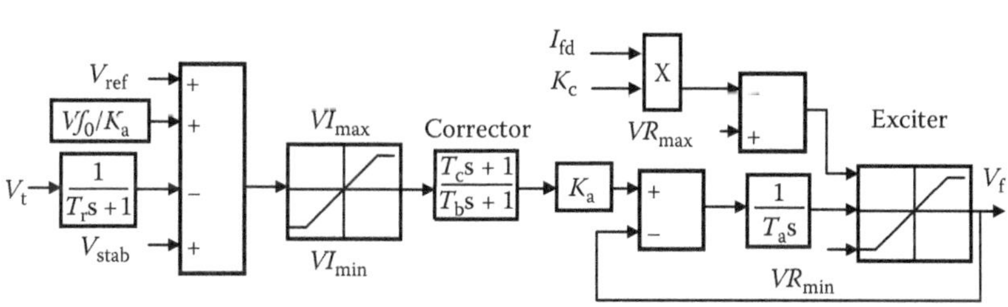

FIGURE 4.16 Diagram of the model AC4A.

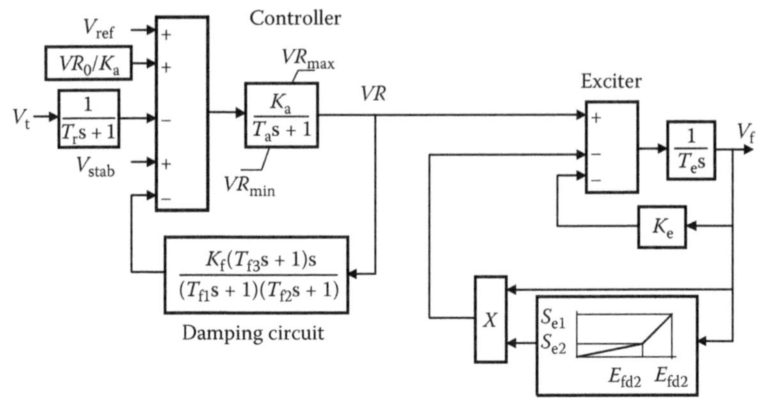

FIGURE 4.17 Diagram of the model AC5A.

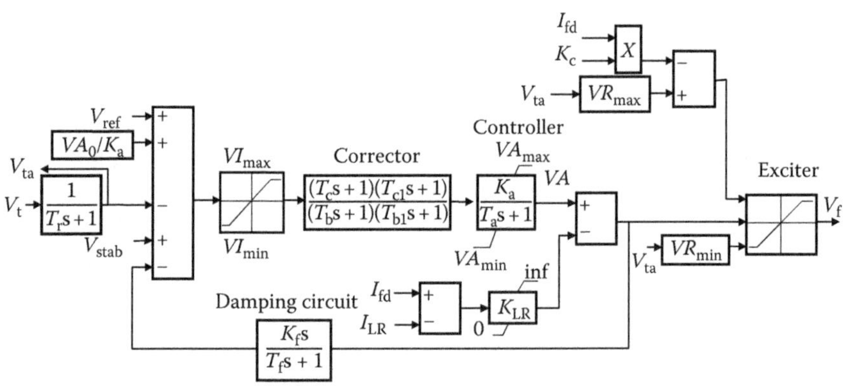

FIGURE 4.18 Diagram of the model ST1A.

The input–output characteristic of the exciter is supposed to be linear. Since in such systems the boosting factor is rather large, the additional feedback with the gain K_{LR} is used, in order to limit the excitation current by the value I_{LR}. Two possibilities are provided to correct the system frequency characteristic: by the series element of the second order or by the vanishing feedback. If the semicontrolled rectifier is utilized, the excitation voltage cannot be negative and $VR_{min} = 0$ must be set.

ST2A models the static exciter, whose powering voltage has a component that is proportional to the SG current; therefore, the excitation current increases with the increase in the SG stator current. The block **Diodes** (Figure 4.19) is the same, as for the model **AC1A**. The winding, which changes the bias current of the transformer, controls the SG excitation current. The limitation E_{fdmax} corresponds to the limit of the exciter voltage caused by saturation of the system magnetic elements. The time constant T_e is determined with the control winding inductance. The value VR_0 is determined by the selected initial conditions.

The parameters of the excitation models are specified in three pages of the dialog box: regulator parameters, rectifier parameters, and initial values. The meaning of

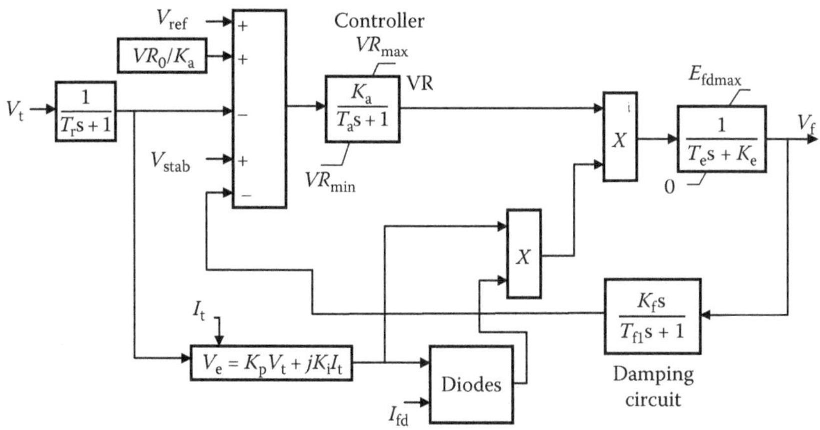

FIGURE 4.19 Diagram of the model ST2A.

the specified parameters is clear from descriptions and diagrams given earlier. The output SG quantities are designated with the symbol t, and the excitation circuit quantities, with the symbol f. The additional information about excitation systems can be found in Refs. [12–14].

The demonstration model **power_machines** is included in SimPowerSystems that contains the subsystem **Excitation**, in which any of the blocks **DC1A**, **DC2A**, **AC1A**, **AC4A**, **AC5A**, **ST1A**, and **ST2A** can be selected. In the model, the 25 kV source powers the 5 MW load and also, with the breaker and the transformer (25 kV/2400 V), the 1 MW load and the induction motor with power of 1680 kVA. The SG of the diesel generator set is connected in parallel with the transformer secondary winding.

The model **Excitation_examples** that is developed for this book is a simpler model, in which the SG supplies the resistive load that changes by step. After simulation execution, it can be seen that all exciters have about the same speed of response: under the accepted parameters, the recovery time of the load voltage is 4–6 s, except for the model **ST1A**, where this time is not more than 0.6 s. This model can be used, when the structure and the parameters of the excitation system are selected during the designing of system of the electric power supply.

4.2.3 Simplified SM Model

The SM model described earlier is rather complicated that can demand much time during the simulation of multibranch circuits with many SG. Therefore, the simplified SM model is included in SimPowerSystems. The round SM is modeled; the damping windings are neglected; and the stator circuits consist of the winding resistance and the inductance $L = L_d = L_q$. It can be written, for instance, for phase A [2] (supposing the generator operation mode and absence of the damping windings) as

$$U_{sa} = E_{fda} - R_s I_{sa} - L\frac{dI_{sa}}{dt} \tag{4.93}$$

$$E_{fda} = \omega_s L_m I_{fd} \sin\theta = E \sin\theta. \tag{4.94}$$

$L = L_m + L_{sl}$; ω_s is the rotor rotation frequency (electrical); θ is the angle between the axes of the rotor and phase A; E is the voltage amplitude that is induced by the rotating rotor. Sin $(\theta - 2\pi/3)$ and $\sin(\theta + 2\pi/3)$ are used for phases B and C, respectively.

The SM parameters can be specified in SI units or in pu. The meaning of all quantities is rather clear, except the damping factor K_d. This factor considers the influence of the damping windings. Its value is supposed to be calculated by

$$K_d = 4\xi\sqrt{\frac{\omega_s H P_{max}}{2}}, \tag{4.95}$$

where ξ is a damping ratio, ω_s is a synchronous frequency (314 rad/s under 50 Hz), H is an inertia constant (s), P_{max} is a maximal power that is equal to $U_s E/X$ (all quantities are in pu), and X is an internal reactance. In order to determine ξ, it is possible to use the simple scheme with the full SM model firstly that is powered with the nominal voltage, to fix the process of the torque angle changing under the load step, and to estimate the value ξ from this process.

The model accelerating torque is computed as

$$T_{din} = \frac{P_m - P_e}{\omega_m}, \tag{4.96}$$

where P_m is the input mechanical power, and P_e is the electrical power that is equal to

$$P_e = E_{fda} I_{sa} + E_{fdb} I_{sb} + E_{fdc} I_{sc}. \tag{4.97}$$

Afterward, the speed changing $\Delta\omega_m$ is calculated as

$$\frac{d\Delta\omega_m}{dt} = \frac{T_{din} - K_d \Delta\omega_m}{2H} \tag{4.98}$$

that is added to the rated speed ω_s (1 pu). The integral of $(\omega_s + \Delta\omega_m)$ determines the angle θ that is utilized for E_{fd} computation. It is seen that this simplified model can be used almost solely under investigation of the processes in the electrical circuits for the simulation of the SG, when the rotation speed slightly deviates from the synchronous.

The model **Simple_SG** helps to estimate the factor K_d. SM is employed with $L_{md} = L_{mq} = 1$ and $L_{sl} = 0.1$. The full and the simplified models of the same SM operate in the motor mode, are supplied in parallel, and are subjected to the same actions: the load increasing from 0 to 0.75 at $t = 2$ s with simultaneous excitation increasing from 1 to 1.5. The plot of the torque angle δ is depicted in Figure 4.20 under $K_d = 9.9$. The angle δ for the simplified model is computed as the difference of the phase angles of phase A internal EMF and of the voltage at phase A terminal measured with the blocks **Fourier Analyser**. The angle values are sufficiently similar.

Let us explain how the value K_d has been determined. The curve δ_1 shows an overshoot $\Delta_{max} = (52 - 33)/33 = 0.57$. From the relationship [15]

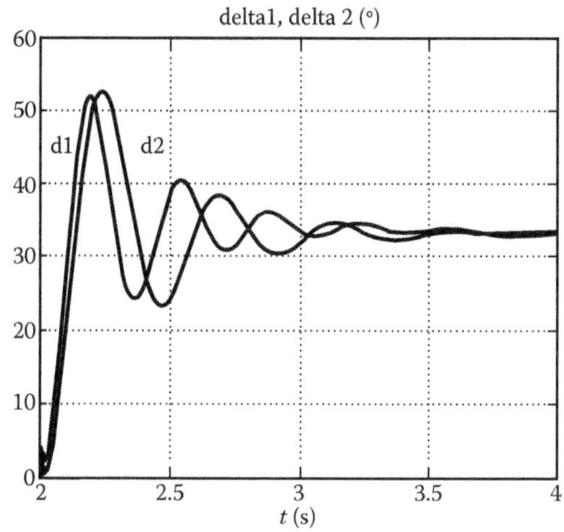

FIGURE 4.20 Torque angle changing for the full and simplified SM models.

$$\Delta_{\max} = \exp\left(-\frac{\pi\xi}{\sqrt{1-\xi^2}}\right), \tag{4.99}$$

it follows that $\xi = 0.18$. According to Equation 4.95, bearing in mind $E = 1.5$, $P_{\max} = 1 \times 1.5/1.1 = 1.36$, it follows that

$$K_d = 4 \times 0.18\sqrt{314 \times 0.89 \times 1.36/2} = 9.9.$$

4.2.4 Model of Six-Phase SG

The high-power SGs, which have two stator windings with a relative shift of 30°, are used for the same reason as the IG with two stator windings considered earlier. Usually, both stator windings are placed in the same slots; therefore, the leakage fluxes exist that embrace both windings, so that they are magnetically coupled. This fact makes the SG equations more intricate. As for the usual SG, the simulation is carried out in the reference frame that rotates in synchrony with the rotor. The equations for the modeling of such six-phase SG are given in Ref. [16].

The three-phase windings are designated as *abc* and *xyz*. It is supposed that the rotor angle θ is counted off from the position of the first winding; the magnetic axis of the second winding leads the first one by 30° (electrical). It is proven in Ref. [16] that the winding magnetic coupling is defined with the parameters

$$\begin{aligned} L_{\text{sm}} &= L_{\text{ax}} \cos g + L_{\text{ay}} \cos(g + 2\pi/3) + L_{\text{az}} \cos(g - 2\pi/3) \\ L_{\text{sdq}} &= L_{\text{ax}} \sin g + L_{\text{ay}} \sin(g + 2\pi/3) + L_{\text{az}} \sin(g - 2\pi/3), \ g = \pi/6, \end{aligned} \tag{4.100}$$

where L_{ax}, L_{ay}, and L_{az} are the mutual leakage inductances of the indicated phases of both windings.

The values of these inductances can be found by the computation of the SG magnetic circuits; some relationships for that are given in Ref. [16]. Let L_{sl} be the leakage inductance of the single winding, $k = L_{sm}/L_{sl}$ be the winding coupling factor, $L_s = (1 - k)\,L_{sl}$, and $L_{sdq} = bL_{sl}$. The voltage equations are as follows:

$$\frac{d\Psi^{r}_{qs1}}{dt} = U_{qs1} - \omega_r \Psi^{r}_{ds1} - R_s I_{q1} \tag{4.101}$$

$$\frac{d\Psi^{r}_{ds1}}{dt} = U_{ds1} + \omega_r \Psi^{r}_{qs1} - R_s I_{d1} \tag{4.102}$$

$$\frac{d\Psi^{r}_{qs2}}{dt} = U_{qs2} - \omega_r \Psi^{r}_{ds2} - R_s I_{q2} \tag{4.103}$$

$$\frac{d\Psi^{r}_{ds2}}{dt} = U_{ds2} + \omega_r \Psi^{r}_{qs2} - R_s I_{d2}. \tag{4.104}$$

The index 1 is related to the *abc* windings; the index 2, with *xyz*. The equations for the damping and exciting winding remain the former.

It is reasonable to introduce the flux linkages Ψ_{mq} and Ψ_{md} as

$$\Psi_{mq} = L_{mq}(I_{q1} + I_{q2} + I_{kq}), \tag{4.105}$$

$$\Psi_{md} = L_{md}(I_{d1} + I_{d2} + I_{kd} + I_{fd}), \tag{4.106}$$

where I_{kq}, I_{kd}, and I_{fd} are the currents of the damping and exciting winding, respectively; L_{mq} and L_{md} are the magnetizing inductances. Then the following relationships for the stator currents can be obtained from the equations given in Ref. [16]:

$$I_{q1} = \frac{\Psi_{q1} - k\Psi_{q2} - (1-k)\Psi_{mq} + b(\Psi_{d2} - \Psi_{md})}{L_{sl}d} \tag{4.107}$$

$$I_{q2} = \frac{\Psi_{q2} - k\Psi_{q1} - (1-k)\Psi_{mq} - b(\Psi_{d1} - \Psi_{md})}{L_{sl}d} \tag{4.108}$$

$$I_{d1} = \frac{\Psi_{d1} - k\Psi_{d2} - (1-k)\Psi_{md} - b(\Psi_{q2} - \Psi_{mq})}{L_{sl}d} \tag{4.109}$$

$$I_{d2} = \frac{\Psi_{d2} - k\Psi_{d1} - (1-k)\Psi_{md} + b(\Psi_{q1} - \Psi_{mq})}{L_{sl} d}, \quad (4.110)$$

where $d = 1 - k^2 - b^2$.

After the substitution of these formulas for currents and also for currents I_{kq}, I_{kd}, and I_{fd} [2,4] in Equations 4.105 and 4.106, the relationships for Ψ_{mq} and Ψ_{md} are written as follows:

$$\Psi_{mq} = L_{aq}\left[\frac{(\Psi_{q1} + \Psi_{q2})(1-k) + b(\Psi_{d2} + \Psi_{d1})}{dL_{sl}} + \frac{\Psi_{kq}}{L_{lkq}}\right] \quad (4.111)$$

$$\Psi_{md} = L_{ad}\left[\frac{(\Psi_{d1} + \Psi_{d2})(1-k) + b(\Psi_{q2} - \Psi_{q1})}{dL_{sl}} + \frac{\Psi_{kd}}{L_{lkd}} + \frac{\Psi_{f}}{L_{lf}}\right] \quad (4.112)$$

$$L_{aq} = \left[\frac{1}{L_{mq}} + \frac{1}{L_{lkq}} + \frac{2(1-k)}{dL_{sl}}\right]^{-1} \quad (4.113)$$

$$L_{ad} = \left[\frac{1}{L_{md}} + \frac{1}{L_{lkd}} + \frac{1}{L_{lf}} + \frac{2(1-k)}{dL_{sl}}\right]^{-1} \quad (4.114)$$

The SG torque is

$$T_e = [(I_{q1} + I_{q2})\,\Psi_{md} - (I_{d1} + I_{d2})\,\Psi_{mq}]. \quad (4.115)$$

The demonstration model **power_6phsyncmachine** is included in SimPower Systems, in which the simplified model of the six-phase SM is employed: the equations of these SMs are (Equations 4.107 through 4.114) $k = b = 0$. In the model **Sixphsynchmacine** developed using these equations, the SG with variable rotation speed is used for active load supplying. The SG power is 1850 kVA, and the rated voltage is 1200 V. The detailed description of the **SM_6Ph** model is given in Ref. [4]. The program, in which SG parameters are specified, as in the program **`SM_MODEL_6ph2`**, has to be executed before simulation start.

In the model **Sixphsynchmacine**, the back-to-back connected inverters are set in the circuit of each three-phase stator winding. The inverter at the SG terminal must keep up the voltage across the capacitor; the inverter connected to the load must control the load voltage. The first task is decided with the subsystems **Control_A** and **Control_X**. The output of the capacitor voltage controller assigns the component I_d of the SG current, an assignation of the component $I_q = 0$. These references are transformed in the phase current references that are reproduced by the controllers of these currents. The load voltage controller of proportional and integral controller (PI) type (subsystem **Voltage_Control**) uses as a feedback the voltage positive sequence amplitude: the controller output is modulated by the three-phase signal with the frequency of 50 Hz, whose amplitude is equal to 1. For simulation speedup, the bridge models are utilized that do not take into consideration the presence of the higher harmonics.

The mechanism that drives SG, for simplicity, is the PI regulator of the speed, whose output acts on the SG torque. At t = 10 s, the speed reference that was equal to 1 pu, initially, changes to 0.5 pu. The changes of the SG rotational speed and of the voltages across the capacitors and across the load are displayed in Figure 4.21. It is seen that the abrupt drop of the SG speed results in an essential sag of the capacitor voltages, but the load voltage changes much less. The SG voltages and currents are shown in Figure 4.22. There is a slight phase shift between voltages and currents that is caused by the errors of the phase current controllers and also by the shift between the rotor position and the phase voltages; the voltage and the current of the *xyz* winding leads the corresponding quantities of the *abc* winding by 30°.

The considered model is idealized somewhat and demonstrates model adequacy. The three-level inverters are used in the more realistic model **SixphSynMach_det**. The additional reactors are set in series with the stator windings with the aim to decrease

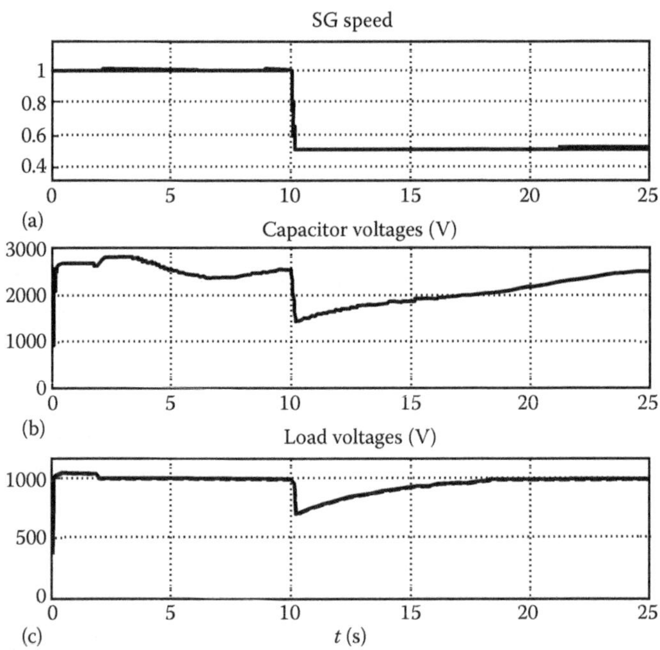

FIGURE 4.21 SG (a) rotational speed, (b) voltages across the capacitors, and (c) load for the model **Sixphsynchmacine**.

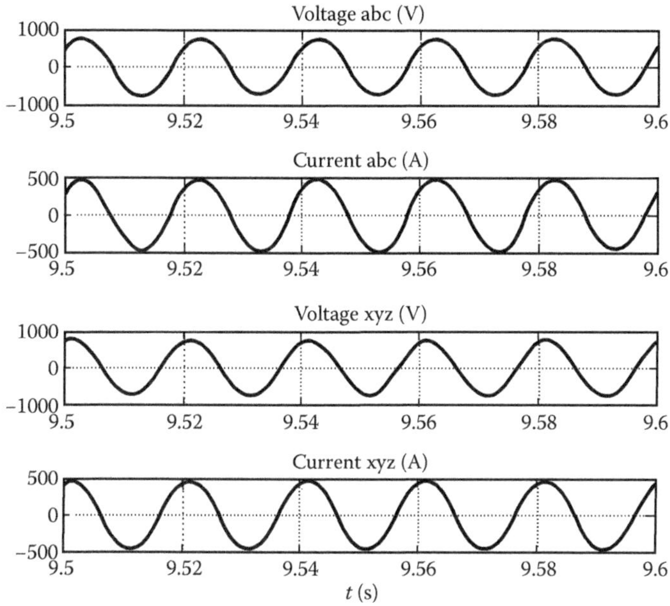

FIGURE 4.22 SG voltages and currents in the model **Sixphsynchmacine**.

current pulsations. The load voltage is filtered with the capacitance filter. The modulation factor of the output inverters is taken as constant; therefore, the load voltage controller is not modeled. The controllers of the quasiconstant currents I_d and I_q are used in the inverters connected to SG, instead of the phase current controllers in the previous model. Since the equations for the flux linkages in both axes are coupled (Equations 4.101 through 4.104), the compensation of the regulator mutual influence is provided (Figure 4.23).

The flux estimations that are obtained with the help of Equations 4.101 through 4.104 are used for compensation, but instead of integrators, the elements with the transfer functions of the first order are utilized, in order to decrease the errors caused by the integration over the infinite time interval. The transfer function time constant T is chosen to be large enough, so that frequency characteristics corresponding to the transfer functions are close to the pure integration. The transfer from the three-phase voltage and current system to d and q components and back is carried out by Park transformation using the rotor angle position θ; at this, for *xyz* winding, the additional shift by $\pi/6$ is applied. The elements for the flux estimation are placed in the subsystem **Est_Fluxes**.

The plots of the SG speed, the voltages across the capacitors, and the load are shown in Figure 4.24. It can be seen on axes 5 and 6 of the scope SG that the stator currents are essentially distorted. To repeat simulation, neglecting the coupling between windings *abc* and *xyz*, it can be seen that the current distortion decreases noticeably. So the simulation of the six-phase SM without consideration of the magnetic coupling between the three-phase stator windings can lead to the wrong conclusions.

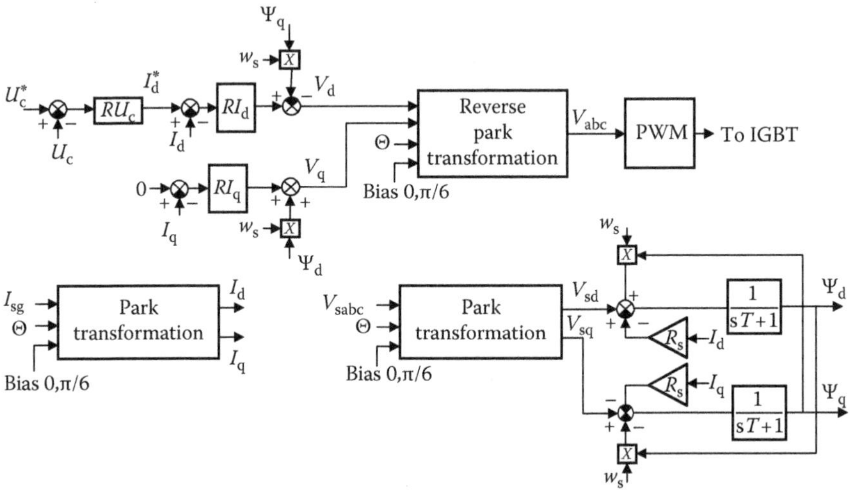

FIGURE 4.23 Block diagram of the control system for the model **SixphSynMach_det**.

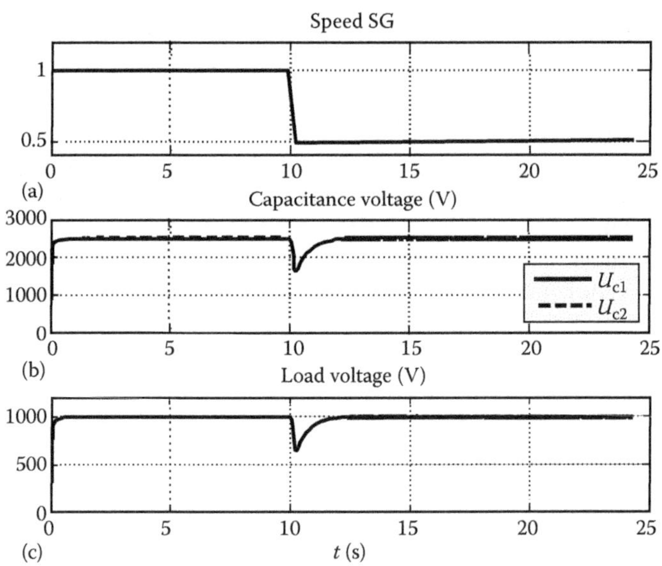

FIGURE 4.24 SG (a) rotational speed, (b) voltages across the capacitors, and (c) load for the model **SixphSynMach_det**.

4.2.5 Model of 12-Phase SG

At present, WGs with power up to 10 MW are being developed; in such WG, the output power halving, as it occurs in six-phase SG, turns out not to be satisfactory, and the employment of the 12-phase SG can solve the problem. A need for such SM appears in industry as well, especially in liquefied gas production [17]; high-speed synchronous machines are used for these sets—to 3000 r/min. Therefore, the development of such SM model that is compatible with the MATLAB/Simulink/SimPowerSystems blocks is of great interest. The modeling of such SM has been already considered in Ref. [18]. The described model differs from the one mentioned earlier and uses the methodology of SimPowerSystems.

Such SM has four three-phase stator windings with a mutual shift of 15° (electrical). The axes of the three-phase windings *abc*, *xyz*, *uvw*, and *efg* with the mutual displacement of 15° are depicted in Figure 4.25. The windings, besides the main flux, produce the leakage fluxes that are linked both with their own and with the other windings but are not linked with the rotor fluxes. It is supposed that the mutual leakage fluxes depend only on the angular distance between the winding axes; it means that it is sufficient to know the mutual leakage fluxes of winding *a* with the nine other windings.

The considered SM usually has the round rotor, so only the mutual inductances of the stator and rotor windings depend on the rotor position.

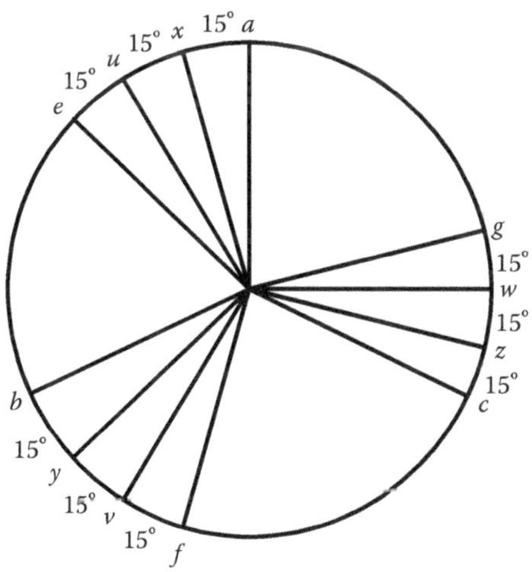

FIGURE 4.25 Arrangement of the windings in 12-phase SM.

Let $\mathbf{I}_a$, $\mathbf{I}_x$, $\mathbf{I}_u$, $\mathbf{I}_v$, and $\mathbf{I}_r$ be the three-dimensional current vectors where $\mathbf{I}_a = [i_a, i_b, i_c]^T$ and the same for $\mathbf{I}_x$, $\mathbf{I}_u$, $\mathbf{I}_v$, $\mathbf{I}_r = [i_{kd}, i_{kq}, i_f]^T$; $\mathbf{I} = [\mathbf{Ia}, \mathbf{Ix}, \mathbf{Iu}, \mathbf{Iv}, \mathbf{Ir}]^T$ is a 15-dimensional vector, and analogous the vectors of the voltages $\mathbf{U}$ and the flux linkages $\boldsymbol{\Psi}$. Then SM equations may be written as

$$\frac{d\boldsymbol{\Psi}}{dt} = \mathbf{U} - \mathbf{RI} \tag{4.116}$$

$$\boldsymbol{\Psi} = \mathbf{LI}, \tag{4.117}$$

where $\mathbf{R}$ is the diagonal matrix of the winding resistances in the order indicated in the previous paragraph, and $\mathbf{L}$ is the matrix of the internal and mutual inductances.

$$\mathbf{L} = \begin{bmatrix} \mathbf{L}_{11} & \mathbf{L}_{12} & \mathbf{L}_{13} & \mathbf{L}_{14} & \mathbf{L}_{1r}; \\ \mathbf{L}_{21} & \mathbf{L}_{22} & \mathbf{L}_{23} & \mathbf{L}_{24} & \mathbf{L}_{2r}; \\ \mathbf{L}_{31} & \mathbf{L}_{32} & \mathbf{L}_{33} & \mathbf{L}_{34} & \mathbf{L}_{3r}; \\ \mathbf{L}_{41} & \mathbf{L}_{42} & \mathbf{L}_{43} & \mathbf{L}_{44} & \mathbf{L}_{4r}; \\ \mathbf{L}_{r1} & \mathbf{L}_{r2} & \mathbf{L}_{r3} & \mathbf{L}_{4r} & \mathbf{L}_{rr} \end{bmatrix} \tag{4.118}$$

All submatrixes have the dimension 3 × 3.

Because of symmetry, it may be written as

$$\mathbf{L} = \begin{bmatrix} \mathbf{L}_{11} & \mathbf{L}_{12} & \mathbf{L}_{13} & \mathbf{L}_{14} & \mathbf{L}_{1r} \\ \mathbf{L}_{12}^T & \mathbf{L}_{11} & \mathbf{L}_{12} & \mathbf{L}_{13} & \mathbf{L}_{2r} \\ \mathbf{L}_{13}^T & \mathbf{L}_{12}^T & \mathbf{L}_{11} & \mathbf{L}_{12} & \mathbf{L}_{3r} \\ \mathbf{L}_{14}^T & \mathbf{L}_{13}^T & \mathbf{L}_{12}^T & \mathbf{L}_{11} & \mathbf{L}_{4r} \\ \mathbf{L}_{1r}^T & \mathbf{L}_{2r}^T & \mathbf{L}_{3r}^T & \mathbf{L}_{4r}^T & \mathbf{L}_{rr} \end{bmatrix}. \tag{4.119}$$

The elements of these matrixes are defined as follows:

$$
\begin{aligned}
\mathbf{L}_{11} = [&l_s + L_1\ L_1\cos(2\pi/3)\ L_1\cos(4\pi/3);\\
&L_1\cos(2\pi/3)\ l_s + L_1\ L_1\cos(2\pi/3);\\
&L_1\cos(4/3)\ L_1\cos(2\pi/3)\ l_s + L_1];\\
\mathbf{L}_{12} = [&l_{ax} + L_1\cos(\pi/12)\ l_{ay} + L_1\cos(\pi/12+2\pi/3)\ l_{az} + L_1\cos(\pi/12-2\pi/3);\\
&l_{az} + L_1\cos(\pi/12-2\pi/3)\ l_{ax} + L_1\cos(\pi/12)\ l_{ay} + L_1\cos(\pi/12+2\pi/3);\\
&l_{ay} + L_1\cos(\pi/12+2\pi/3)\ l_{az} + L_1\cos(\pi/12-2\pi/3)\ l_{ax} + L_1\cos(\pi/12)];\\
\mathbf{L}_{13} = [&l_{au} + L_1\cos(\pi/6)\ l_{av} + L_1\cos(\pi/6+2\pi/3)\ l_{aw} + L_1\cos(\pi/6-2\pi/3);\\
&l_{aw} + L_1\cos(\pi/6-2\pi/3)\ l_{au} + L_1\cos(\pi/6)\ l_{av} + L_1\cos(\pi/6+2\pi/3);\\
&l_{av} + L_1\cos(\pi/6+2\pi/3)\ l_{aw} + L_1\cos(\pi/6+2\pi/3)\ l_{au} + L_1\cos(\pi/6)];\\
\mathbf{L}_{14} = [&l_{ae} + L_1\cos(\pi/4)\ l_{af} + L_1\cos(\pi/4+2\pi/3)\ l_{ag} + L_1\cos(\pi/4+4\pi/3);\\
&l_{ag} + L_1\cos(\pi/4+4\pi/3)\ l_{ae} + L_1\cos(\pi/4)\ l_{af} + L_1\cos(\pi/4+2\pi/3);\\
&l_{af} + L_1\cos(\pi/4+2\pi/3)\ l_{ag} + L_1\cos(\pi/4+4\pi/3)\ l_{ae} + L_1\cos(\pi/4)];\\
\mathbf{L}_{1r} = 1.5L_1\,[&\cos(t) \quad -\sin(t) \quad \cos(t);\\
&\cos(t-2\pi/3) \quad -\sin(t-2\pi/3) \quad \cos(t-2\pi/3);\\
&\cos(t+2\pi/3) \quad -\sin(t+2\pi/3) \quad \cos(t+2\pi/3)];\\
\mathbf{L}_{2r} = 1.5L_1\,[&\cos(t-\pi/12) \quad -\sin(t-\pi/12) \quad \cos(t-\pi/12);\\
&\cos(t-2\pi/3-\pi/12) \quad -\sin(t-2\pi/3-\pi/12) \quad \cos(t-2\pi/3-\pi/12);\\
&\cos(t+2\pi/3-\pi/12) \quad -\sin(t+2\pi/3-\pi/12) \quad \cos(t+2\pi/3-\pi/12)];\\
\mathbf{L}_{3r} = 1.5L_1\,[&\cos(t-\pi/6) \quad -\sin(t-\pi/6) \quad \cos(t-\pi/6);\\
&\cos(t-2\pi/3-\pi/6) \quad -\sin(t-2\pi/3-\pi/6) \quad \cos(t-2\pi/3-\pi/6);\\
&\cos(t+2\pi/3-\pi/6) \quad -\sin(t+2\pi/3-\pi/6) \quad \cos(t+2\pi/3-\pi/6)];\\
\mathbf{L}_{4r} = 1.5L_1\,[&\cos(t-\pi/4) \quad -\sin(t-\pi/4) \quad \cos(t\text{-}\pi/4);\\
&\cos(t-2\pi/3-\pi/4) \quad -\sin(t-2\pi/3-\pi/4) \quad \cos(t-2\pi/3-\pi/4);\\
&\cos(t+2\pi/3-\pi/4) \quad -\sin(t+2\pi/3-\pi/4) \quad \cos(t+2\pi/3-\pi/4)];\\
\mathbf{L}_{rr} = [&l_{skd} + 1.5L_1 \quad 0 \quad 1.5L_1;\\
&0 \quad l_{skq} + 1.5L_1 \quad 0;\\
&L_1 \quad 0 \quad l_{sf} + 1.5L_1]
\end{aligned}
\tag{4.120}
$$

At this, according to Ref. [2], the rotor winding currents are referred to the stator as $i'(k_d, k_q, f) = 2/3N(k_d, k_q, f)/N_{stator}i(k_d, k_q, f)$; N is the number of the turns of the corresponding windings.

Here l_s is the internal leakage inductance of the single stator winding, l_{ax} and so on are the mutual leakage inductances of the indicated stator windings, l_{skd}, l_{skq}, and l_{sf} are the leakage inductances of the proper rotor windings, and L_1 is the inductance connected with the main flux.

SM is modeled in the rotating reference frame connected with the rotor. The transformation from the stationary reference frame to the rotating one is carried out with the aid of the Clarke transformation abc—αβ and consecutively of the Park transformation αβ—dq that in the case under consideration, may be written as

$$\mathbf{T} = \begin{vmatrix} \mathbf{T}_0 & \mathbf{0} & \mathbf{0} & \mathbf{0} & \mathbf{0} \\ \mathbf{0} & \mathbf{T}_1 & \mathbf{0} & \mathbf{0} & \mathbf{0} \\ \mathbf{0} & \mathbf{0} & \mathbf{T}_2 & \mathbf{0} & \mathbf{0} \\ \mathbf{0} & \mathbf{0} & \mathbf{0} & \mathbf{T}_3 & \mathbf{0} \\ \mathbf{0} & \mathbf{0} & \mathbf{0} & \mathbf{0} & \mathbf{I} \end{vmatrix}, \tag{4.121}$$

where $\mathbf{I}$ is the unit matrix 3 × 3,

$$\mathbf{T}_i = \mathbf{P}_i\,\mathbf{C},\ i = 0\ldots3\ (i = 0, 1, 2, 3), \tag{4.122}$$

$$\mathbf{C} = \frac{1}{3}\begin{vmatrix} 2 & -1 & -1 \\ 0 & \sqrt{3} & -\sqrt{3} \\ 1 & 1 & 1 \end{vmatrix} \tag{4.123}$$

$$\mathbf{P}_i = \begin{vmatrix} \cos(t - is) & \sin(t - is) & 0 \\ -\sin(t - is) & \cos(t - is) & 0 \\ 0 & 0 & 1 \end{vmatrix}, \tag{4.124}$$

where $s = \pi/12$.

Let us apply this transformation to Equation 4.117 bearing in mind Equation 4.120. The variables in the rotating reference frame are designated with the index dq. We have $\Psi_{dq} = \mathbf{T}\Psi = \mathbf{TLI} = \mathbf{TLT}^{-1}\mathbf{TI} = \mathbf{L}_{dq}\mathbf{I}_{dq}$ and $\mathbf{L}_{dq} = \mathbf{TLT}^{-1}$. To expel the zero sequence components, the matrix $\mathbf{L}_{dq}$ will have dimension 11 × 11; at this, the matrixes for the stator windings have the dimension 2 × 2; and for the rotor windings, 2 × 3. Designate $1.5L_1 = L_m$. Because the formulas for the matrix rows are too long, the elements of the rows are separated with the commas and the rows with the semicolon in the following relationships:

$$
\begin{aligned}
\mathbf{L}_{dq11} &= [L_m + l_s,\ 0;\ 0,\ L_m + l_s] \\
\mathbf{L}_{dq12} &= [L_m + \cos(\pi/12)l_{ax} - \cos(\pi/4)l_{ay} - \sin(\pi/12)l_{az}), \\
&\quad \sin(\pi/12)l_{ax} + \cos(\pi/4)l_{ay} - \cos(\pi/12)l_{az}); \\
&\quad -\sin(\pi/12)l_{ax} - \cos(\pi/4)l_{ay} + \cos(\pi/12)l_{az}), \\
&\quad L_m + \cos(\pi/12)l_{ax} - \cos(\pi/4)l_{ay} - \sin(\pi/12)l_{az}] \\
\mathbf{L}_{dq13} &= [L_m + \cos(\pi/6)l_{au} - \cos(\pi/6)l_{av},\ \ \sin(\pi/6)l_{au} + \sin(\pi/6)l_{av} - l_{aw}; \\
&\quad -\sin(\pi/6)l_{au} - \sin(\pi/6)l_{av} + 1_{aw},\ L_m + \cos(\pi/6)l_{au}) - \cos(\pi/6)l_{av})] \\
\mathbf{L}_{dq14} &= [L_m + \cos(\pi/4)l_{ae} - \cos(\pi/12)l_{af} + \sin(\pi/12)l_{ag}, \\
&\quad \cos(\pi/4)l_{ae} + \sin(\pi/12)l_{af} - \cos(\pi/12)l_{ag}); \\
&\quad -\cos(\pi/4)l_{ae} - \sin(\pi/12)l_{af}) + \cos(\pi/12)l_{ag}, \\
&\quad L_m + \cos(\pi/4)l_{ae} - \cos(\pi/12)l_{af} + \sin(\pi/12)l_{ag}] \\
\mathbf{L}_{dq1r} &= [L_m, 0, L_m; 0, L_m, 0] \\
\mathbf{L}_{r1} &= (2/3)\mathbf{L}_{1r}^{T},
\end{aligned}
$$

and so on. Therefore, it may be written as follows:

$$
\begin{aligned}
\mathbf{L}_{dq} = [&\mathbf{L}_{dq11},\ \mathbf{L}_{dq12},\ \mathbf{L}_{dq13},\ \mathbf{L}_{dq14},\ \mathbf{L}_{dqr}; \\
&\mathbf{L}_{dq12}^{T},\ \mathbf{L}_{dq11},\ \mathbf{L}_{dq12},\ \mathbf{L}_{dq13},\ \mathbf{L}_{dqr}; \\
&\mathbf{L}_{dq13}^{T},\ \mathbf{L}_{dq12}^{T},\ \mathbf{L}_{dq11},\ \mathbf{L}_{dq12},\ \mathbf{L}_{dqr}; \\
&\mathbf{L}_{dq14}^{T},\ \mathbf{L}_{dq13}^{T},\ \mathbf{L}_{dq12}^{T},\ \mathbf{L}_{dq11},\ \mathbf{L}_{dqr}; \\
&\mathbf{L}_{dqr}^{T},\ \mathbf{L}_{dqr}^{T},\ \mathbf{L}_{dqr}^{T},\ \mathbf{L}_{dqr},\ \mathbf{L}_{rr}] \\
\mathbf{L}_{rr} = [&l_{skd} + L_m, 0, L_m; 0, l_{skq} + L_m, 0; L_m, 0, l_{sf} + L_m]
\end{aligned}
\tag{4.125}
$$

The rotor winding leakage inductances are increased in the ratio of 1.5 relatively real values, because the rotor currents are decreased in the same ratio.

The matrixes of the mutual leakage inductances may be written as follows:

$$\mathbf{L}_{dq1j} = [L_{1jd}, L_{1jq}; -L_{1jq}, L_{1jd}],\ j = 2, 3, 4 \tag{4.126}$$

so that there are only six quantities that define the mutual leakage inductances:

$$\begin{aligned}
L_{12d} &= \cos(\pi/12)l_{ax} - \cos(\pi/4)l_{ay} - \sin(\pi/12)l_{az};\\
L_{12q} &= \sin(\pi/12)l_{ax} + \cos(\pi/4)l_{ay} - \cos(\pi/12)l_{az};\\
L_{13d} &= \cos(\pi/6)l_{au} - \cos(\pi/6)l_{av};\\
L_{13q} &= \sin(\pi/6)l_{au} + \sin(\pi/6)l_{av} - l_{aw};\\
L_{14d} &= \cos(\pi/4)l_{ae} - \cos(\pi/12)l_{af} + \sin(\pi/12)l_{ag};\\
L_{14q} &= \cos(\pi/4)l_{ae} + \sin(\pi/12)l_{af} - \cos(\pi/12)l_{ag}.
\end{aligned} \tag{4.127}$$

Therefore, the fluxes of both axes are coupled, as explained in the previous paragraph. The mutual leakage inductances are maximal when the winding axes are aligned, and are equal to zero when they are perpendicular. Let us suppose that these inductances are proportional to the cosine of the angle between the winding axes with the proportional factor l_{ss}. Then $L_{1jd} = 1.5l_{ss}$ and $L_{1jq} = 0$; the matrix $\mathbf{L}_{dq}$ is decomposed in two independent ones (or matrixes). By designating $1.5l_{ss}/l_s = k$, one obtains vectors $\Psi_{sd} = [\Psi_{d1}, \Psi_{d2}, \Psi_{d3}, \Psi_{d4}]$ and $\Psi_{sq} = [\Psi_{q1}, \Psi_{q2}, \Psi_{q3}, \Psi_{q4}]$ for the stator flux linkages and vectors $\mathbf{I}_d = [I_{d1}, I_{d2}, I_{d3}, I_{d4}, I_{kd}, I_f]$ and $\mathbf{I}_q = [I_{q1}, I_{q2}, I_{q3}, I_{q4}, I_{kq}]$ for the stator and rotor current:

$$\begin{aligned}
\boldsymbol{\Psi}_{sd} = [&L_m + l_s, L_m + kl_s, L_m + kl_s, L_m + kl_s, L_m, L_m;\\
&L_m + kl_s, L_m + l_s, L_m + kl_s, L_m + kl_s, L_m, L_m;\\
&L_m + kl_s, L_m + kl_s, L_m + l_s, L_m + kl_s, L_m, L_m;\\
&L_m + kl_s, L_m + kl_s, L_m + kl_s, L_m + l_s, L_m, L_m]\mathbf{I}_d
\end{aligned} \tag{4.128}$$

$$\begin{aligned}
\boldsymbol{\Psi}_{sq} = [&L_m + l_s, L_m + kl_s, L_m + kl_s, L_m + kl_s, L_m;\\
&L_m + kl_s, L_m + l_s, L_m + kl_s, L_m + kl_s, L_m;\\
&L_m + kl_s, L_m + kl_s, L_m + l_s, L_m + kl_s, L_m;\\
&L_m + kl_s, L_m + kl_s, L_m + kl_s, L_m + l_s, L_m]\mathbf{I}_q
\end{aligned} \tag{4.129}$$

To take

$$\Psi_{md} = L_m(I_{d1} + I_{d2} + I_{d3} + I_{d4} + I_{kd} + I_f) \tag{4.130}$$

$$\Psi_{mq} = L_m(I_{q1} + I_{q2} + I_{q3} + I_{q4} + I_{kq}), \tag{4.131}$$

then the following equations are obtained:

$$\Psi_{sd} - \Psi_{md}\begin{vmatrix}1\\1\\1\\1\end{vmatrix} = l_s\begin{vmatrix}1 & k & k & k\\ k & 1 & k & k\\ k & k & 1 & k\\ k & k & k & 1\end{vmatrix}\mathbf{I}_{sd} \tag{4.132}$$

$$\Psi_{sq} - \Psi_{mq}\begin{vmatrix}1\\1\\1\\1\end{vmatrix} = l_s\begin{vmatrix}1 & k & k & k\\ k & 1 & k & k\\ k & k & 1 & k\\ k & k & k & 1\end{vmatrix}\mathbf{I}_{sq} \tag{4.133}$$

The solution of these equations yields the following:

$$I_{d1} = \frac{F_{d1} - k(F_{d2} - 2F_{d1} + F_{d3} + F_{d4})}{dl_s} \tag{4.134}$$

$$d = (1 - k)\,(3k + 1) \tag{4.135}$$

$$F_{di} = \Psi_{sdi} - \Psi_{md} \tag{4.136}$$

$$I_{d2} = \frac{F_{d2} - k(F_{d1} - 2F_{d2} + F_{d3} + F_{d4})}{dl_s} \tag{4.137}$$

$$I_{d3} = \frac{F_{d3} - k(F_{d1} + F_{d2} - 2F_{d3} + F_{d4})}{dl_s} \tag{4.138}$$

$$I_{d4} = \frac{F_{d4} - k(F_{d1} + F_{d2} + F_{d3} - 2F_{d4})}{dl_s} \tag{4.139}$$

By substituting these relationships for the currents in Equation 4.130, we receive

$$\Psi_{md} = L_{ad}\left[\frac{\Psi_f}{l_{sf}} + \frac{\Psi_{kd}}{l_{skd}} + \frac{(\Psi_{d1} + \Psi_{d2} + \Psi_{d3} + \Psi_{d4})(1-k)}{dl_s}\right] \tag{4.140}$$

$$L_{ad} = \left[\frac{1}{L_m} + \frac{1}{l_{sf}} + \frac{1}{l_{skd}} + \frac{4(1-k)}{dl_s}\right]^{-1}. \tag{4.141}$$

The same for Ψ_{mq}:

$$\Psi_{mq} = L_{aq}\left[\frac{\Psi_{kq}}{l_{skq}} + \frac{(\Psi_{q1} + \Psi_{q2} + \Psi_{q3} + \Psi_{q4})(1-k)}{dl_s}\right] \tag{4.142}$$

$$L_{aq} = \left[\frac{1}{L_m} + \frac{1}{l_{skq}} + \frac{4(1-k)}{dl_s}\right]^{-1} \tag{4.143}$$

The formulas for the currents I_q are obtained by replacing index d for q. The voltage equations are as follows:

$$\begin{aligned}
s\Psi_{qs1} &= U_{qs1} - \omega_r\Psi_{ds1} - R_sI_{q1}\\
s\Psi_{ds1} &= U_{ds1} + \omega_r\Psi_{qs1} - R_sI_{d1}\\
s\Psi_{qs2} &= U_{qs2} - \omega_r\Psi_{ds2} - R_sI_{q2}\\
s\Psi_{ds2} &= U_{ds2} + \omega_r\Psi_{qs2} - R_sI_{d2}\\
s\Psi_{ds3} &= U_{ds3} + \omega_r\Psi_{qs3} - R_sI_{d3}\\
s\Psi_{qs3} &= U_{qs3} - \omega_r\Psi_{ds3} - R_sI_{q3}\\
s\Psi_{qs4} &= U_{qs4} - \omega_r\Psi_{ds4} - R_sI_{q4}\\
s\Psi_{ds4} &= U_{ds4} + \omega_r\Psi_{qs4} - R_sI_{d4}\\
s\Psi_{kq} &= \frac{R_{kq}}{l_{skq}}(\Psi_{mq} - \Psi_{kq})\\
s\Psi_{f} &= \frac{R_{f}}{l_{sf}}(\Psi_{md} - \Psi_{f}) + U_f\\
s\Psi_{kd} &= \frac{R_{kd}}{l_{skd}}(\Psi_{md} - \Psi_{kd}),
\end{aligned} \tag{4.144}$$

where s is an operator of the differentiation. Substituting the expressions found for the currents in the first eight equations (Equation 4.144), taking into account formulas for Ψ_{md} and Ψ_{mq}, we receive the system of differential equations for the stator flux linkages, from which the SM flux linkages can be found, and afterward the SM currents can be calculated.

The torque may be expressed as

$$T_e = 1.5Z_p[(I_{q1} + I_{q2} + I_{q3} + I_{q4})\Psi_{md} - (I_{d1} + I_{d2} + I_{d3} + I_{d4})\Psi_{mq}], \tag{4.145}$$

where Z_p is the pole pair number.

The utilization of the developed SM model is demonstrated with some examples. The model of 12-phase SM is given in the model **SG_12ph**. The SM model consists of four blocks for connection with the elements of SimPowerSystems **Power_connect**, the electrical and mechanical subsystems **Sub_Electr** and **Sub_mech**. The former contain two voltage sensors for the measurement of the phase-to-phase voltages and two current sources for output of SM currents in the power circuits. The model of the mechanical part is the same as in the previous models. The model of the electrical part consists of a number of subsystems.

The subsystem **Vabc_Vdq** transforms two phase-to-phase voltages in the voltages V_d and V_q in the rotating reference frame, whose angle position θ is determined by the rotor position. The signals θ for each input have the mutual shift by 15°. The voltages V_d and V_q are computed in pu.

The subsystem **Stator_Rotor** is the main one, in which relationships 4.134–4.144 are realized. The input quantities are four pairs of the voltages V_d and V_q, the excitation voltage V_f, and the frequency of the supplying voltage ω_e. The output quantities are four pairs of the currents I_d and I_q, the flux linkage Ψ_{md} and Ψ_{mq}, and also a 13-dimensional vector of the measured variables arranged as I_{q1}, I_{d1}, I_{q2}, I_{d2}, I_{q3}, I_{d3}, I_{q4}, I_{d4}, I_{fd}, I_{kq}, I_{kd}, Ψ_{mq}, Ψ_{md}. The torque is computed in the subsystem **Te** by Equation 4.145 (all computation in pu); the active and reactive SM powers are calculated in the subsystem **PQ**, at that,

$$P = \sum_{i=1}^{4} (V_{qi} I_{qi} + V_{di} I_{di}) \tag{4.146}$$

$$Q = \sum_{i=1}^{4} (V_{qi} I_{di} - V_{di} I_{qi}). \tag{4.147}$$

The transformation of the computed winding currents I_d and I_q into the three-phase currents of the same windings is carried out by means of the reverse Park transformation with the employment of the earlier-used angles θ. These currents control the controlled current sources in the blocks **Power_connect**.

Twenty-seven output signals are united at the output *m*1: 12 stator phase currents, 13 aforesaid signals, and *P* and *Q* SM powers. The SM rotational speed, the mechanical power, the angle rotor position (electrical grad), the SM torque, and the torque angle δ are added to 27 signals at the output *m*.

The model **SG_12ph** is intended for the confirmation of the model validity and for the study of its main characteristics. The SG speed is kept equal to 1 with the help of the PI speed controller. The parameters of the nonsalient pole SG are 10 MW, 2400 V, 50 Hz, and $Z_p = 2$.

The rest of the parameters are given in the program `MODEL_SM_12ph1` that is executed before the model start. Each winding is loaded with the active load of 2.5 MW. The excitation voltage is 1.2 (after forcing completion) that is computed by the following way. For nonsalient pole SG, the load voltage is roughly

$$U_n = \frac{R_n E_{fd}}{R_n + jX_d}, \tag{4.148}$$

where R_n is the load resistance. In the considered case, $R_n = (2400)^2/(2.5 \times 10^6) = 2.3\Omega$; the base resistance is $R_b = U_{nom}^2/P_{nom} = 2400^2/10^7 = 0.576\Omega$; the equivalent SG load in pu is 2.3/4/0.576 = 1; the SG inductance impedance (supposing all four windings are connected in parallel, without coupling) $X_d = X_{md} + X_{sl}/4 = 0.6272 + 0.1473/4 = 0.664$. Therefore, in order to obtain the rated voltage, it must be

$$E_{fd} = \frac{\sqrt{R_n^2 + X_d^2}}{R_n} = 1.2.$$

To run the simulation under the stipulation that $k = 0$, one can see that after transients, the torque, the power, and the voltage are very close to 1 pu. The winding currents exhibit the mutual shift by 15°. The torque angle can be found from the relationship

$$P = \frac{U_s E_{fd}}{X_d} \sin\delta, \tag{4.149}$$

from which, under $P = 1$, $U_s = 1$, and $E_{fd} = 1.2$, we have $\delta = 33.6°$ that corresponds to display the **Delta** reading. Therefore, this model operates according to the theoretical premises.

To repeat the simulation with $k = 0.5$, it can be seen that the torque, the power, and the load voltage decrease by some percentage; the angle δ increases to 35.5°; in order to reach the rated values, one must take $E_{fd} = 1.23$.

The model **SG_12ph1** is an extension of the model **Sixphsynchmacine** for 12-phase SG. Moreover, SG supplies not the resistive load, but the network via the 1.6 kV/25 kV transformer. The reference capacitor voltage is 5 kV. The output inverters control the current component I_d, which is sent to the network; this component aligns with the network voltage; the reference for the component I_q is zero. The current reference that was initially equal to 2 kA increases to 3 kA at $t = 10$ s; at $t = 20$ s, the SG rotational speed decreases by 20%. The SG speed, the capacitor voltages, and the network power are shown in Figure 4.26. The reactive power is equal to zero. Although the capacitor voltage response is relatively slow, the power sent in the network grows up rather quickly.

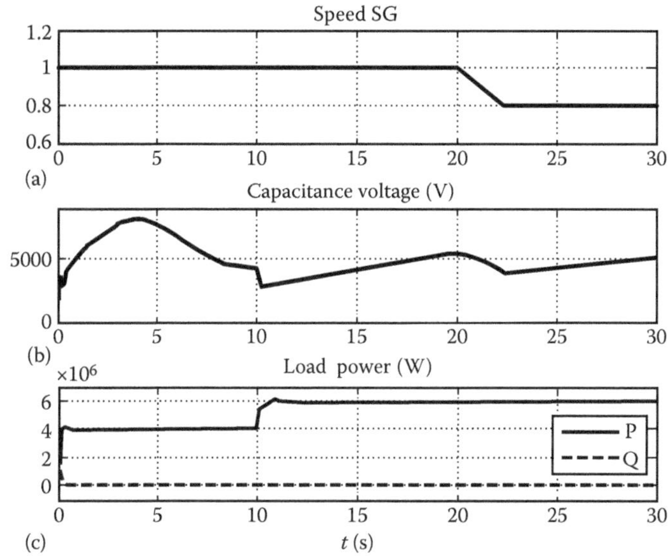

FIGURE 4.26 SG (a) speed, (b) capacitor voltages, and (c) network power for the model **SG_12ph1**.

The model **SG_12ph1_det** has the same structure as the previous one, but the three-level inverters are utilized for the rectifiers and the inverters. The additional L–C filters are set in series with the SG windings. The control system structure remains. The current reference is set as 4 kA. The network power and the current of the first SG winding are shown in Figure 4.27; THD < 10%. To execute the simulation with $k = 0$, the THD is seen to essentially decrease; so in order to receive the correct results, the magnetic coupling of the windings has to be taken into account.

For completeness, the model of the high-power electrical drive with 12-phase SM is given: **SM_12phase_drive**. As was said earlier, SMs of such type are intended for operation in sets for the production of liquefied gas. The peculiarities of such drives are the huge powers: the dozens of megawatt; the high rotational speed: up to 3000 r/min; and the large inertia, because the motor is built in the kinematic chain of the whole big assembly. Because a number of low-frequency natural oscillations can arise in the kinematic chain, the torque pulsations have to be low too, in order to not to cause the elastic vibrations and not to damage the turbine blades. Because the four-pole rotor is better balanced dynamically than a two-pole rotor, SM has four poles, and the supplying voltage has the frequency of 100 Hz.

The 40 MW and 7200 V SM is modeled. Its parameters are specified in the program **MODEL_SM_12ph.m**. SM is supplied from four three-level VSIs; the reactors with the inductance of 1.8 mH are set in the circuits of the stator windings. The fabrication of the 2 × 7500 V DC voltage is not modeled. The PWM frequency is taken as 900 Hz; therefore, the ratio of the commutation frequency to the voltage frequency is 9, which is less than what is usually accepted.

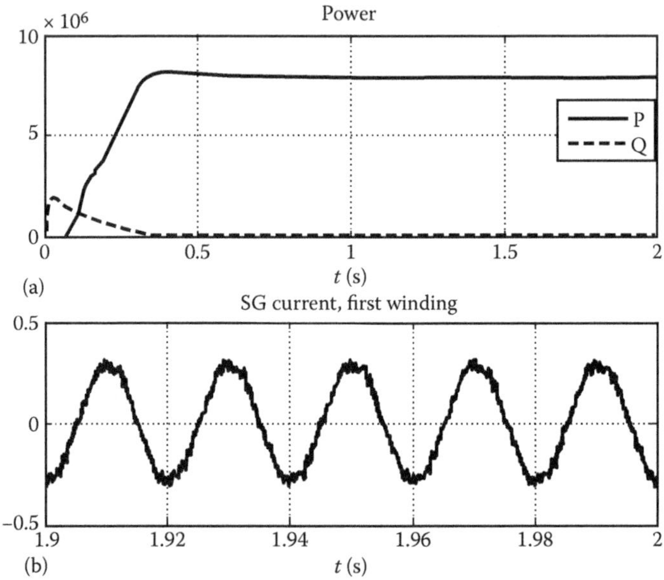

FIGURE 4.27 Network (a) power and (b) current of the SG winding for the model **SG_12ph1_det**.

The electrical drive control system uses the field-oriented control principle. The components of the full SM current I_T and I_M are controlled; the former component is aligned in the direction of the supplying voltage space vector and is responsible for the active power; the latter is aligned perpendicular to the former and is responsible for the reactive power; to secure a better effectiveness, the reference $I_M^* = 0$. As for the reference I_T^*, it is defined by the output of the speed controller (or torque reference) divided by the stator flux module.

For the computation of the components I_T and I_M, knowing the full SM current I_s, it is necessary to know the angle position α of the supplying voltage space vector V_s. At this, $\alpha = \theta + \delta$, where θ is the rotor angle and δ is the torque angle that can be calculated as

$$\delta = \text{arctg}(V_d/V_q), \tag{4.150}$$

where V_q and V_d are the projections of the space vector V_s on the axis of the rotor and are perpendicular to it. The angle θ is supposed to be measured; the angle δ is calculated using the voltage applied to the smoothing reactor and differs from the angle δ computed using the voltage at the SM terminals.

The control system contains the PI regulator of the stator flux F_{smod} affecting on SM excitation voltage. The calculation of F_{smod} is carried out by following relationships [19]:

$$\Psi_{s\alpha} = \int (U_{s\alpha} + \lambda U_{s\beta} - \lambda\omega_s \Psi_{s\alpha})\,dt$$

$$\Psi_{s\beta} = \int (U_{s\beta} - \lambda U_{s\alpha} - \lambda\omega_s \Psi_{s\beta})\, dt$$

$$U_{s\alpha} = V_{s\alpha} - R_s I_{s\alpha},\ U_{s\beta} = V_{s\beta} - R_s I_{s\beta}$$

$$\Psi_{s\,\mathrm{mod}} = \sqrt{\Psi_{s\alpha}^2 + \Psi_{s\beta}^2} \tag{4.151}$$

The arrangement of the control system is displayed in Figure 4.28. The three-phase output signals of four current controllers $U_1^* - U_4^*$ come to the PWM blocks.

The modules of the flux linkages and the angles δ are computed for each winding with the use of relationships 4.150 and 4.151, and the average value is utilized afterward. The current controllers I_T and I_M are separate for each winding as well. The transformation of their outputs in the three-phase signals for PWM is executed with the employment of the angles $\theta + \delta + i\pi/12$ and $i = 0...3$ ($i = 0, 1, 2, 3$); the calculation of the synchronizing signals is carried out in the subsystem **sin_cos** with the help of the following relationships:

$$\sin(\theta + i\,\pi/12) = \sin(\theta)\cos(i\,\pi/12) + \cos(\theta)\sin(i\,\pi/12)$$

$$\cos(\theta + i\,\pi/12) = \cos(\theta)\cos(i\,\pi/12) - \sin(\theta)\sin(i\,\pi/12)$$

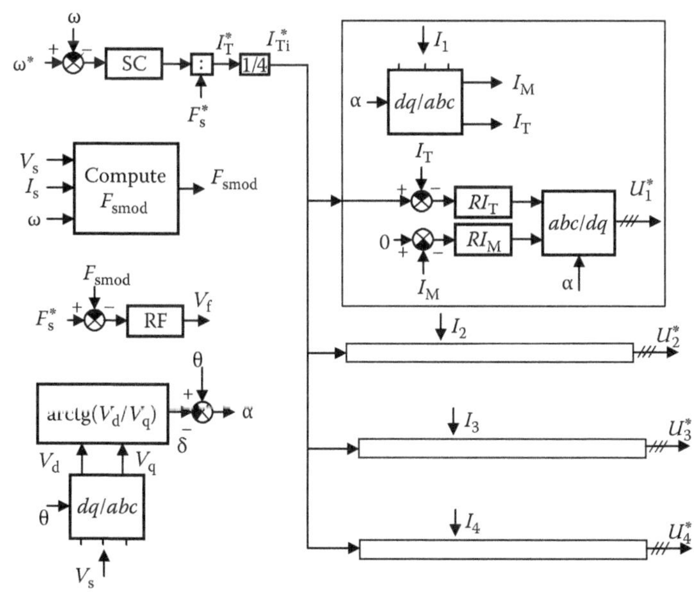

FIGURE 4.28 Control system block diagram of the power drive with 12-phase SM.

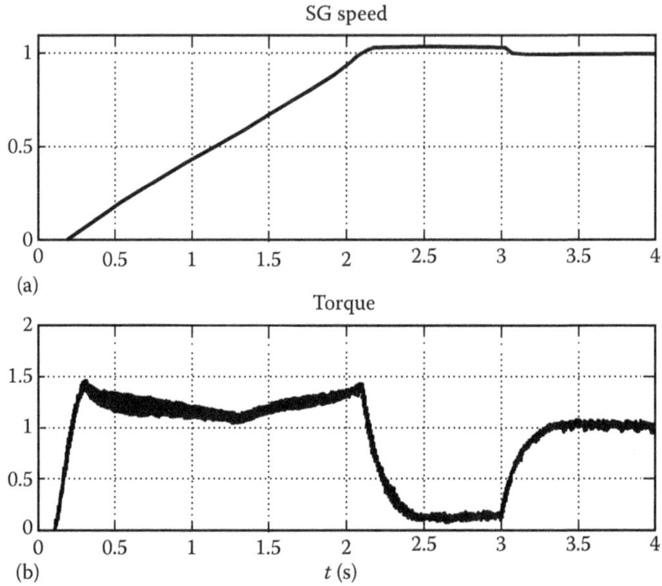

FIGURE 4.29 (a) Speed and (b) torque of power drive with 12-phase SM.

The process of acceleration to the rated speed with the load torque 0.2 and increasing the latter quantity to 1 at $t = 3$ s is simulated. At this, the inertia constant H was essentially decreased in order to speed up the simulation. The SM speed and torque are shown in Figure 4.29. To investigate the plots of the voltages and the currents in the circuits of the windings (scopes in the subsystem **Measurements**), with the help of the option *Powergui/FFT*, it can be seen that the phase shift between voltages of the neighboring windings is close to 15°, and the phase shift between the voltage and the current of the same winding is not more than 1°.

4.3 SYNCHRONOUS GENERATORS WITH PERMANENT MAGNETS

4.3.1 Generator Model

The synchronous machine with permanent magnets (PM) has only a stator winding; an excitation is carried out with the permanent magnets mounted on the rotor. One distinguishes the surface magnet machines (SPM) and the interior magnet machines (IPM). The former have magnets that are glued on the rotor surface, whereas the latter have magnets that are mounted inside the rotor. Since the relative permeability of PM material is very close to 1, the presence of PM does not have influence on the stator winding inductance. Therefore, $L_d = L_q = L_s$ for SPM, whereas $L_d < L_q$ for IPM, because of the bigger air gap in the direction of axis d, owing to the necessity to place magnets. Synchronous machine with permanent magnets (PMSM) is described by four electromagnetic parameters: the stator winding resistance R_s, the inductances L_d and L_q, the rotor flux Ψ_r, and also the pole pair number Z_p.

PMSM is modeled in the reference frame that rotates in synchrony with the rotor. The PMSM equations may be obtained from the equations of the usual SM as follows:

$$v_{sd} = R_s i_{sd} + L_d \frac{di_{sd}}{dt} - \omega_s L_q i_{sq} \tag{4.152}$$

$$v_{sq} = R_s i_{sq} + L_q \frac{di_{sq}}{dt} + \omega_s (L_d i_{sd} + \Psi_r) \tag{4.153}$$

because the components of the stator flux Ψ_s are

$$\Psi_{sq} = L_d i_{sd} + \Psi_r,\ \Psi_{sq} = L_q i_{sq}. \tag{4.154}$$

Using Equations 4.75 and 4.154, the PMSM torque may be written as

$$Te = 1.5Z_p[(L_d i_{sd} + \Psi_r)i_{sq} - L_q i_{sq} i_{sd}] = 1.5Z_p[\Psi_r i_{sq} + (L_d - L_q)i_{sd} i_{sq}]. \tag{4.155}$$

On the first page of the PMSM dialog box, the construction of the investigated machine can be selected: the phase number N_{ph}: 3 or 5; with sinusoidal or trapezoidal EMF waveform (only the former is considered in the following); IPM or SPM (only for $N_{ph} = 3$ and for sinusoidal EMF waveform). One of the preset PMSM models may be chosen as well.

The stator winding resistance, the inductances L_d and L_q, and the rotor flux Ψ_r are specified on the second page. The flux value can be specified in three ways: by direct indication of this value in Wb; by specification of the maximum phase-to-phase voltage *V* that is induced at the PMSM terminal when it rotates with the speed of 1000 r/min under no-load condition; as the proportional factor *TC* between the torque and the current amplitude. All these values are connected to one another. For example, for the PMSM in the model PMSMFLY described in the following, $\Psi_r = 0.45$ Wb, $V = 1.732 \times 1000/30$, $\pi Z_p \Psi_r = 81.62$ V, and $TC = 1.5Z_p\Psi_r = 0.675$ Nm/A.

The moment of inertia *J* (kgm^2), the viscous damping factor *F*, and the pole pair number $p = Z_p$, together with the initial conditions, are defined on the second page of the dialog box too. On the third page, the rotor flux direction, relatively phase *A*, is assigned, when the rotor is in the position taken for zero; in fact, it is fixed, from which vector component, *d* or *q*, the angle for Park transformation is counted off.

The electrical part of the PMSM model consists of four blocks: transformation of the phase-to-phase voltages in the components V_d and V_q as for SM with wounded rotor, computation of i_{sq} and i_{sd} by the solution of Equations 4.152 and 4.153, transformation of i_{sq} and i_{sd} in the phase currents, and torque computation by Equation 4.155. The mechanical part is described by Equations 4.16 and 4.17. The blocks Simulink **Bus Selector** are used for PMSM variable measurement; the stator currents i_a, i_b, and i_c (A), the current and voltage components i_{sq} and i_{sd} (A), V_q and V_d (V), the rotational speed ω_m (rad/s), the rotor angle θ_m (rad), and the torque T_e (Nm) are available for measurement. Also the signal *Hall effect signal* is formed, which can be utilized for the logical indication of the position of the EMF space vector.

The main principles for PMSM control are the following: For SPM, the torque is independent on i_{sd}; therefore, it is reasonable to have $i_{sd} = 0$, in order to decrease the PMSM heating and to increase the load-carrying capability. It follows from Equation 4.155 for IPM that under $i_{sd} < 0$, its torque will be bigger, because of $(L_d - L_q) < 0$; but at this, the maximal allowed value i_{sq} decreases, because it has to be

$$i_s = \sqrt{i_{sd}^2 + i_{sq}^2} \le i_{s0}, \tag{4.156}$$

where i_{s0} is the maximal permissible current of the PMSM or inverter. Therefore, the optimal value of i_{sd} exists.

As an example of PMSM model employment, the system of the electric energy storage with the flywheel is considered. These systems are environmentally friendly and have fast response when it is necessary to transfer the stored energy to the consumers, but, with reasonable flywheel size, they can operate for a short time, so that they are used, mainly, for smoothing the source power oscillations and for load supplying during a short power interruption. The flywheel unit consists of the electrical machine and the flywheel; when the source possesses more power than it is in demand, the machine operates as a motor and the flywheel speed rises; under power shortage, the machine changes over in the generator mode, and the energy stored in the flywheel returns to the load.

The energy stored by a flywheel is

$$E = \frac{J\omega^2}{2} = \frac{mR^2\omega^2}{2} = \frac{\pi R^4 h\rho\omega^2}{2}, \tag{4.157}$$

where J is the moment of inertia, m is the mass, R is the radius, h is the height of the spinning cylinder, ρ is the material density, and ω is the rotational speed.

Therefore, in order to increase the energy storage, the flywheel radius and its rotational speed have to be increased, but the mechanical factors limit these possibilities. If the electrical drive provides the speed control range $k = \omega_{min}/\omega_{max}$, the maximal extracted energy is

$$\Delta E = E_{max}\,(1 - k^2). \tag{4.158}$$

Under the flywheel unit design, the measures are taken for the loss decrease, specifically, the loss in the bearings and the windage loss. For instance, the flywheel is covered with an airtight housing, from which the air is evacuated.

In the model **PMSMFLY**, the subsystem **Source** models the three-phase source with the rated phase-to-phase voltage of 400 V, whose frequency and voltage are distorted in a random way. The source of the white noise and a shaper of the correlation function are used for the fabrication of the random perturbations. The source is connected to the load with an active power of 20 kW. The flywheel system consists of a two-pole IPM with the maximal speed of 900 rad/s; of the flywheel, with J = 50 kgm^2; and of two VSI, with the capacitor between them. The direct torque control system is utilized for PMSM control.

The reference values are the wanted PMSM torque and the stator flux. The subsystem **DTC_Reg** that is placed in the subsystem **DTC** contains two relays, which perceive the errors between the references and the actual values of the torque and the flux. The block **Lookup Table1** converts the relay states into the number 0.3 (Table 4.1).

The next inverter state depends on the output of this table and on the angular position θ of the stator flux linkage space vector, which is defined by the number of the sector having 60° extent. One of the six possible states corresponds to each output of **Lookup Table1**; these states depend on the sector number, in which vector Ψ_s are placed.

In Figure 4.30, the vector Ψ_s in the system of the inverter voltage space vectors is shown. Every inverter state is defined by numbers from 1 to 6. In the binary code, the first bit describes the state of phase *A* devices; the second, phase *B*; and the low, phase *C*, under which this inverter state is implemented. At this, 1 defines the state when the device, connected to the DC positive pole, conducts; correspondingly, 0 defines the state when the device, connected to the DC negative pole, conducts. For example, Vk = 5 (101) means that this inverter state takes place when the devices of

TABLE 4.1
Relay State Coding

DF\DM	0	1
0	0	1
1	2	3

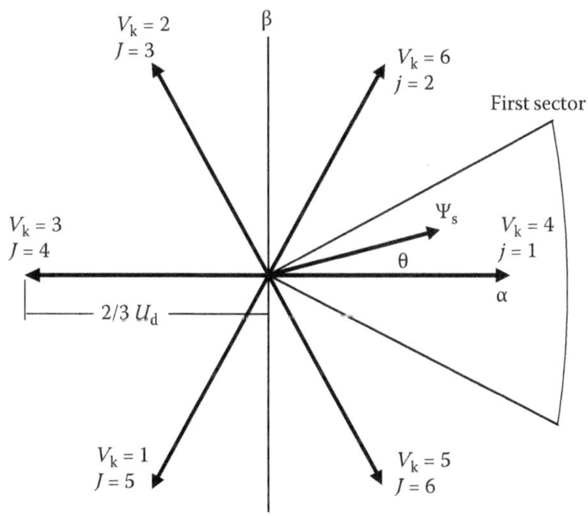

FIGURE 4.30 Vector Ψ_s and VSI voltage space vectors.

phases A and C that connect to the positive pole and the device of phase B that connects to the negative pole conduct.

If vector Ψ_s is in the first sector, as shown in Figure 4.30, the torque increases for the states $Vk = 6$ and $Vk = 2$ and decreases for $Vk = 1$ and $Vk = 5$. The stator flux increases for the states $Vk = 6$ and $Vk = 5$ and decreases for the states $Vk = 2$ and $Vk = 1$. The described program of actions is programmed in the block **Lookup Table0**. It has two inputs: from the output of **Lookup Table1** and from the block **Round** that determines the sector number, in which vector Ψ_s is placed. The tables **Table_a**, **_b**, and **_c** convert the output of the **Lookup Table0** into inverter gate pulses.

The components of stator flux linkages $\Psi_{s\alpha}$, and $\Psi_{s\beta}$ are calculated in the subsystem **Flux**: at first, Ψ_{sd} and Ψ_{sq} are computed by Equation 4.154, afterward Ψ_{sa}, Ψ_{sb}, and Ψ_{sc} as a result of the Park transformation, and further, $\Psi_{s\alpha} = \Psi_{sa}$ and $\Psi_{s\beta} = (\Psi_{sb} - \Psi_{sc})/\sqrt{3}$.

The reference values of the torque and the flux linkage are computed in the subsystem **Control_1**.

The first value is determined by the source power deviation from the average value that corresponds to the load power. The computation of the average value takes much time; therefore, although such a possibility is intended in the considered model, under the simulation, the constant value of 20 kW is used. A power deviation, divided by the rotation speed, determines the torque reference. Its value is limited if the module of the current space vector i_s (Equation 4.156) is getting more than I_0. When the speed reaches 900 rad/s, its subsequent rise is limited by the torque decrease. The torque is also limited when the source power essentially reduces, in order to prevent the penetration of the PMSM power into the grid (the PMSM power is employed only for the load power).

The reference for Ψ_s is computed in the following way [4].

Let $I_q = L_d i_{sq}/\Psi_r$, $I_d = L_d i_{sd}/\Psi_r$, $I_0 = L_d i_{s0}/\Psi_r$, $\mathrm{d}l = (L_q - L_d)/L_d$, $m = T_e L_d/1.5Z_p\Psi_r^2$; and $f_s = \Psi_s/\Psi_r$, then it follows from Equations 4.154 and 4.155 that

$$f_s = \sqrt{(1+I_d)^2 + (1+\mathrm{d}l)^2 I_q^2} \tag{4.159}$$

$$m = I_q(1 - I_d\mathrm{d}l). \tag{4.160}$$

The quantities I_q and I_d are chosen so that the torque would be maximal under the given current i_s (Equation 4.156). After substitution of $I_q = \left(I_0^2 - I_d^2\right)^{0.5}$ in Equation 4.160, maximization by I_d, and the substitution in the final expression $I_0 = \left(I_q^2 + I_d^2\right)^{0.5}$, the optimal relationship will be received as

$$I_q^2 = I_d^2 - \frac{I_d}{\mathrm{d}l}. \tag{4.161}$$

From Equations 4.159 through 4.161, the optimal dependence f_s on m can be obtained. It is preferable to have this relationship in the explicit form. For this aim, the wanted dependence is approximated by

$$f_s = am^2 + bm + c + 1, \tag{4.162}$$

where the coefficients a, b, and c depend on dl. If the approximated value $f_s < 1$, it is taken as $f_s = 1$.

Therefore, before simulation start, the MATLAB program `Contr_PMSM_DTC1.m` has to be executed:

```
Contr_PMSM_DTC1
Ld = 0.0023;
Lq = 0.007;
Fr = 0.45;
dl = (Lq-Ld)/Ld;
I0 = 120;
p = 1;
Km = Ld/(1.5*p*Fr*Fr);
for j = 1:30,
Id(j) = -j*0.01;
M1(j) = (1-dl*Id(j))*sqrt(Id(j)^2-Id(j)/dl);
fs(j) = sqrt((1+Id(j))^2+(1+dl)^2*(Id(j)^2-Id(j)/dl));
end;
p1 = polyfit(M1,fs-1,2)
a = p1(1)
b = p1(2)
c = p1(3)
```

At first, the PMSM parameters are specified. Furthermore, for a number of I_d values, with I_q according to Equation 4.161, the values m from Equation 4.160 and f_s from Equation 4.159 are computed; in this way, the wanted dependence of the stator flux linkage on the torque is received. Afterward, this dependence is approximated by the second-order polynomial using the standard MATLAB function `p1 = polyfit (M1, fs-1, 2)`; the polynomial coefficients a, b, and c get into the MATLAB Workspace and can be employed by the model. The block **Fcn** in the subsystem **Control_1** realizes the computed dependence, fabricating the flux reference: first, the torque reference T_{ref} is converted to m by multiplication by K_m; afterward, the value f_s is calculated, which, by multiplication by Ψ_r, converts to the reference stator flux.

It is supposed that the position sensor is jointed with PMSM that can be used as the speed sensor too. For model simplification, it is also supposed that the torque can be measured.

The converter connected to the load controls the power flow that is transferred from the PMSM to the load or is consumed from the grid. The equilibrium state is characterized by the constancy of the voltage across the capacitor in the DC link, so that the voltage controller is placed in the subsystem **Control**, the regulator of the three-phase converter current is the inner part of this controller. The reference of the component I_d is determined by the voltage controller output; the reference for the component I_q is zero.

The process that lasts 100 s is fixed in Figure 4.31. The source power is seen to change ranging from –20 to +60 kW, but owing to the smoothing action of the flywheel unit, the load power slightly deviates from the rated value of 20 kW. The rotational speed of PMSM changes not much as well, so that there is sufficient reserved energy, but

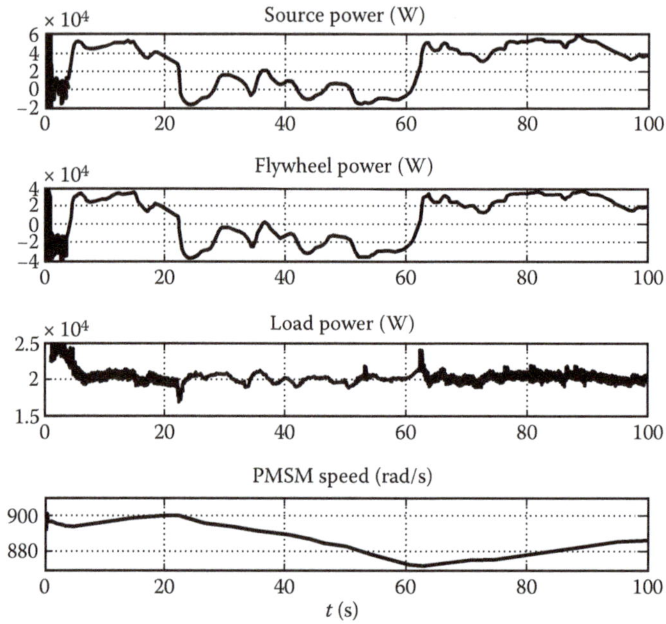

FIGURE 4.31 Smoothing of the grid voltage perturbations by the flywheel.

it is necessary to have in mind that under simulation, the frequency of the disturbances was increased with the aim to decrease simulation time. When the real processes are under investigation, the data that are gained during the field tests must be utilized.

4.3.2 Six-Phase PMSG

PMSG with two windings that are mutually shifted by 30° are employed in the same cases as the six-phase SG with wounded rotor. The voltage equations are the same as Equations 4.101 through 4.104; the equations that connect flux linkages and currents may be written as

$$\begin{aligned}
\Psi_{q1} &= L_q I_{q1} + L_{q12} I_{q2} - L_{qd} I_{d2} \\
\Psi_{d1} &= L_d I_{d1} + L_{d12} I_{d2} + L_{qd} I_{q2} + \Psi_r \\
\Psi_{q2} &= L_q I_{q2} + L_{q12} I_{q1} + L_{qd} I_{d1} \\
\Psi_{d2} &= L_d I_{d2} + L_{d12} I_{d1} - L_{qd} I_{q1} + \Psi_r.
\end{aligned} \tag{4.163}$$

Here L_{d12} and L_{q12} are the mutual inductances of the windings in the direct and quadrature axes; L_{qd} is the mutual inductance of the windings in different axes. It is supposed that the mutual inductances in the same axis are proportional to the internal inductances: $L_{d2} = kL_{d2}$ and $L_{q12} = kL_q$. It was said earlier that $L_{qd} = 0$, when assuming that the mutual inductances of the windings are proportional to the cosine of the angle between them.

From the equations that are obtained by this way,

$$\begin{aligned}\Psi_{q1} &= L_q I_{q1} + kL_q I_{q2} \\ \Psi_{d1} &= L_d I_{d1} + kL_d I_{d2} + \Psi_r \\ \Psi_{q2} &= L_q I_{q2} + kL_q I_{q1} \\ \Psi_{d2} &= L_d I_{d2} + kL_d I_{d1} + \Psi_r,\end{aligned} \tag{4.164}$$

the expressions for currents may be found as

$$\begin{aligned}I_{q1} &= \frac{\Psi_{q1} - k\Psi_{q2}}{dL_q} \\ I_{q2} &= \frac{\Psi_{q2} - k\Psi_{q1}}{dL_q} \\ I_{d1} &= \frac{\Psi_{d1} - \Psi_r - k(\Psi_{d2} - \Psi_r)}{dL_d} \\ I_{d2} &= \frac{\Psi_{d2} - \Psi_r - k(\Psi_{d1} - \Psi_r)}{dL_d}, \\ d &= 1 - k^2.\end{aligned} \tag{4.165}$$

To substitute Equation 4.165 in Equations 4.101 through 4.104, the equations for flux linkages can be solved; afterward, the currents may be found by Equation 4.165. The torque may be expressed as

$$T_e = 1.5p(\Psi_{d1}I_{q1} - \Psi_{q1}I_{d1} + \Psi_{d2}I_{q2} - \Psi_{q2}I_{d2}). \tag{4.166}$$

Be reminded that the second winding leads the first one by 30°.

The model of the six-phase PMSM is used in the model **SixphPMSM**. The electrical part solves the equations given earlier; the model of the mechanical part is the same as in the previous models. The following signals are collected at the output *m*: three currents of the first winding (A), three currents of the second winding (A), the rotational speed (r/min), the mechanical power, the rotor angle (grad), and the torque (Nm). The program `Contr_PMSM_T1` has to be executed before the model start, in which the optimal dependence I_d on the torque is calculated.

PMSG in the considered model has 20 pole pairs, voltage of 1200 V, and power of about 2 MW. In WG sets, the converter at the generator side controls its speed in such a way that the WG output power would be maximal; it means that the speed control system is implemented. The block diagram of the controller is depicted in Figure 4.32.

The SC output is considered as the value *m* in Equation 4.160. With the aid of the tables that was computed under the execution of the program `Contr_PMSM_T1`, the relative values of the references for I_d and I_q are found, which are converted to

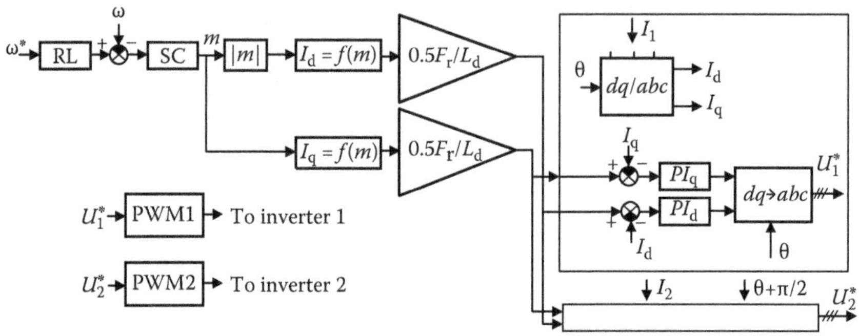

FIGURE 4.32 Block diagram of the control system of six-phase SG.

the absolute quantities and are reproduced by means of the current controllers. The gains of all four controllers are the same and are specified in the option *File/Model Propeties/Callbacks/InitFcn.*

The dependences of the currents I_d and I_q on m are displayed in Figure 4.33. It is seen that, for example, $I_0 = 0.415$ under $m = 0.5$, whereas I_0 would be equal to 0.5 under $I_d = 0$, that is, the current decreases by 17%.

In the model under consideration, the fabrication of DC voltages, which is usually carried out with the converter at the grid side, is not modeled; the DC value is accepted as 2800 V.

The following process is simulated: the steady state in the no-load condition at the rotational speed of $300/Z_p$ rad/s, the increase in the generator load at $t = 1$ s to −100 kNm with a rate of −200 kNm/s, the decrease in the rotational speed to $150/Z_p$ rad/s at $t = 3$ s, and the increase in the rotational speed to $250/Z_p$ rad/s at $t = 4$ s.

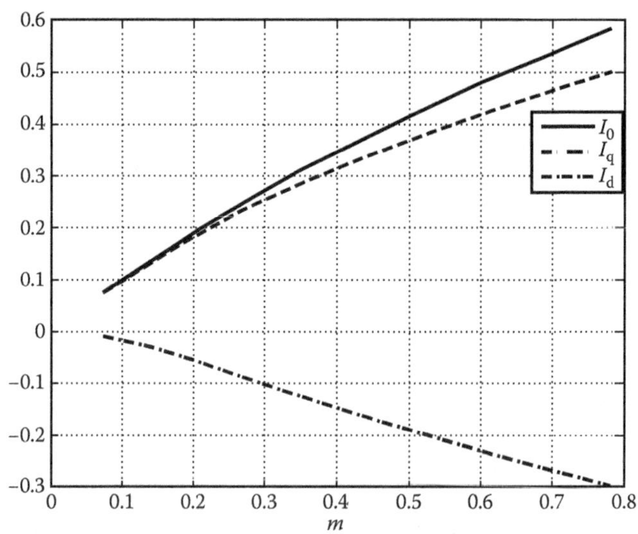

FIGURE 4.33 Optimal dependencies I_d and I_q on m.

The rotational speed, the torque, the power, and the current of one of the stator windings are shown in Figure 4.34; the currents of both windings are depicted in Figure 4.35: THD ≈ 15%; $I_{rms} = 357$ A.

To repeat simulation with $k = 0$, the current THD decreases to 8% under $I_{rms} =$ 377 A.

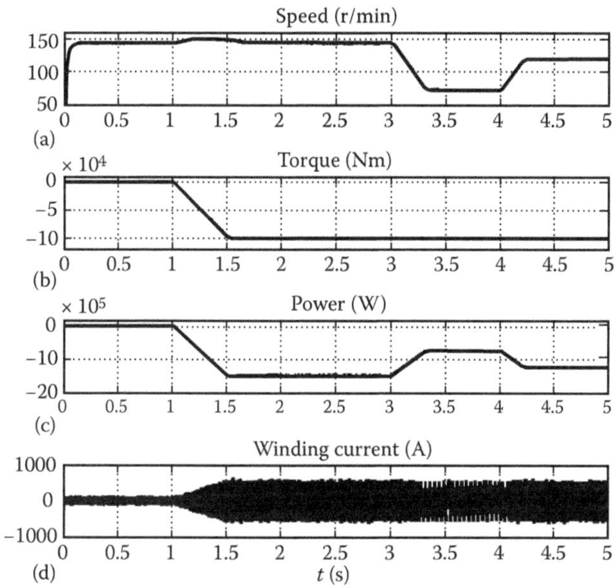

FIGURE 4.34 (a) Rotational speed, (b) torque, (c) power, and (d) current of one of the stator windings in six-phase SG.

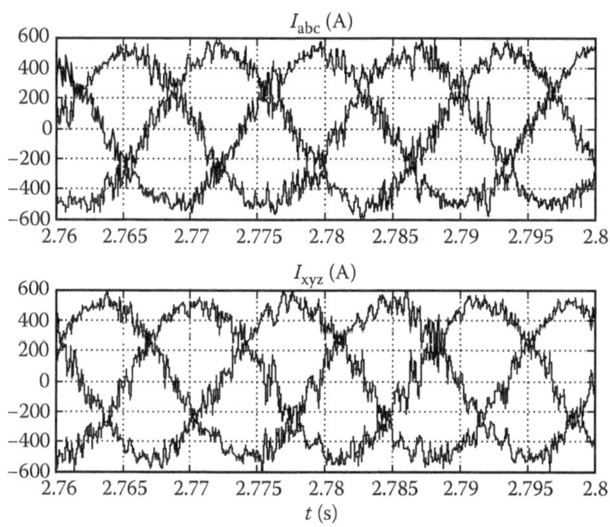

FIGURE 4.35 Currents in six-phase SG.

4.4 SWITCHED RELUCTANCE GENERATOR SIMULATION

It follows from Equation 4.155 that the SM torque is not equal to zero under $L_d \neq L_q$ even in the case when the excitation source on the rotor is absent ($\Psi_r = 0$). The electrical machines that use this peculiarity are called *reluctance machines*. They are subdivided into sinusoidal and switched machines. The former is in reality the usual SM without windings on the rotor; special steps are undertaken during the design of these machines for obtaining $L_d \gg L_q$.

The switched reluctances machines (SRM) have a completely different design. The machine shown in Figure 4.36 has six poles on the stator. Each pole carries a winding; the opposite windings (a and a', b and b', c and c') are connected in series, so that these windings may be considered as a three-phase system. The rotor has four poles. The current pulses are sent in the stator windings in the instants that depend on the rotor position; for that, the position sensor is mounted on the machine shaft.

Under the neglect of saturation, the torque applied to the rotor may be expressed as [5]

$$T_e = 0.5 I^2 \frac{dL}{d\beta}, \tag{4.167}$$

where L is the phase inductance, I is its current, and β is the rotor angle position. To send the current into phase b when the rotor is in the position depicted in Figure 4.36, the produced torque will be positive, under the indicated direction of the rotation, because the phase inductance is rising. The rotor pole 1 moves away from the stator pole; hence, to send the current in phase a, the torque will be negative, because phase a inductance is decreasing. After sending the current in phase b and the rotor rotation by 30°, pole 1 of the rotor takes about phase c the same position as pole 2 about

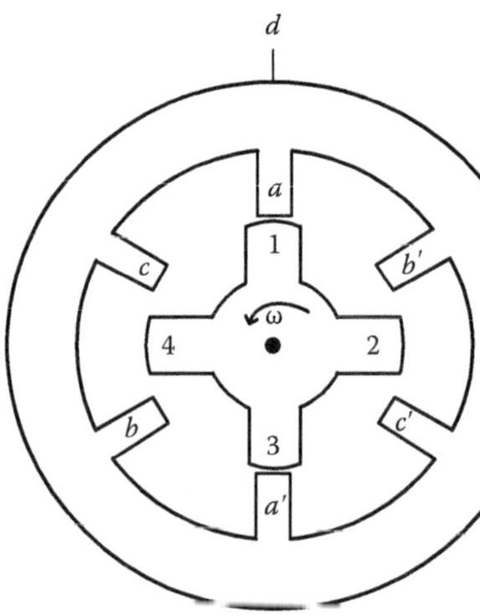

FIGURE 4.36 SRM 6/4.

the phase *b* had occupied. Now, the current pulse may be sent in phase *c*; when the rotor rotates by 30° the current pulse may be sent in phase *a* and so on.

To count off angle β from axis *d* and to send the current pulse to phase *a* when β is small, the torque will be negative, but to do the same when angle β is sufficiently big, for example β > 45°, the torque will be positive; therefore, by changing the angle, at which the current pulses are sent to a certain phase, the operation mode can be changed.

In order to improve the SRM technical-and-economic indexes, its parameters are selected in such a way that its poles are essentially saturated in the aligned position. At this, the machine flux depends not only on the current but also on the rotor position: $\Psi = \Psi(I, \beta)$. The typical curves are shown in Figure 4.37 that are made with the help of the means intended in Ref. [1]. Accordingly, the torque is the function of I and β: $T_e = T(I, \beta)$. These nonlinearities make both the computation and the realization of the SRM control system difficult.

The SRM can have the numbers of the stator phases and the rotor poles that are different from that shown in Figure 4.36. Usually, the stator has from 3 to 6 phases; and the rotor, from 4 to 10 poles, respectively. The SRM type is defined by the fraction, whose numerator is double the number of the phases, and the denominator is the number of the rotor poles; therefore, the SRM of the types 6/4 (as in Figure 4.36), 8/6, 10/8, and 12/10 exist.

SRM has a number of merits: the high reliability and the high rotational speed because there are no windings and other mounted elements on the rotor, and a small moment of inertia. The model **Switched Reluctance Motor** is included in

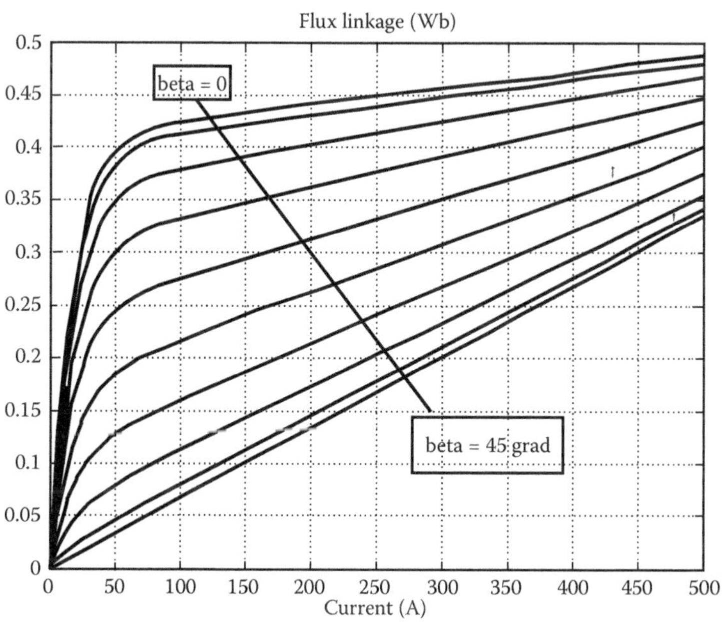

FIGURE 4.37 SRM typical curves $\Psi = \Psi(I, \beta)$.

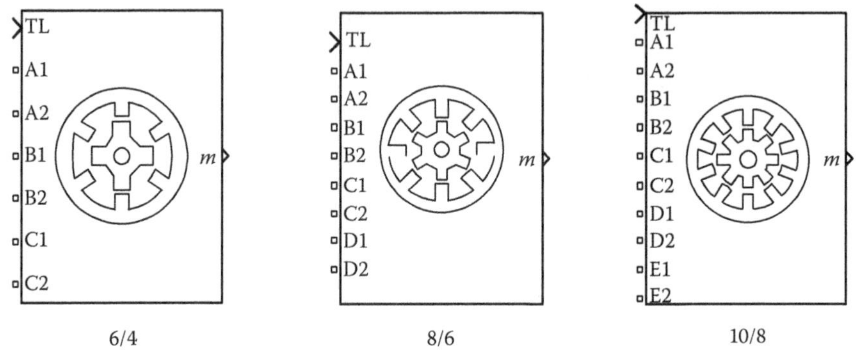

FIGURE 4.38 Pictures of different SRMs.

SimPowerSystems that can operate as a generator of the types 6/4, 8/6, and 10/8; their pictures are shown in Figure 4.38.

The connection ports A1, A2, B1, B2, C1, C2, and so on are the terminals of the three, four, or five stator windings. The driving or load moment comes to the input *TL*; the output *m* contains the vector of the SRM measured variables in the following order: *V* is the stator voltage (V); *flux* is the flux linkage (Vs); *I* is the stator current (A); *Te* is the electromagnetic torque (Nm); *w* is the rotor speed (rad/s); and *teta* is the rotor position (rad).

The block diagram of the SRM model is depicted in Figure 4.39. It consists of three (for SRM 6/4) identical circuits. The phase flux linkages are calculated by Equation 4.1; these quantities come to the identical tables **ITBL** for the phase current computation. These tables also have the inputs for the rotor angular position

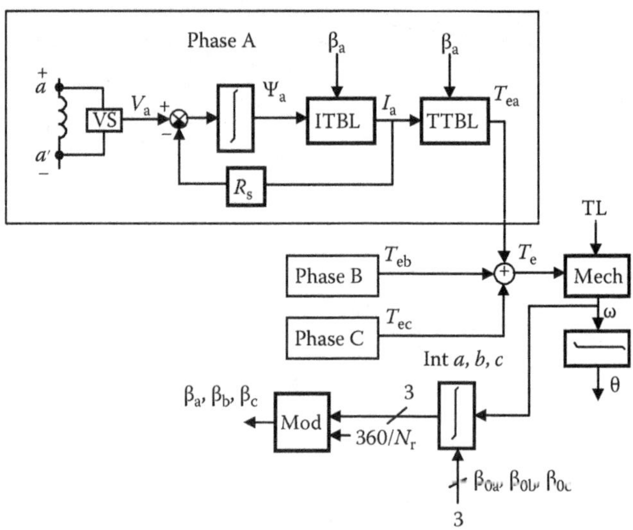

FIGURE 4.39 Block diagram of the SRM model.

about the phases. The tables contain the data that are analogous to those displayed in Figure 4.37. The table outputs are the current values, which, together with the angles, come to the tables **TTBL**, whose outputs are the electromagnetic torques of the phases. The resulting torque is equal to the sum of the phase torques. The block **Mex** solves the equation of motion in Equation 4.16: the rotational rotor speed and rotor angle are founded. The value of the speed comes to the three identical integrators **Inta**, **Intb**, and **Intc** with different initial values; for the SRM under consideration, Inta = 0, Intb = −30°, and Intc = −60° (Figure 4.39). The outputs of the integrators come to the blocks, which find the remainder after division by $m = 360/Z_p$; in the case under consideration, ($Z_p = 4$) $m = 90$; therefore, the angles β_a, β_b, and β_c change from 0° to 90°.

In the first line of the SRM dialog box on the first page, the SRM type is selected: 6/4, 8/6, or 10/8; the second line allows choosing either the generic model or the specific one. The former uses some typical curves $\Psi = \Psi(I, \beta)$ inherent in the model; they are displayed in Figure 4.37. These curves can be modified with the employment of the parameters defined on the second page, including the maximum current (500 A for Figure 4.37) and the maximum flux linkage (0.486 Wb in Figure 4.37). The utilized curves $\Psi = \Psi(I, \beta)$ can be seen if the option *Plot magnetization curves* is selected.

Under the choice of the specific model, the curves $\Psi = \Psi(I, \beta)$ have to be determined as the table; the data for this table are received either by experiment when a prototype exists or by computation with the help of the finite-element analysis.

SRG models the system with the switched reluctance generator. The SRG parameters are specified in Figure 4.40. The subsystem **Converter** fabricates a three-phase voltage system for powering the generator windings and consists of the three units shown in Figure 4.41. When the enabling pulses come to the transistors, the DC voltage is applied to the phase winding; when the pulses go away, the circuit for the current is formed containing the diodes and the voltage source; at that, the current flow direction is opposite the voltage polarity, so that the current dies rapidly. The current regulation is carried out with the hysteresis controllers **Relay**. The phase current reference comes to the current regulator of this phase only in the case when the subsystem **Position Sensor** has the logical 1 at the corresponding output.

The same circuits for forming the angles β as in the SRM model are used in this subsystem (Figure 4.39). The values of the angles are compared with the references of the turn-on and turn-off angles, which are different for speed-up and speed-down modes. The references are accepted as constants, although for the optimization of the operation, it is reasonable to change them in some way [5]. The switching over of the modes is fulfilled according to the sign of the signal at the output of the limiter 500 A. When the angle β for the certain phase is in the proper range, the logical 1 appears at the corresponding output *sig* of the subsystem.

In the generator mode, the turn-on angle is equal to zero. Under switching on, the energy of the DC source transfers in the stator winding. Under switching off, the energy stored in the winding returns in the DC source. Since the mechanical energy of the rotating rotor is added to the stored energy, the energy that returns to the grid is more than the energy that is expended for excitation; owing to that, generation of the electrical energy and transfer in the grid take place.

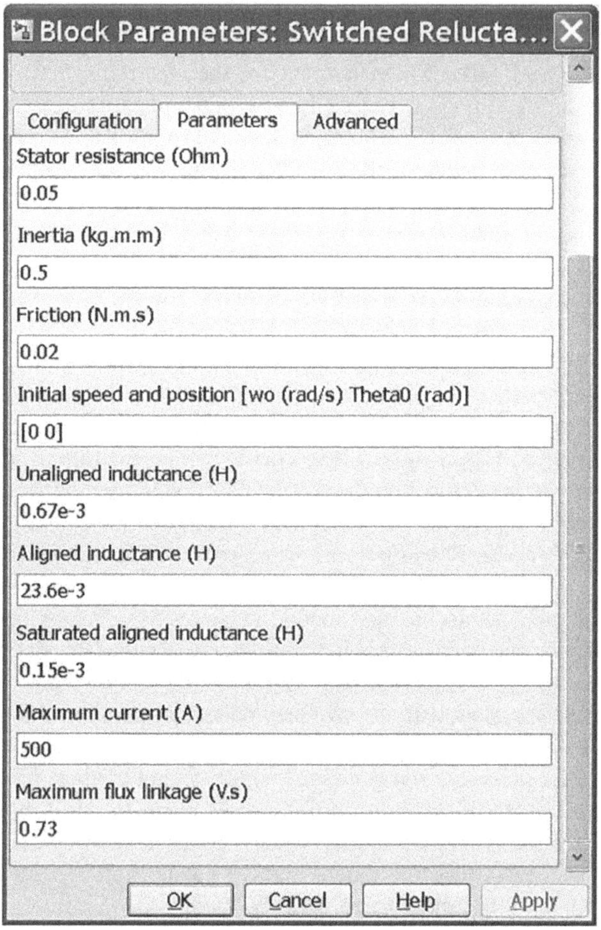

FIGURE 4.40 Dialog box of SRM in the model SRG.

The DC source consists of the active front-end rectifier with a capacitor that is connected to the 380 V grid with the aid of the 380/110 V transformer. The block diagram of the control system of this converter is displayed in Figure 4.42, where V_{sabc} and I_{sabc} are the voltage and the current on the low-voltage transformer side, respectively. The reference DC voltage is 300 V.

The scope **Gen** fixes the phase flux linkages, phase currents, SRG torque, and its rotational speed. The scope **Torque** shows the mean torque value, and the scope **Scope** records the DC voltage and the active and reactive power in the grid.

The following process is simulated. First, the SRG reaches the steady speed of 2000 r/min without applied torque (as a motor). At t = 0.3 s, the active torque of −300 Nm is applied to the SRG that corresponds to the mechanical power of 62.8 kW. At t = 1 s, the speed reference decreases to 1000 r/min; since in WG sets,

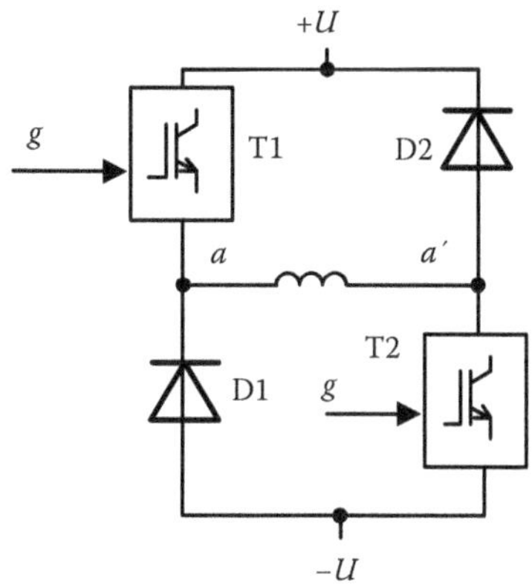

FIGURE 4.41 Diagram of the SRM stator phase.

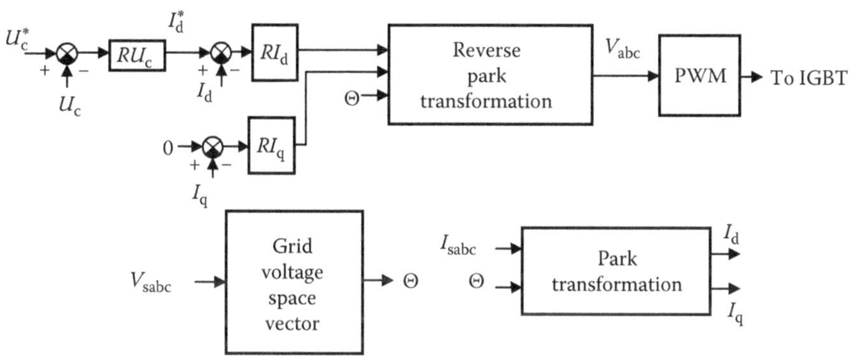

FIGURE 4.42 Block diagram of the control system of the front-end converter.

the optimal torque is proportional to the square of the speed, the active torque decreases to −75 Nm that corresponds to the power of 7.85 kW.

The speed, the SRG torque mean value, the grid active and reactive power, and the voltage across the capacitor are shown in Figure 4.43. It is seen that the reactive power is close to zero; the active power takes the values of 48 and 4 kW in the steady state, respectively. Therefore, the efficiency is correspondingly 48/62.8 = 0.76 and 4/7.85 = 0.51.

The pulse fabrication for the stator phases is shown in Figure 4.44.

It is not difficult to check that, accepting that the turn-off angle is equal to 30° (but not 35°), under the low speed, the power sent to the grid increases to 4.4 kW, by 10%. Therefore, the modification of the turn-off angle in dependence on the operation mode can increase the SRG efficiency to some extent.

The other system with SRG is simulated in Chapter 5 (the model **SRGFLY**).

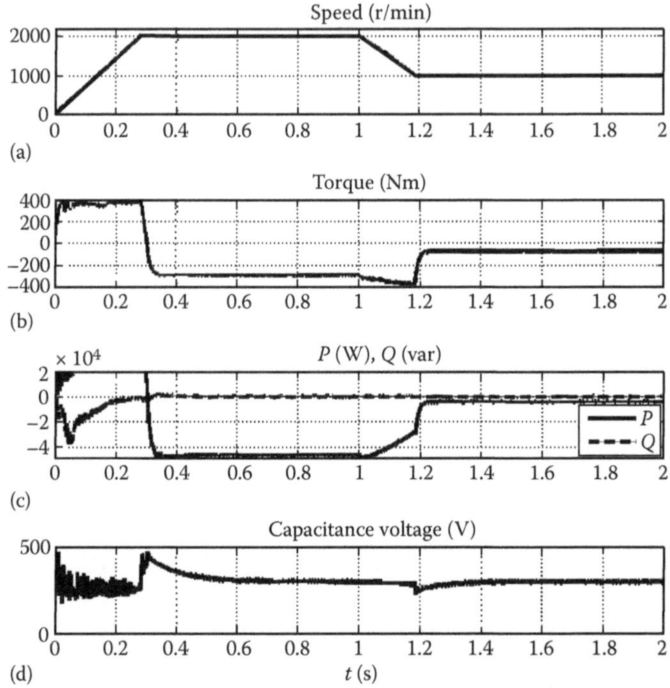

FIGURE 4.43 SRG (a) speed, (b) torque mean value, (c) grid active and reactive power, and (d) voltage across the capacitor in the model **SRG**.

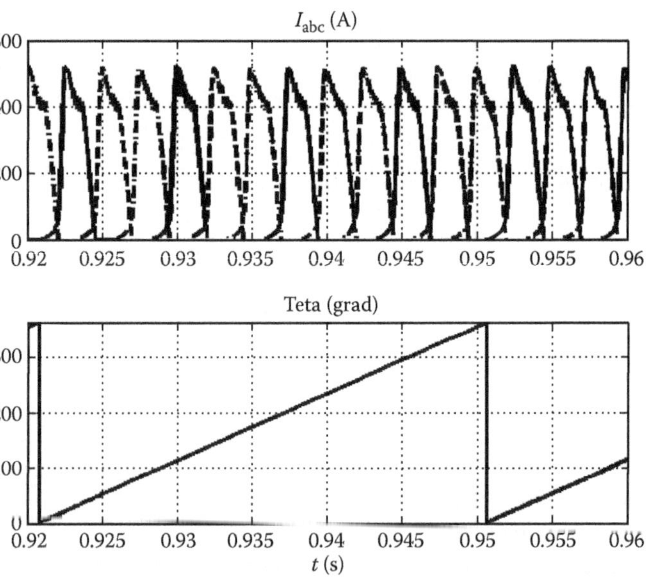

FIGURE 4.44 Pulse fabrication in the model **SRG**.

4.5 MECHANICAL COUPLING SIMULATION

It was supposed in the previous models that the generator and the device, which sets the former in motion (the prime mover), form the united mechanical system (the so-called single-mass system). In reality, the driving torque is transferred from the prime mover to the generator by means of a number of mechanical elements—shafts, gears, couplings—which possess certain elasticity. Therefore, the torque can be transferred only at the expense of the deformation of the mechanical parts. When the deformation is not large, the transferred torque is proportional to the deformation and, in some extent, its velocity. Therefore, the torque T_y transferred through the shaft may be defined as

$$T_y = C\int(\omega_1 - \omega_2)\,dt + B(\omega_1 - \omega_2), \tag{4.168}$$

where ω_1 is the rotational speed of the shaft input point, ω_2 is the same for the shaft output point, and C and B are the factors. These speeds may be found as

$$\frac{d\omega_1}{dt} = \frac{T_e - T_y}{J_1} \tag{4.169}$$

$$\frac{d\omega_2}{dt} = \frac{T_y - T_l}{J_2}, \tag{4.170}$$

where T_e is the torque applied to the shaft input (the torque of the prime mover); J_1 is the moment of inertia of the prime mover; T_l is the torque of the resistance at the output shaft end; and J_2 is the moment of inertia of the rotating mass connected to the shaft output end.

In order to make the simulation of such a system easier, SimPowerSystems have the models **Mechanical Shaft** and **Speed Reducer**. The former realizes the relationship 4.168 and has two parameters: C (stiffness; Nm) and B (damping; Nms); its inputs are the rotational speeds in r/min. The latter contains two models, **Mechanical Shaft** and the model **Reduction Device**, which may be used independently. Block **Reduction Device** realizes the relationship as in Equation 4.170 for the gear unit; the diagram of the block is depicted in Figure 4.45. Here T_h is the torque at the gear input shaft, T_l is the torque at the gear output shaft, i is the gear ratio, ef is its efficiency, J is the gear moment of inertia referred to the input shaft, N_{rdh} are N_{rdl} are the rotational speeds of the shafts (r/min), and T_s is the sampling time.

The employment of these blocks depends on the peculiarity of the prime mover model. The block **Reduction Device** models the reducer, assuming that all the mass is concentrated in the gear. The input and output shafts are massless, but each of them is elastic, with different, in the general case, coefficients C and B. If the moments of inertia of the prime mover and of the load are known, and they both are connected to the elastic shaft, the diagram depicted in Figure 4.46 is suitable. It will be shown

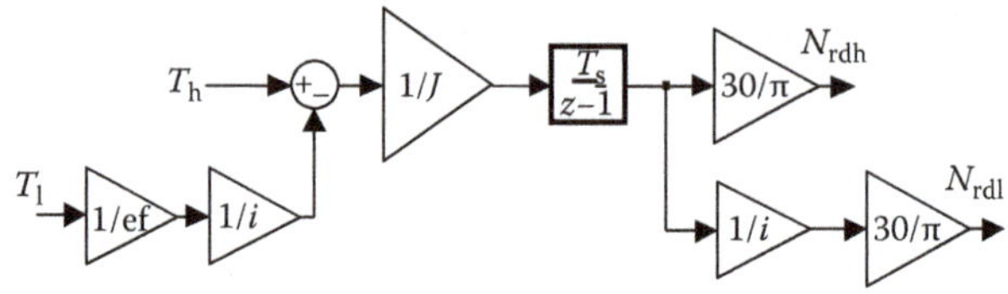

FIGURE 4.45 Block diagram of the block **Reduction Device**.

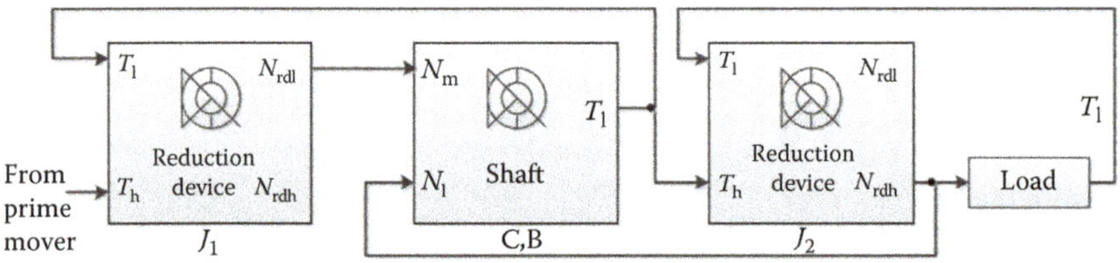

FIGURE 4.46 Model of two mass set connected to the elastic shaft.

in Chapter 6 that the wind turbine model operates with the parameters, referring to the generator shaft, so that it is necessary to take $i = 1$ in the elastic coupling model, and to refer the other parameters to this axis. With reference to the considered model **SRG**, such elastic coupling model is used in the model **SRG1**, when the tumbler **Switch** is in the top position. The turbine moment of inertia, referring to the SRG rotational speed, is equal to the SRG moment of inertia; the rest of the parameters

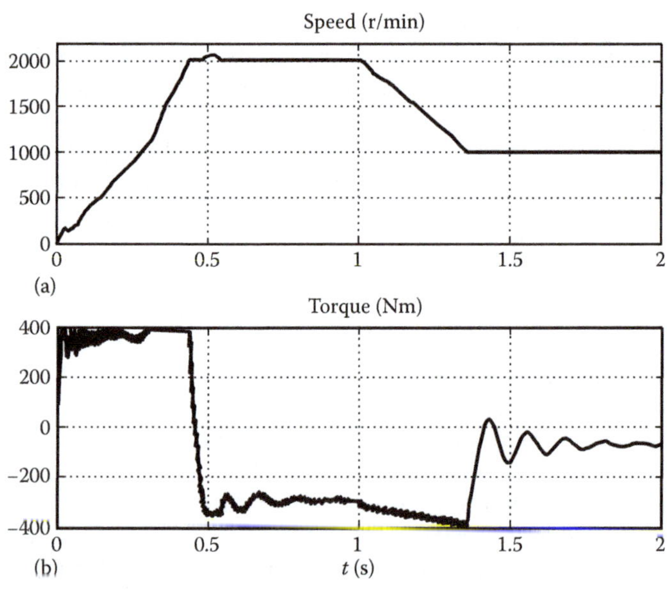

FIGURE 4.47 SRG (a) rotational speed and (b) torque with elastic coupling.

can be seen in the model dialog boxes. The SRG rotational speed and its torque are shown in Figure 4.47; the perceptible oscillations are seen.

It is possible to also presume that SRG is rotated with low-speed diesel with the maximum speed of 500 r/min through the speed-up gear 4:1. Then, taking $i = 1/4$ and increasing the moment of inertia of the first mass by 16 times, we receive the model that is built with the tumbler **Switch** in the low position. The processes in this system are the same as those displayed in Figure 4.47.

REFERENCES

1. MathWorks. *SimPowerSystems™: User's Guide*. MathWorks, Natick, MA, 2011–2016.
2. Krause, P. C., Wasynczuk, O., and Sudhoff, S. D. *Analysis of Electric Machinery*. IEEE Press, Piscataway, NJ, 2002.
3. Singh, B., Murthy, S. S., and Gupta, S. STATCOM-based voltage regulator for self-excited induction generator feeding nonlinear loads. *IEEE Transactions on Industrial Electronics*, 53(5), 2006, October, 1437–1452.
4. Perelmuter, V. M. *Electrotechnical Systems, Simulation with Simulink and SimPowerSystems*. CRC Press, Boca Raton, FL, 2013.
5. Boldea, I. *Variable Speed Generators*, Boca Raton, FL, CRC Press, 2006.
6. Ammasaigounden, N., Subbiah, M., and Krishnamurthy, M. R. Wind-driven self-excited pole-changing induction generators. *Electric Power Applications, IEEE Proceedings B*, 133(5), 1986, 315–321.
7. Zhao, Y., and Lipo, T. Space vector PWM control of dual three-phase induction machine using vector space decomposition. *IEEE Transactions on Industry Applications*, 31(5), 1995, September/October, 1100–1109.
8. Bojoi, R., Lazzari, M., Profumo, F., and Tenconi, A. Digital field-oriented control for dual three-phase induction motor drives. *IEEE Transactions on Industry Applications*, 39(3), 2003, May/June, 752–769.
9. Barati, F., Shao, S., Abdi, E., Oraee, H., and McMahon, R. Generalized vector model for the brushless doubly-fed machine with a nested-loop rotor. *IEEE Transactions on Industrial Electronics*, 58(6), 2011, June, 2313–2321.
10. Pozal, J., Oyarbide, E., Sanrasola, I., and Roriguez, M. Vector control design and experimental evaluation for the brushless doubly fed machine. *IET Electric Power Applications*, 3(4), 2009, 247–256.
11. Institute of Electrical and Electronics Engineers (IEEE). IEEE Std. 421.5: Recommended practice for excitation system models for power system stability studies. IEEE, Piscataway, NJ. 2005 (Revision of *IEEE Std 421.5*-1992).
12. Kundur, P. *Power System Stability and Control*. McGraw-Hill, New York, 1994.
13. Boldea, I. *Synchronous Generators*. CRC Press, Boca Raton, FL, 2006.
14. Berets, T. Model development and validation of brushless exciters. *Power Tech, 2015, IEEE Eindhoven*. 2015, pp. 1–6.
15. Franklin, G. F., Powell, J. D., and Emami-Naeini, A. *Feedback Control of Dynamic Systems*, third edn. Addison-Wesly Publishing Company, Reading, MA, 1994.
16. Schiferl, R. F., and Ong, C. M. Six phase synchronous machine with AC and DC stator connections: Part I: Equivalent circuit representation and steady-state analysis. *IEEE Transactions on Power Apparatus and Systems*, 102(8), 1983, August, 2685–2693.
17. Tessarolo, A., Zocco, G., and Tonello, C. Design and testing of a 45-MW 100-Hz quadruple-star synchronous motor for a liquefied natural gas turbo-compressor drive. *SPEEDAM 2010, International Symposium on Power Electronics, Electrical Drives, Automation and Motion*, 2010, pp. 1754–1761.

18. Contin, A., Grava, A., Tessarolo, A., and Zocco, G. Novel modeling approach to a multi-phase, high power synchronous machine. *SPEEDAM 2006, International Symposium on PowerA., Electronics, Electrical Drives, Automation and Motion*, 2006, pp. 19-7–19-12.
19. Hinkkanen, M., and Luomi, J. Modified integrator for voltage model flux estimation of induction motors. *IEEE Transactions on Industrial Electronics*, 50(4), 2003, August, 818–820.

5 Simulation of the Renewable Electrical Sources

5.1 SIMULATION OF SYSTEMS WITH BATTERIES

The model of the **Battery** is included in SimPowerSystems that models in detail the charge–discharge processes in different types of batteries: lead acid, lithium ion, nickel–cadmium, and nickel metal hydride. The battery nominal voltage V_{nom}, its rated capacity Q_{rated} (Ah), and the initial charge for simulation *SOC* (%) are specified in the block dialog box. At this, the nominal voltage is the voltage value in the end of the discharge linear zone. A number of additional characteristics must be defined that can be done in two ways: if the option *Use parameters based on Battery type and nominal values* is chosen, all the necessary parameters will be defined automatically; if this option is not checked, the necessary parameters have to be specified manually. These parameters are the maximum capacity Q_{max} (Ah) that is usually equal to 105% of the rated capacity; the fully charged voltage (V) V_{full}; the nominal discharge current, which the discharge characteristic is calculated for; the internal resistance that, for the model on default, is determined as the resistance, across which 1% of the power is dissipated: $R = 0.01\ V_{nom}/Q_{rated}$; the capacity under nominal voltage, which is the capacity that can be extracted from the battery until the voltage is kept not less than the nominal value; and exponential zone—the values V_{exp} and Q_{exp} correspond to the end of this zone.

The characteristics of the battery, whose data are fixed in Figure 5.1, are depicted in Figure 5.2 for the discharged currents 20 and 100 A.

The voltage response time under the step current change may be specified on the third page of the dialog box; by default, it is taken as 30 s. More detailed information about the described model is given in Ref. [1].

The model **Battery_Example** demonstrates the employment of the block. The load of 4 Ω is supplied from the source with the voltage of 500 V. At this, the **Diode** is cut off, because the battery voltage in less than 471 V. In the time interval 100–700 s, the source voltage decreases to 100 V, so that the **Diode** begins to conduct; the load voltage is determined by the battery voltage, which is discharging. When the source voltage is restored, the process of the battery recharge begins that comes to the end when either the rise of the voltage stops or this voltage reaches 475 V. The simulated process is shown in Figure 5.3. It is seen that, at the instant when the source voltage is restored, the battery voltage has dropped from 470 to 410 V, and the charge has decreased by 15%. The battery state returns to the initial level at about 2500 s.

A more complicated system is simulated in the model **Battery_Example1**. The two-level VSI is utilized in this model, which contains the AC voltage source of

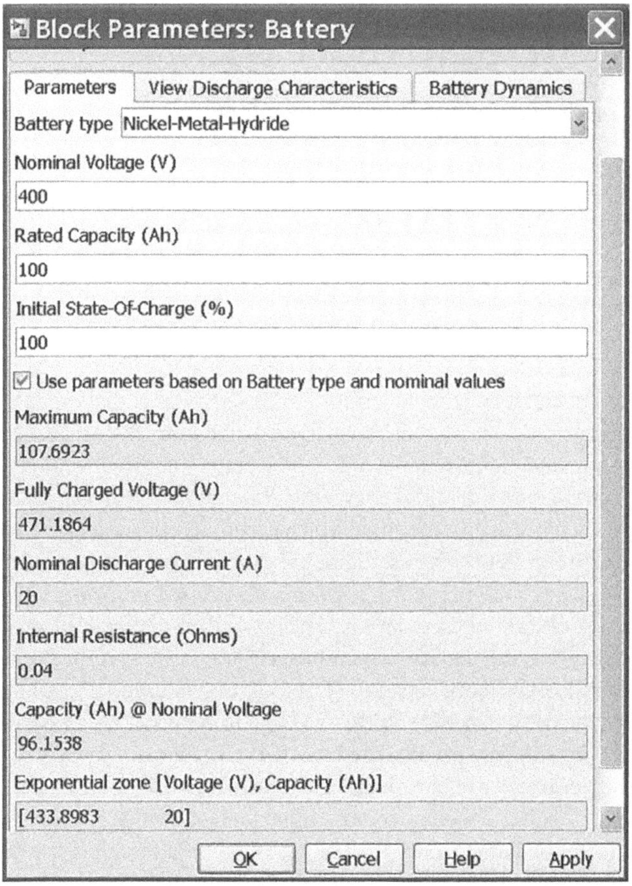

FIGURE 5.1 Battery dialog box.

380 B that is stepped up with the 720/380 V transformer, and is rectified and converted into the AC voltage with controlled amplitude and frequency. The wanted load voltage is kept by means of the voltage controller. In the case when the supply voltage fails or drops essentially, the voltage controller cannot keep the reference voltage up, and the normal operation stops. Such a situation can have a place in WG under wind speed decrease. In order to prevent load supply failure, the battery is used. It is connected to the circuits in the DC link; the DC boost/buck converter is utilized for this aim that has a possibility to transfer the electric energy in both directions. The diagram of this converter is shown in Figure 5.4. The transistors are controlled with the pulses with variable relative duration $d = t_1/T_0$, where t_1 is the pulse duration, and T_0 is the pulse repetition period. When the battery is charging from the source V_a (DC voltage), the pulses come to the transistor T_1, and the current flows in the direction I_{b2}; when the pulse comes, the current in the inductance rises, and when the pulse is removed, the current flows through the battery and the diode D_2, and it decreases. The mean voltage across the capacitor C_2 is $V_b = dV_a$; therefore,

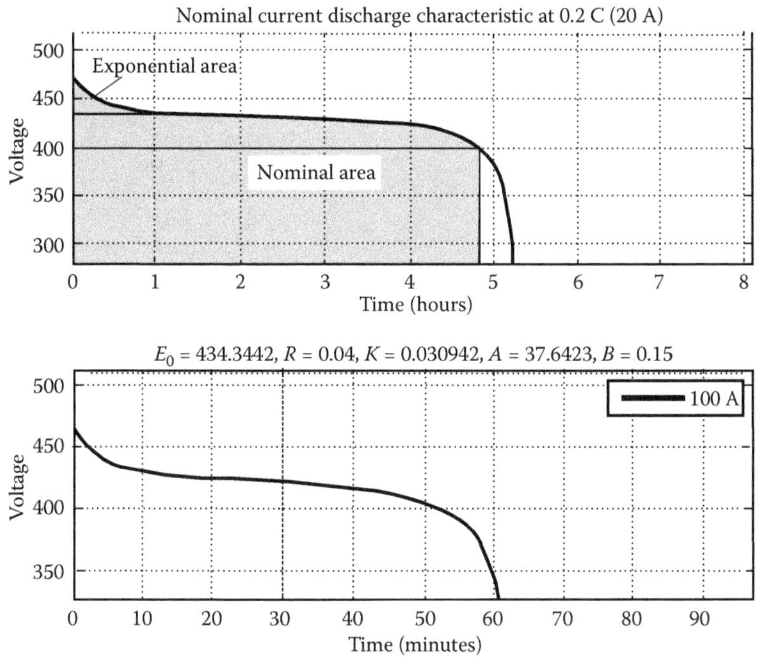

FIGURE 5.2 Characteristics of the battery with the data in Figure 5.1.

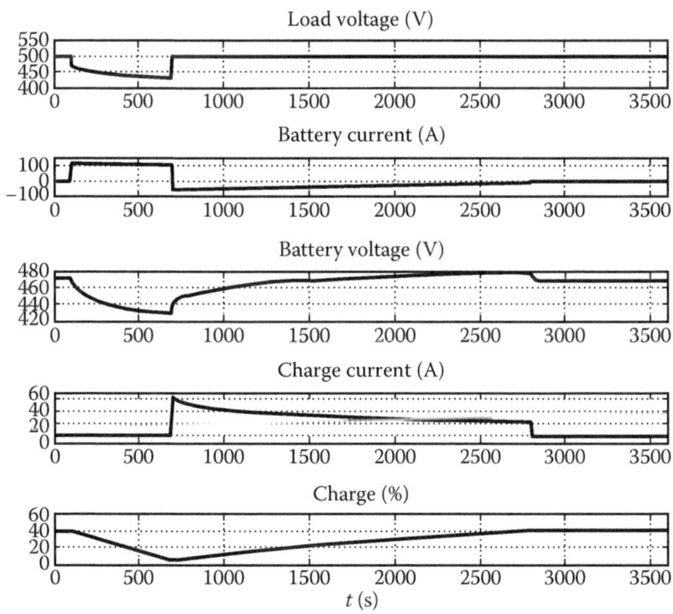

FIGURE 5.3 Battery discharge–charge in the model **Battery_Example**.

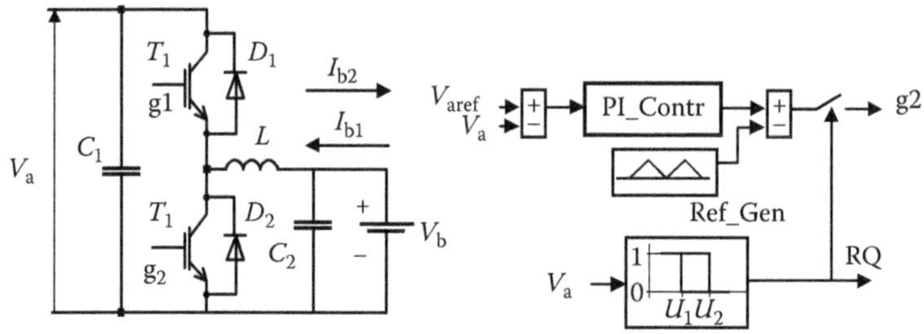

FIGURE 5.4 Diagram of DC boost/buck converter.

the converter steps down the voltage (buck mode). When the battery supplies DC link, the pulses come to the transistor T_2, the current flows in direction I_{b1}, and the battery discharges; when the pulse comes, the current in the inductance rises rapidly; when the pulse is removed, the current flows through the capacitor C_1 and the diode D_1. The mean voltage across the capacitor C_1 is

$$V_a = \frac{V_b}{1-d}; \tag{5.1}$$

therefore, the converter steps up the voltage (boost mode) [2].

In Figure 5.4, the diagram of the transistor T_2 control is displayed. The control of the DC voltage V_a is fulfilled. The regulator output is compared with the unipolar triangle carrier waveform; while the regulator output exceeds the carrier, the gate pulse is sent to the transistor. The regulator begins to operate, when DC voltage is getting less than U_1 that can have a place under essential supply voltage sag, and stops to actuate, when this voltage is getting more than U_2 that has a place when the supply voltage recovers.

The diagram of the transistor T_1 control is shown in Figure 5.5. This scheme fulfills the battery-charging current control. The regulator operates when T_2 control is not active ($RQ = 0$), the battery voltage V_b is less than value U_3, and the battery charge is less than Q_1. The regulator is deactivated when either $V_b > U_4$ or $Q > Q_2$. Moreover, the regulator is turned off before these conditions will be reached, if the signal $RQ = 1$ appears.

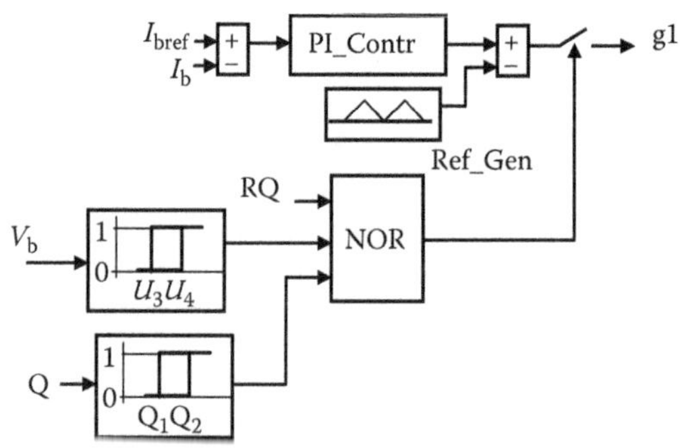

FIGURE 5.5 Diagram of the transistor T_1 control in boost/buck converter.

The described units are placed in the subsystem **Battery**. The possibility to give a pulse relative duration manually for system checking and investigation is intended.

With the nominal DC voltage of 800 V, the control system that regulates the voltage in that link is turned on under the voltage sag to 650 V, and it is turned off, when this voltage increases to 775 V. The charge of the battery that has the nominal voltage of 400 V and the fully charged voltage of 471 V begins with its drop to 465 V and stops either with the increase to 475 V or with the charge rise to 96%.

The supply voltage halving at $t = 1$ s following the restoration to the nominal value of 380 V at $t = 40$ s is simulated. The DC voltage, the load voltage amplitude, the battery voltage, and the charge are shown in Figure 5.6. It is seen that, under essential decrease in the supply voltage, the DC voltage, owing to the battery, is kept at the level of 750 V, giving the possibility to keep the load voltage up. The battery voltage that was initially equal to 471 V decreases to 434 V to the end of the discharge time; the charge decreases for this time to 92.4%. When the supply voltage is restored, the battery charge with the current of 200 A takes place; at the beginning of the process, the voltages of the battery and the DC oscillate, owing to switching over of the device that connects the current controller to the transistor T_1. These oscillations do not affect the accuracy of the load voltage control and disappear when the battery gains the full charge. As a result, the battery voltage of 457 V is set. We note that, to save time, the battery capacity (Ah) is essentially decreased relative to the real values.

When the battery must interact with the other sources of electric energy, it is often desirable to isolate the circuits of the battery from the other circuits,

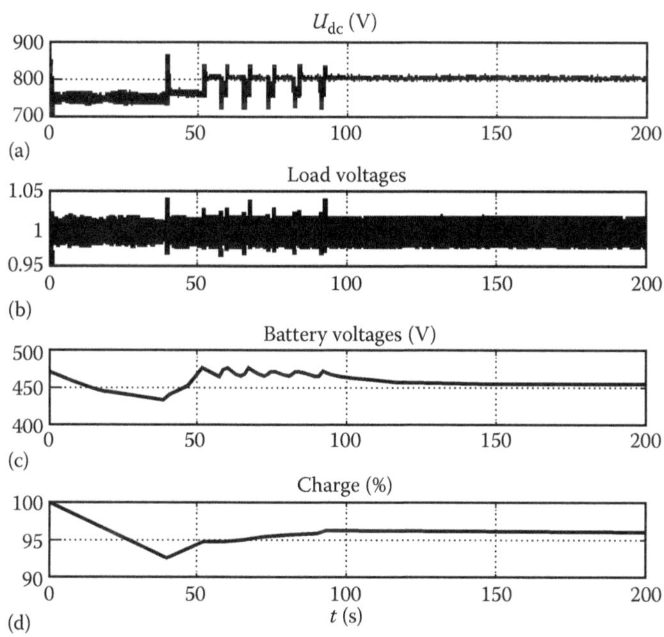

FIGURE 5.6 (a) DC voltage, (b) load voltage amplitude, (c) battery voltage, and (d) charge for the model **Battery_Example1**.

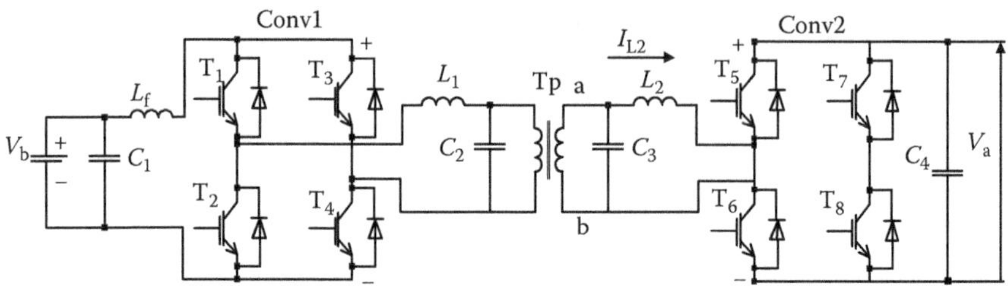

FIGURE 5.7 Diagram of DC converter with galvanic isolation.

for instance, when the DC link voltage is much more than the battery voltage. In this case, the scheme that consists of two single-phase bridge rectifiers connected through the high-frequency transformer (Figure 5.7) can be used. There are many articles (for example, Ref. [3]), in which it is proposed to form the square wave voltages in both transformer windings by means of a proper gate pulse control; these voltage waves are shifted to one another by some angle; at this, the angle sign and value define the direction and the value of the power transferred through the transformer. The frequency of 10 kHz and higher is utilized. Here such converters are not considered.

In the model **Battery_Example2**, the transformer operates with the moderate frequency of 1 kHz, and the winding voltage waves are close to the sinusoidal. The transformation ratio is 1:3 (400/1200 V), and the power is 100 kW. The diverse converter control systems are possible. In the considered model, converter 1 operates in the mode of the sinusoidal modulation with the modulation factor equal to 1; converter 2 controls the current of the reactor L_2.

The control system diagram is displayed in Figure 5.8. The reference current with the amplitude I^* is in-phase or antiphase, depending on the power transfer direction, relative to the winding voltage V_{ab}. The hysteresis current controller is used. Let, for example, the reference current be in-phase with V_{ab} and $I < I^*$. Then, as it follows from Figure 5.8, transistors T_6 and T_7 are turned on, the voltage V_a is added in the circuit that consists of the transformer winding and the reactor, and the current increases. If $I > I^*$, transistors T_5 and T_8 are turned on, V_a is opposite to V_{ab}, and the current decreases. $V_a > V_{abmax}$ has to be in this scheme.

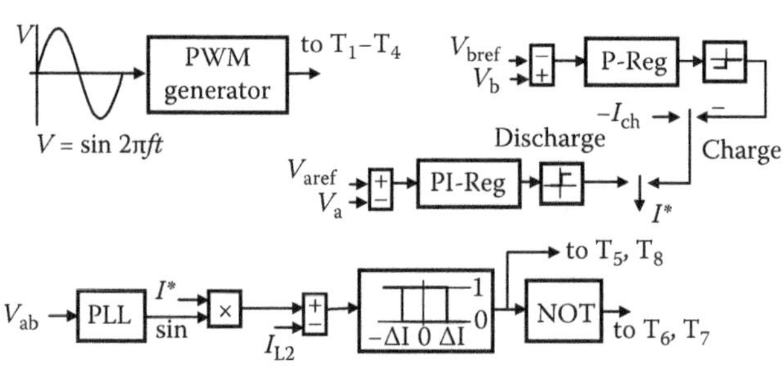

FIGURE 5.8 Diagram of the control system for DC converter with galvanic isolation.

The current controller is inner for the voltage controller. When the battery discharges, it is the regulator of the voltage V_a at the output. When the battery charges, the current reference is either the constant value I_{ch} or the output of the battery voltage proportional controller with limitation; the regulator gain is taken big enough, so that at most part of the control range, the system operates in the current control mode, and only close to the voltage reference value transfers in the voltage control mode.

The simulation of the full system has a number of problems. Since the transformer voltage frequency is rather big, the simulation sampling time has to be small; at the same time, the charge–discharge processes are very slow, so that the simulation time turns out to be very large. Therefore, the main converters are not modeled (unlike in the previous model); the source of the rectifier voltage is modeled with the DC source $V_a = 2$ kV. Moreover, the simulation proved that even with the sampling time of 5 μs, the essential distortions can be seen in the current waves; therefore, the simulation was fulfilled with the sampling time of 1 μs for the plots given in the following.

The battery discharge for the time interval of 10 s is simulated (the **Breaker** is initially turned off and is turned on at $t = 10$ s). When V_a drops below 1700 V, the regulator of this voltage, with the inner current controller, is activated, and the reference voltage is 1800 V. When V_a is restored ($V_a > 1950$ V), the battery discharge stops, the charge process begins either with the constant current of 150 A or with the help of the voltage control, how it was described earlier. The charge ends when the battery voltage reaches the reference value (may be with a control of the battery charge state).

Some results of the simulation are shown in Figures 5.9 through 5.11 when both tumblers **Manual Switch** and **Manual Switch1** are in the top positions (the discharge process is shown in Figure 5.10, and the charge process is shown in Figure 5.11).

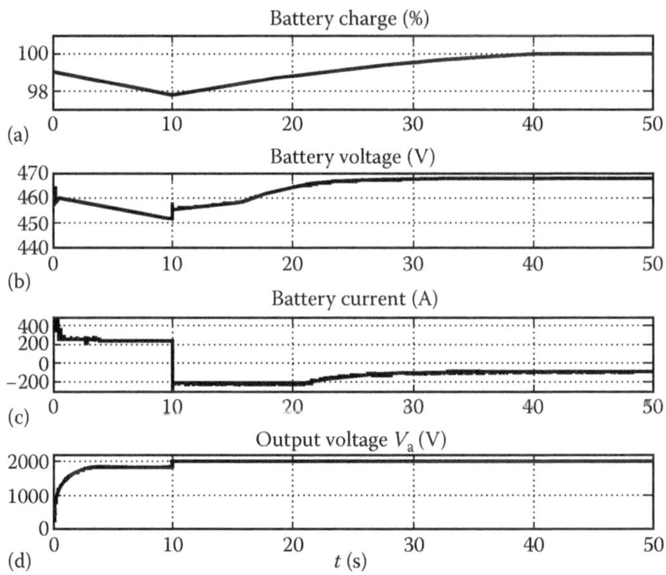

FIGURE 5.9 (a) Battery charge, (b) voltage, (c) current, and (d) output voltage in the model **Battery_Example2**.

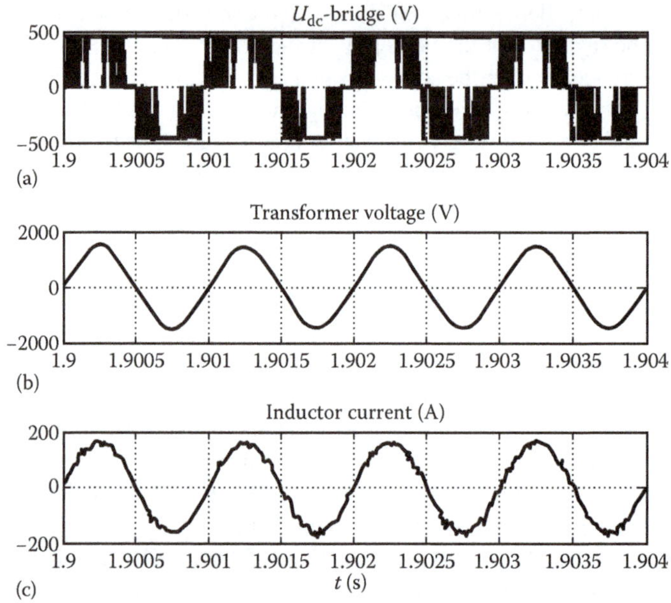

FIGURE 5.10 (a) DC, (b) transformer voltages, and (c) inductor current in the discharge mode.

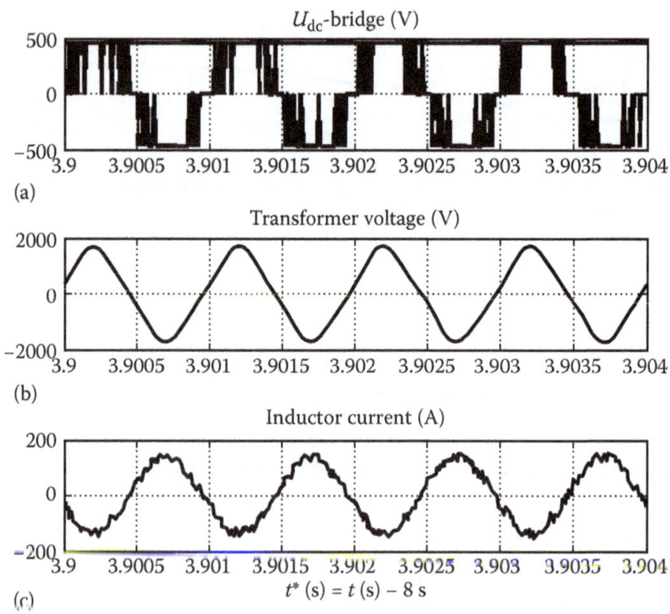

FIGURE 5.11 (a) DC, (b) transformer voltages, and (c) inductor current in the charge mode.

5.2 PHOTOVOLTAIC SYSTEM SIMULATION

Photocells (PV cells) are widely employed for powering various devices, independently or together with the other sources. For instance, it is reasonable to use them with WG sets: when the weather is bad, and it is cloudy, the wind blows and the WG operates; when it is sunny, under light breeze, the PV units provide electric energy. The equivalent PV cell circuit may be diagrammed as shown in Figure 5.12 [4,5]. Without light, the PV cell characteristic corresponds to the diode characteristic; when the PV cell is lit up, the characteristic shifts by the value of photocurrent I_{ph}. This current is equal to the load current if the output terminals of PV cell are shortened, $V_1 = 0$ (under the neglect of the inner loss of PV). If $V_1 \neq 0$, then [6]

$$I_d = I_{d0}\left\{\exp\left[\frac{qV_1}{K_b T_c A(1 - b_t(T_c - T_{cr}))}\right] - 1\right\}, \tag{5.2}$$

where I_{d0} is the diode saturation current (A), $q = 1.602 \times 10^{-19}$ C is the electron charge, $K_b = 1.38 \times 10^{-23}$ J/K is Boltzmann's constant, T_c is the PV absolute temperature, T_{cr} is the nominal temperature, A is the adjusting factor, and b_t is the voltage temperature coefficient.

The magnitude of the photocurrent depends on the irradiance level and on the temperature; this dependence may be taken as

$$I_{ph} = I_{sr}[1 + a_k(T_c - T_{cr})]\frac{Q}{100}, \tag{5.3}$$

where I_{sr} is the photocurrent under the nominal conditions, when $T_c = T_{cr}$ (usually 298 K) and the irradiance level is $Q = 100\%$, and a_k is the current temperature coefficient.

Resistors R_p and R_s take into account the losses in the PV cell. It is necessary to say that different relationships can be found in the literature.

The volt-ampere characteristics of PV cell under various conditions, and also the power curves ($P = IV$) are given in Figures 5.13 and 5.14. The losses are ignored. The values that are obtained under simulation with the developed model (see further) are marked with dots. The PV cell parameters are specified in Figure 5.15. By some voltage V_m, the power reaches the maximum P_m. If I_0 is the output current

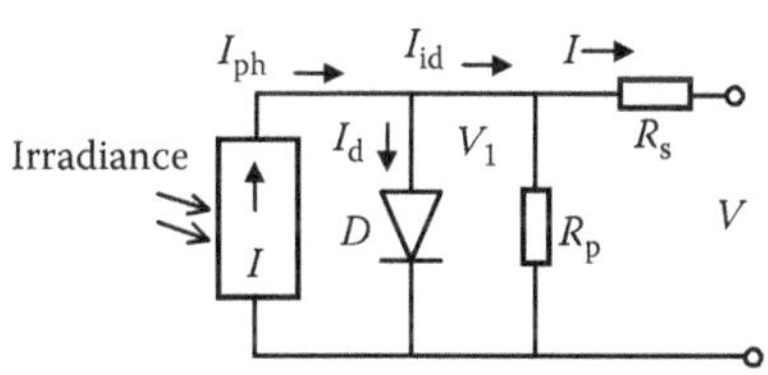

FIGURE 5.12 Equivalent circuits of the photocell.

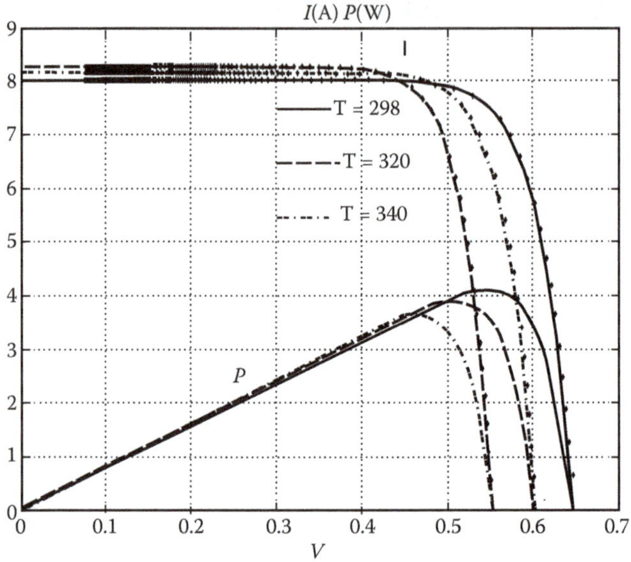

FIGURE 5.13 PV cell characteristics under different temperatures.

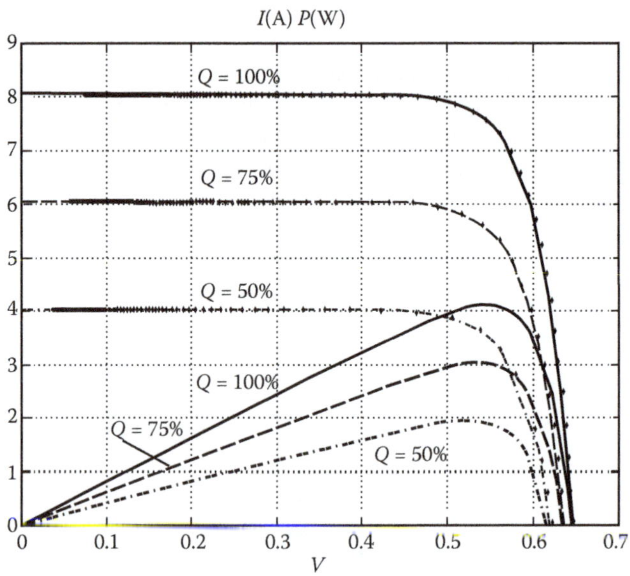

FIGURE 5.14 PV cell characteristics under different irradiance.

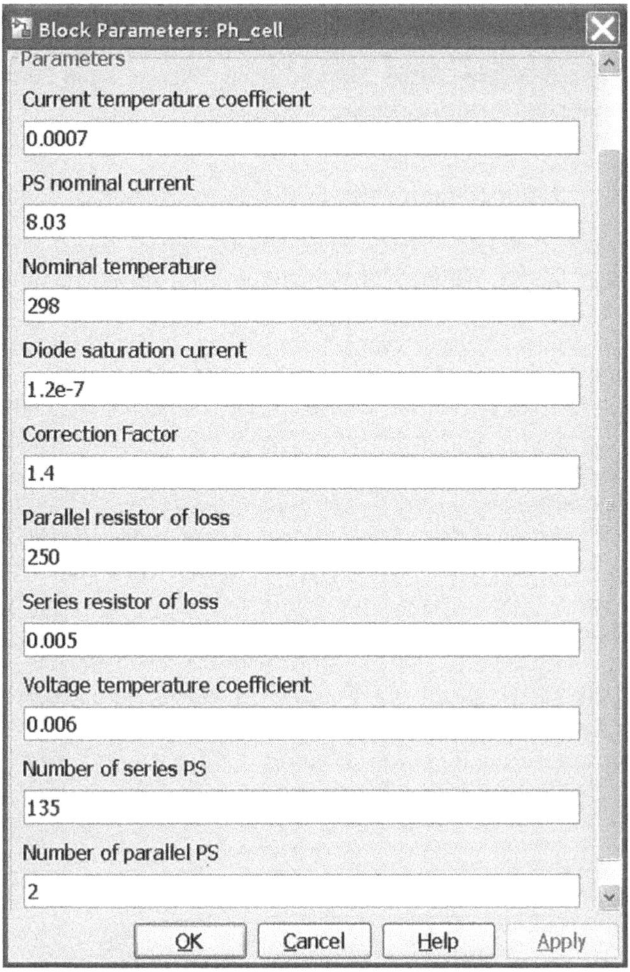

FIGURE 5.15 Dialog box of the developed PV cell model.

under zero voltage, V_0 is the voltage under the zero current and then the fill factor is defined as

$$FF = \frac{P_{\mathrm{m}}}{V_0 I_0}. \tag{5.4}$$

For example, for $T_{\mathrm{c}} = 298$ K, $Q = 100\%$ we have

$$FF = \frac{4.15}{0.645 \times 8.03} = 0.8.$$

In order to obtain the greater values of the voltage and the current, the individual PV cells are connected in series and in parallel. If all PV cells are in the same conditions, the relationships given earlier must be changed as shown in Equation 5.3; the right side is multiplied by n_p, and Equation 5.2 obtains the form

$$I_d = n_p I_{d0}\left\{\exp\left[\frac{qV_1}{n_s K_b T_c A(1 - b_t(T_c - T_{cr}))}\right] - 1\right\}, \tag{5.5}$$

where n_p and n_s are the numbers of the parallel- and series-connected PV cells, respectively. The resistances of R_p and R_s increase in the ratio of n_s/n_p. According to Ref. [4], the procedure for the calculation of the equivalent cell circuit parameters is given, knowing the nameplate data of the utilized PV cells.

The PV cell model **Ph_cell** that is developed by the author and realizes the relationships given earlier is used in the model **Photo_m1**. Its dialog box is displayed in Figure 5.15. In the model **Photo_m1**, the PV cells supply the variable resistor, whose resistance smoothly changes from $0.0039 \times Rn_s/n_p$ Ω to Rn_s/n_p Ω by 255 levels and afterward to infinity. The resistance block consists of eight resistors, whose resistances change as a power of two, and which are switched in parallel. The coding unit converts the signal 0—255 in the binary code that controls the switches. The volt-ampere characteristics that are obtained with $R_s = 0$ and $R_p = \text{inf}$ are marked with dots in Figures 5.13 and 5.14 and are tallied with the calculations by the formulas given earlier. The characteristics of PV unit (array) with $n_p = 2$ and $n_s = 46$ are shown in Figure 5.16; at this, $FF = 379/(16.06 \times 29.78) = 0.79$.

The demonstration model **power_PVarray_grid_det** is included in SimPower Systems that contains the model **PV Array**; the latter models the PV arrays that

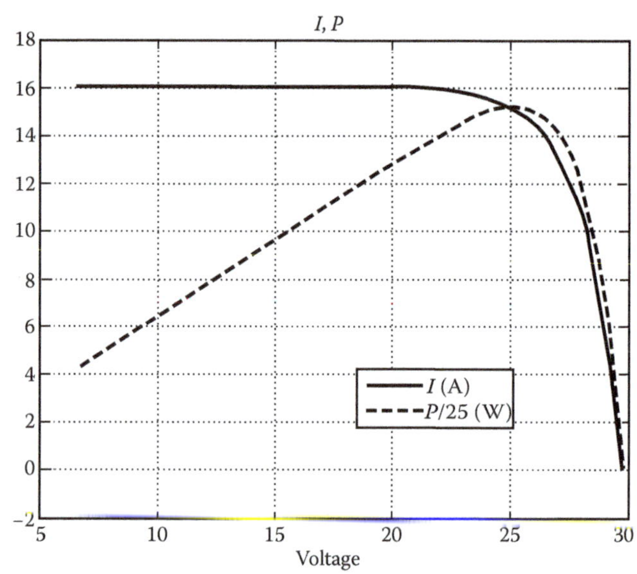

FIGURE 5.16 Characteristic of PV array with $n_p = 2$, $n_s = 46$.

are produced by a number of manufacturers. Under simulation, the type of array is selected, and its parameters appear in the model dialog box. Only the numbers of the series and parallel arrays must be specified manually. Of course, the employment of this model makes the preparation for the simulation much easier, but it is necessary to keep in mind that the array parameters cannot be changed, so the new developments and modification of the old ones cannot be taken into consideration. Furthermore, we shall use the model presented in Figure 5.12.

The condition of the equal illumination intensity of all PV cells in the array is observed far from always, especially for the large sets, because of the partial shading with the elements of construction, with clouds passing by, and so on. The shaded PV cells produce the lesser current, and they define the current of the series string of PV cells; the higher voltage applies to these PV cells that can cause overheating and their failure [4]. For failure prevention, the individual PV cells or the groups of PV cells are shunted with the bypass diodes D_1 and D_2, as displayed in Figure 5.17, so that the voltage across the slightly illuminated diode is limited by the diode voltage drop. If the voltage of some parallel string decreases, the other strings will try to power this string that can destroy their PV cells. In order to prevent this, the blocking diodes (D_{B1} and D_{B2} in Figure 5.17) are set in every parallel PV string.

Because SimPowerSystems is intended, mainly, for the investigation of the systems, not of the separate semiconductors, the other software, for instance, PSpice that can operate together with Simulink blocks [6], may be used; however, with the employment of the developed PV cell model, the notion of influence of the partially shaded PV cells can be obtained. Five PV cells are connected in series in the model **Photo_m2**; at t = 3 s, the illumination of the third PV cell decreases by 25%. To execute simulation without bypass diodes, it can be seen that the current of all PV cells decreases from 8.04 to 6.04 A, the voltage across badly the illuminated PV cell is becoming –2 V, and the voltage across the other four PV cells increases to 0.6 V.

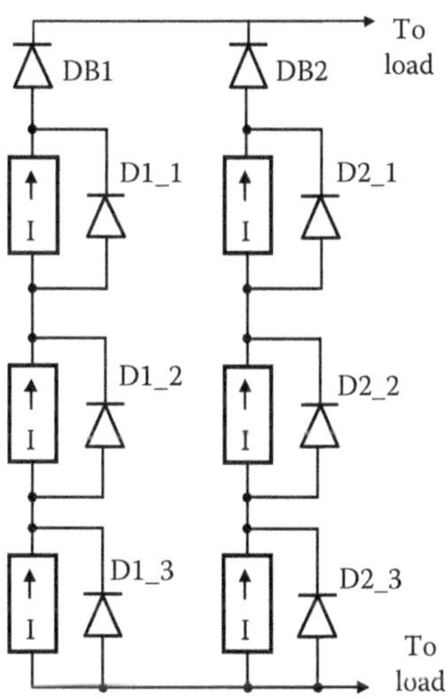

FIGURE 5.17 Integration of the individual PV cells in the array.

To repeat the simulation with the bypass diodes, one can observe that the voltage across the badly illuminated PV cell is limited by the value of −0.29, and the output current is kept as 8.04 A; the output characteristics are distorted: the local maximum appears in the power curve (Figure 5.18).

As can be seen from the displayed figures, the dependence of P on V has a maximum, and for gaining the maximum efficiency from a PV cell, its voltage must be V_m. In order to realize this requirement, the following conditions have to be fulfilled: a—presence of the device that can change V; b—availability of the unit that can measure the efficiency; c—availability of the algorithm for V adjusting. As the device by item *a*, the DC boost converter is usually employed, whose diagram is shown in Figure 5.19. The relative duration d of the closed state of the switch K can be assigned independently or can be the output of the voltage or the current controller (see further). In this scheme,

$$V_0 = \frac{V_i}{1-d}. \tag{5.6}$$

Since

$$P_0 = \frac{V_0^2}{R} = \frac{V_i^2}{R(1-d)^2} = P_i, \tag{5.7}$$

(neglecting the loss in the converter), it has to be in the maximum power point

$$V_m^2 = R(1-d)^2 P_m. \tag{5.8}$$

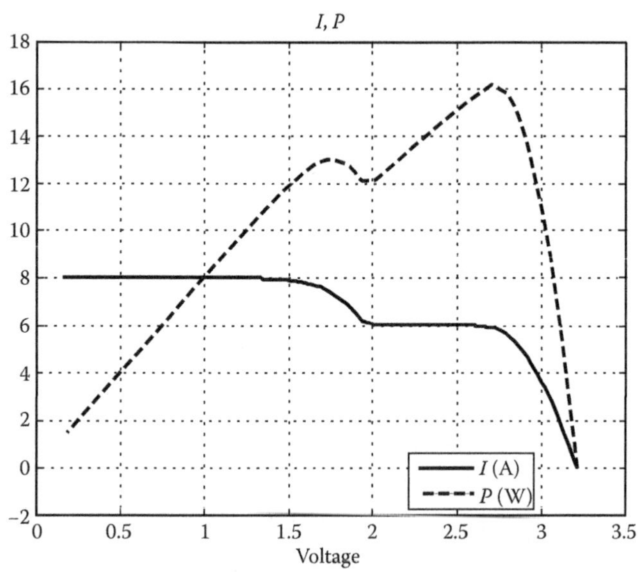

FIGURE 5.18 Characteristics of PV array under partial shadowing.

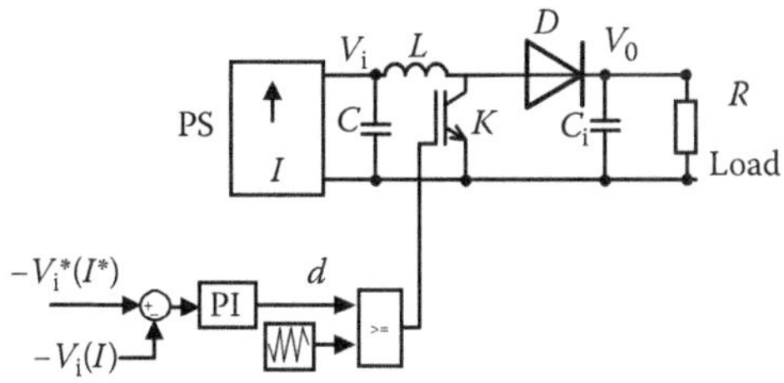

FIGURE 5.19 Diagram of the DC boost converter.

As an example, the model **Photo_m3** is considered. The PV array consists of two parallel strings; each of them has 135 PV cells. To put these values into the model **Photo_m1**, it can be seen after simulation that $V_m = 69.5$ V, $I_m = 14.86$ A, and $P_m = 1032.5$ W. In the model **Photo_m3**, $R = 20\ \Omega$, from Equation 5.8, it follows that $d = 0.516$. The process of d changing in the model **Photo_m3** is shown in Figure 5.20. The power is seen to reach the maximum that is equal to 1028 W under $d \approx 0.528$: at this, the load voltage is $V_{0m} = 143$ B. In Figure 5.20, V_i and P_i are the quantities at the PV unit output; V_o and P_o—at the load.

The measurement of the temperature and the irradiance of the PV array with the subsequent recalculation of the received values in the parameter d are rarely employed, because of the system complexity; therefore, other methods are used that

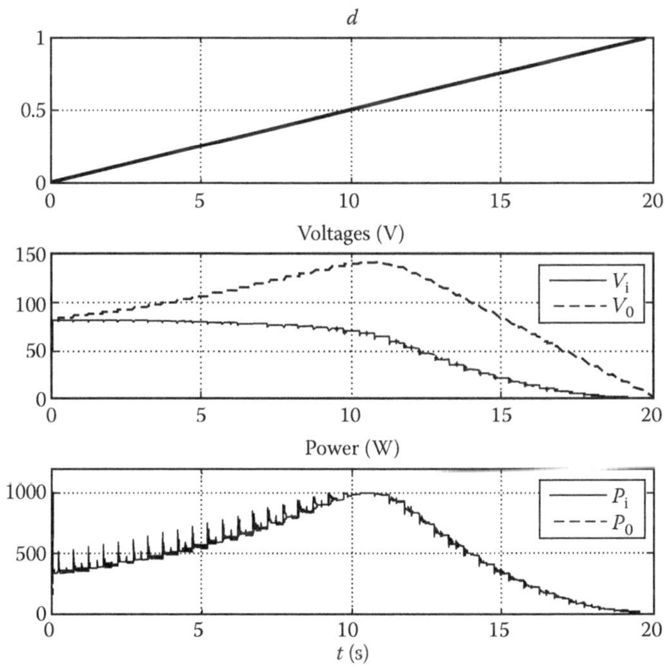

FIGURE 5.20 Dependencies of voltages and powers on d.

can be designated as the direct and indirect ones [4]. The latter is based on the fact that the ratios $K_v = V_m/V_0$ and $K_i = I_m/I_0$ for the certain PV cell type are almost constant under environment change. If the values K_v or K_i are known by means of computations, simulation, or experiment, and the PV cell open circuit voltage or the PV cell output short-circuit current can be measured, the reference quantities of the voltage or the current may be obtained, which can be realized with the aid of the feedback control system as shown in Figure 5.19 (voltage and current changing under d changing have the opposite directions).

In Table 5.1, which is calculated with the use of the data that are given in demonstration model **power_PVarray_grid_det** mentioned earlier, the coefficients K_v and K_i are given for the nominal condition and under the temperature increasing by 50°C. The influence of the temperature can be seen, so it may be reasonable to correct these coefficients as a function of the temperature, even if it is only approximate.

The values V_0 and I_0 can be obtained in two ways: either to place inside PV array the special measuring PV cells of the same type as the working ones or to switch over the output array circuits. The additional PV cells are necessary in the first case, and the switching devices in the second case; moreover, the load loses supply in the measuring moments. The first variant, with V_0 measurement, is realized in the model **Photo_m4V**. Four PV cells connected in series are used; these PV cells are loaded with the resistors having sufficiently large resistance. Usually, these PV cells are placed in the different areas of the unit in order to receive an average value. It has been found, with the help of the model **Photo_m1**, that under changing Q in the range of 25–100% and T in the range of 290–350 K, the coefficient K_v changes in the range 0.78–0.82 (the average value is 0.8), so that the amplifier gain is $K_v = 0.8 \times 135/4$ (135 is, as in the previous model, the number of series-connected PV cells). The process of the response to the irradiance and the temperature change is shown in Figure 5.21. The steady-state power values are 1010, 250, and 197 W. It was found, with the help of the model **Photo_m1**, that P_m = 1053, 251, and 205 W, respectively, but it is without consideration of the losses in the converter; therefore, one may assume that the tracking maximum point is carried out precisely enough.

TABLE 5.1
Optimal Coefficients of the Voltage and the Current

NN PV Array	K_v	K_{v50}	K_i	K_{i50}
1	0.7941	0.7425	0.92	0.882
2	0.8153	0.7651	0.893	0.8835
3	0.7461	0.7428	0.8560	0.8221
4	0.8009	0.7514	0.9116	0.8788
5	0.8012	0.7518	0.9222	0.8891
6	0.8019	0.7534	0.9368	0.9076
7	0.811	0.7894	0.93	0.9113
8	0.852	0.8202	0.9366	0.8925
9	0.7865	0.7346	0.9402	0.9119

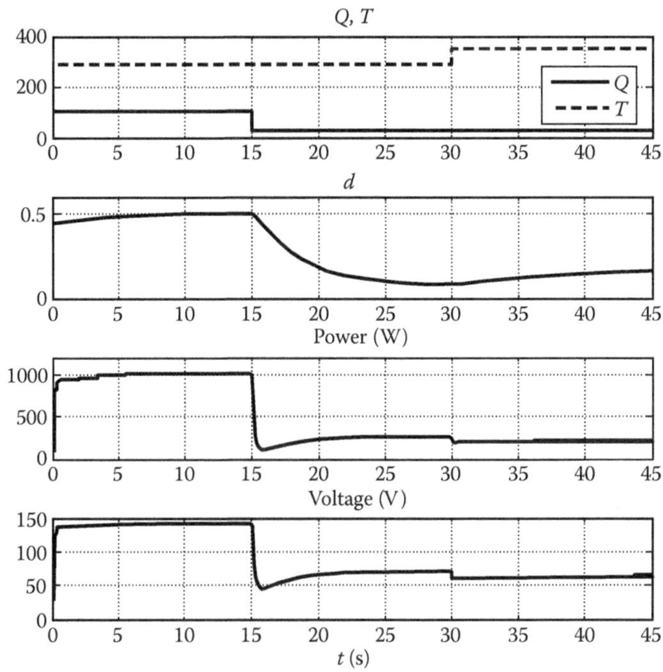

FIGURE 5.21 Response to the irradiance and temperature changing in the model **Photo_m4V**.

In the model **Photo_m4I**, the current control, instead of the voltage control, is fulfilled; at this, the average value $K_i = 0.91$ was found. The processes in this system are displayed in Figure 5.22.

The second variant, with the load circuit disconnected for a short time, is utilized in the model **Photo_m4V1**. The switch that is set at the input of the boost converter opens for the time, defined by the pulse generator. The duration of disconnection is 20 ms, and the period is 2 s. When the switch is open, the voltage at the PV array output is stored in the sample-and-hold unit **S/H**; this voltage, with the proper gain, is the reference for the voltage controller at the PV array output when the switch is closed. The resulting process is shown in Figure 5.23; the power steady-state values are about the same as in the previous models (1050, 250, 206 W).

In the direct methods, the measurement of the PV array output power $P(k)$ under some value of the system input parameter, for instance, $d(k)$, is performed, and afterward it is decided in which direction and how much the value d has to be changed, in order to increase the value P. The algorithm may be written as

$$d(k + 1) = d(k) + [d(k) - d(k - 1)] \text{ sign } [P(k) - P(k - 1)]. \tag{5.9}$$

If the magnitude of d changing is constant and is equal to g, then Equation 5.9 may be written as

$$d(k + 1) = d(k) + g \text{ sign}[d(k) - d(k - 1)] \text{ sign } [P(k) - P(k - 1)]. \tag{5.10}$$

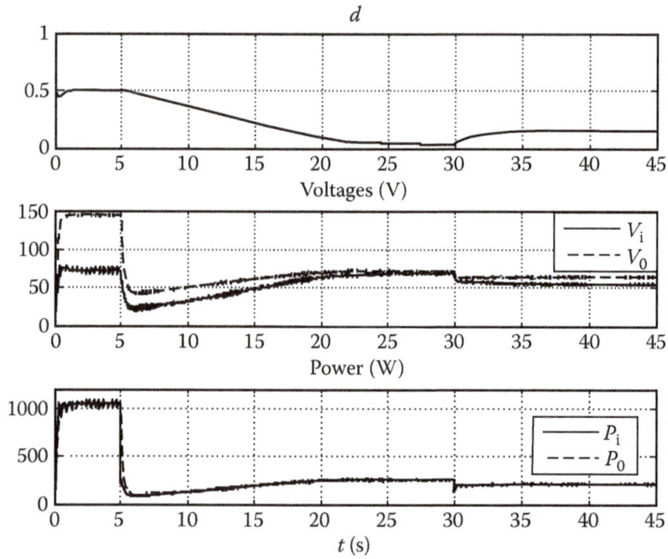

FIGURE 5.22 Response to the irradiance and temperature changing in the model **Photo_m4I**.

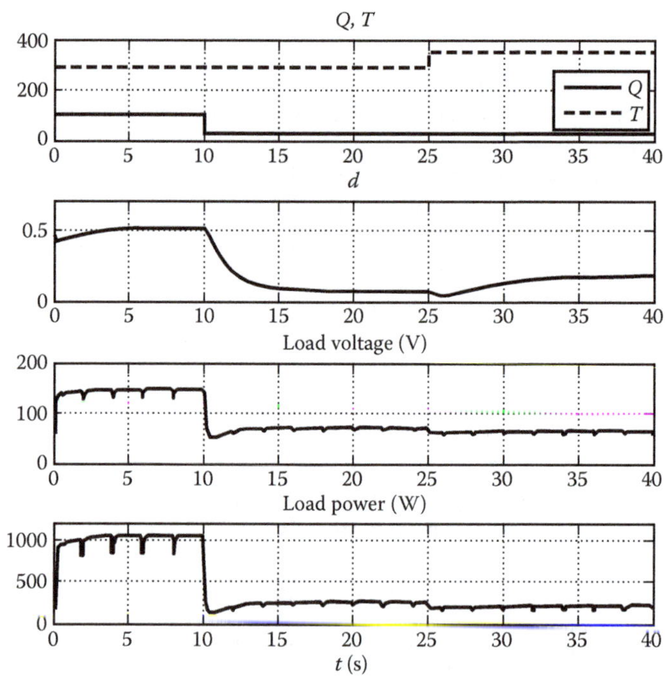

FIGURE 5.23 Processes in the model **Photo_m4V1**.

The realization of this algorithm that is called *Perturb&Observe Approach* is demonstrated in the model **Photo_m4S**. The limitations for the rate of irradiance and temperature changes are included. It is usually accepted [4] that the rate of irradiance change must be not more than 100 W/m^2/s under the nominal value of 1000 W/m^2, that is, $|dQ/dt| \leq 0.1$.

The subsystem **Control** is made with the employment of the blocks **Sample&Hold**. When input $S = 1$, the output follows input In; when $S = 0$, the output is held. The system operations are synchronized with the pulse generator; the pulse frequency is equal to 1 Hz, and the relative pulse duration is 0.5. With the pulse leading edge, the values $P(k)$ and $d(k)$ and also sign $[d(k) - d(k-1)]$ and sign $[P(k) - P(k-1)]$ are stored. The value $g \times$ sign $[d(k) - d(k-1)] \times$ sign $[P(k) - P(k-1)]$ is formed. With the pulse trailing edge, the value $d(k+1)$ is fabricated and stored that will be realized during the next sampling period. By this edge, with some delay, the values $P(k)$ and $d(k)$ are rewritten in the blocks $P(k-1)$ and $d(k-1)$ for utilization for the next calculation.

The possibility to change (to increase) the iteration step g when the rate of power change is rather large is intended in the system; in the considered case, the value g doubles when the absolute value of the rate of the measured power change is becoming more than 60 W/s.

The processes in the system are shown in Figure 5.24. It is seen that the system follows the environment conditions because the power steady-state values are successively equal to 1025, 510, 185, 718, and 600 W; at this, the power values that are obtained with the use of the model **Photo_m1** under the same values of irradiances and temperatures are 1037, 512, 191, 722, and 600 W.

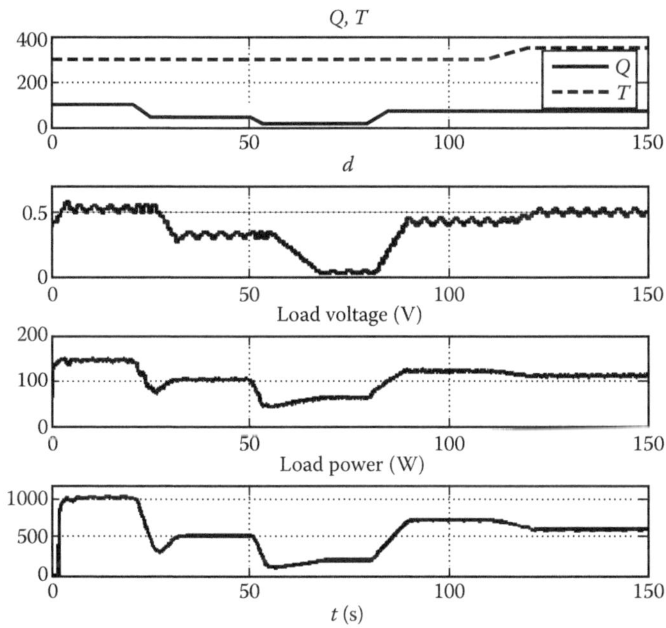

FIGURE 5.24 Processes in the model **Photo_m4S**.

The diverse modifications of the considered method are described in the literature [4]. So, *The Incremental Conductance Method* gains widespread popularity. This method is based on the fact that in the maximum power point,

$$\frac{dP}{dV} = 0 = \frac{d(VI)}{dV} = I + V\frac{dI}{dV}. \tag{5.11}$$

Therefore, in the optimal point,

$$e(k) = \frac{I(k)}{V(k)} + \frac{I(k) - I(k-1)}{V(k) - V(k-1)} = 0. \tag{5.12}$$

If $e(k) > 0$, the value V must be increased; when $e(k) < 0$, it must be decreased. The method merit is the theoretical possibility to avoid the oscillations of the voltage V in the point $e(k) = 0$, but in practice, it is difficult to realize [4]. The method is used in the model **Photo_m5**; the same blocks **Sample&Hold** as in the previous model are utilized in the control system. The iterations (perturbations) are carried out not with direct d changing, as in the previous model, but with alteration of the reference for the voltage controller. The opportunity to exclude the perturbations about the point $e(k) = 0$ with the help of the blocks **Abs1**, **e0**, and **Switch2** is intended in the system. The simulated processes are shown in Figure 5.25. The steady-state values (1050,

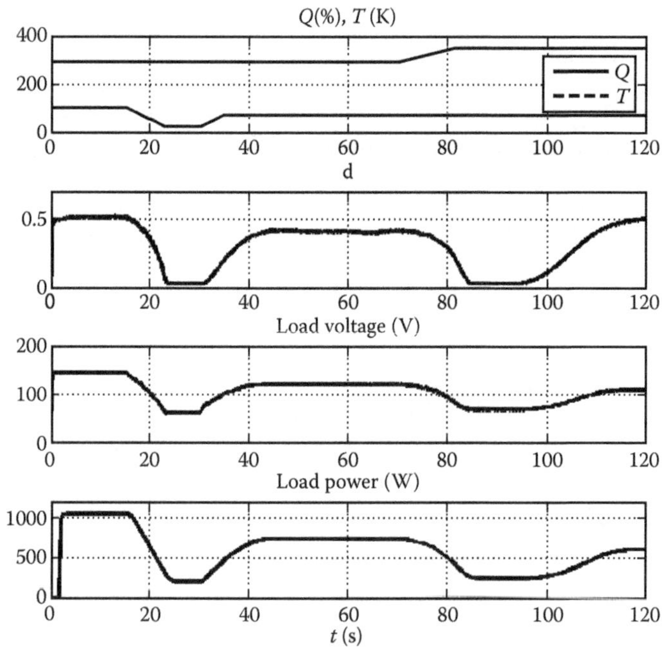

FIGURE 5.25 Processes in the model **Photo_m5**.

190, 732, and 600 W) are rather close to the values that are obtained with the aid of the model **Photo_m1** at the same environmental conditions.

The amount of electric energy produced by PV cells depends on the time of day and the weather conditions; therefore, for maintaining the supply, the PV units operate together with one or other device for electric energy accumulation, more often, with a battery. Such a system is considered in the model **Photo_m6**.

The outputs of the subsystems **PV** and **Battery** are connected in parallel to the load resistor of 10 Ω, and the rated load voltage is V_a = 220 V. The load resistor is set in the DC circuit, in order to decrease the simulation time essentially, which is large as it is. The load resistance is specified in the option *File/Model Properties/Callbacks*. The PV control system provides the maximum output power of PV array under existing environment conditions; the battery control system has to keep the load voltage up. The PV unit model from the model **Photo_m4V** is used, in which an optimization is reached with the employment of the measuring PV cells that operate without output current. The number of the series-connected PV cells is increased to 430, and of the parallel connected to the PV cells to 4, so that the rated load power of 4840 W is obtained under Q = 80%. The power circuits of the **Battery** are the same as shown in Figure 5.4; the fully charged voltage is 160 V; the control system is modified and displayed in Figure 5.26. It contains the load voltage controller with the inner battery current controller. The gate pulses are sent to transistors T_1 or T_2, depending on the sign of the $e_a = V_a^* - V_a$: when $e_a > 0$, the pulses come to T_2, the load voltage increases, and the battery discharges; when $e_a < 0$, the pulses are sent to T_1, the load voltage decreases, and the battery charges.

However, the capability to convert the PV surplus energy into the battery charge energy is limited by the battery capacity, and when the battery is fully charged, further charging must stop. For this, when the battery voltage reaches the value U_2, the gate pulses are not sent for T_1 (it is set U_2 = 162 V). If this PV unit has the energy surplus, the load voltage increases and can reach inadmissible values; the additional loads can be connected as useful (***R*~1~**) or ballast (**Ballast**), in order to avoid a failure. The choice is carried out with the tumblers **Tumbler1** and **Tumbler2**. The useful load is equal to 25% of the constant one; it is connected under V_a = 245 V and disconnected under V_a = 220 V. As for the ballast load, the same load unit that is used in the model **Photo_m1** is utilized, with some alterations. The ballast load begins to increase under V_a = 245 V, keeps constant under decreasing V_a to 235 V, and begins to decrease under V_a < 230 V.

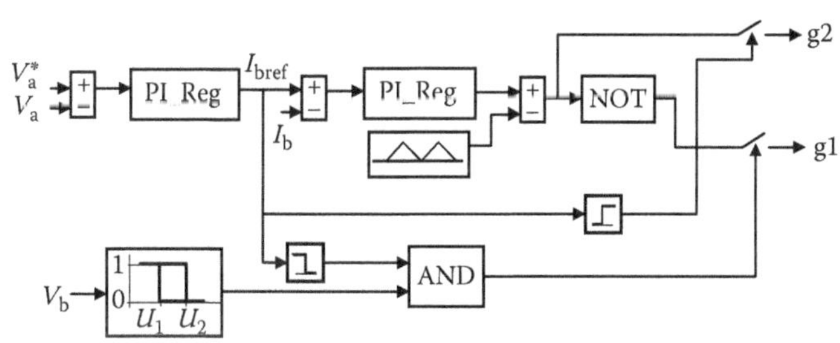

FIGURE 5.26 The diagram of the control system of the model **Photo_m6**.

The following schedule is simulated: start under $Q = 0.8$; decreasing Q to 0.25 at $t = 6$ s; increasing Q to 1 at $t = 16$ s; decreasing Q to 0.8 at $t = 40$ s. The rate of irradiance changing is taken as 10%/s. We note once more that the battery capacity and some time parameters are decreased, in comparison with real quantities, in order to have the reasonable simulation time.

The processes with the closed tumblers **Tumbler2** or **Tumbler1** are shown in Figures 5.27 and 5.28, respectively. The load voltage is kept up, after transient, in the given limits.

Another opportunity to limit the load voltage, when the PV unit has a surplus power, is investigated in **Photo_m6_1**, namely, the decrease in the fabricated power by the deviation of the PV output voltage from the optimal value. It is seen from the model diagram that, when the load voltage exceeds 240 V, the signal dV appears at the output of the additional controller (**V-P_Controller**) in the **PV** subsystem that decreases the reference of the output voltage relative to the optimal value, which is produced by the measuring PV cells. The processes in the system are shown in Figure 5.29.

The model **Photo_m6_2** is a modification of the previous model, in which the load resistor is replaced with a single-phase regulated rectifier with PWM that fabricates the output voltage with the frequency of 50 Hz. The resistance of the resistive–inductive load is 2.5Ω, and the time constant is 2 ms. The load voltage controller is

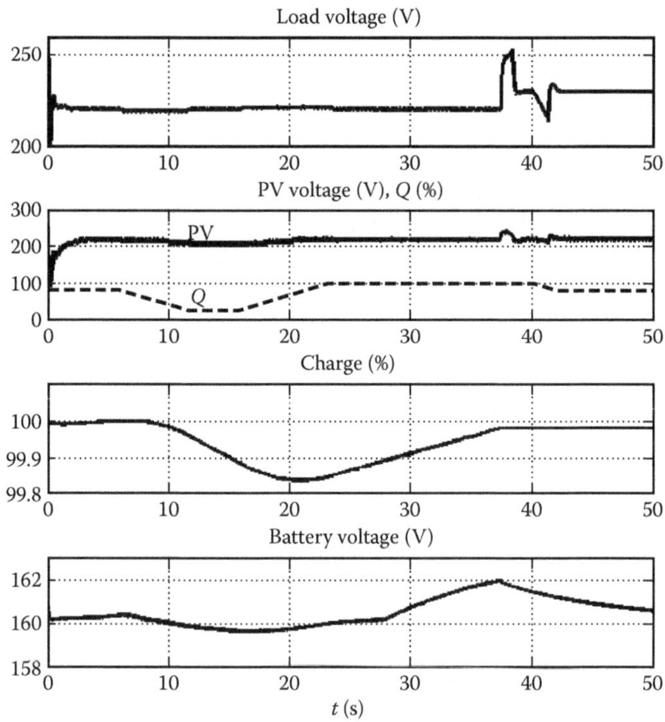

FIGURE 5.27 Processes in the model **Photo_m6** when **Tumbler2** is closed.

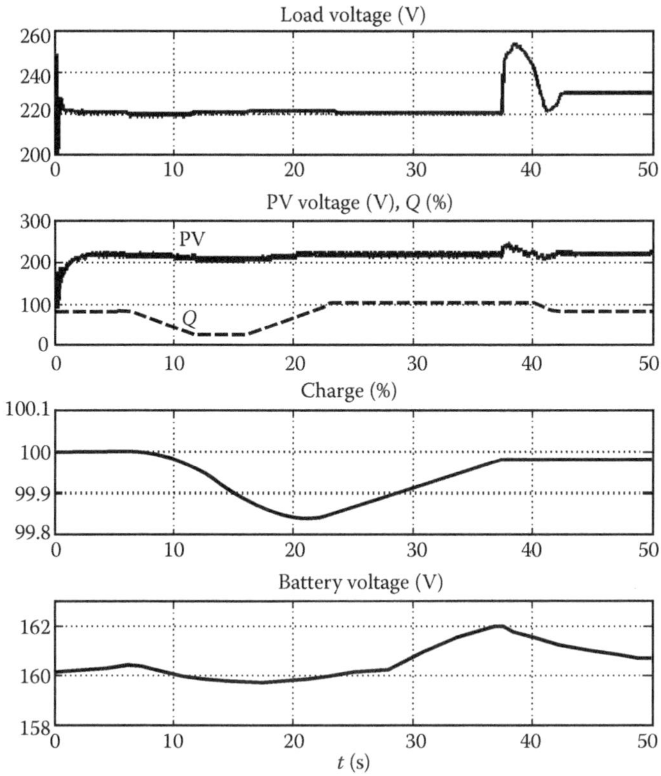

FIGURE 5.28 Processes in the model **Photo_m6** when **Tumbler1** is close.

employed, whose reference is 200 V. It follows from Figure 5.30 that this quantity is maintained with a high degree of accuracy.

When the PV units are situated not far from the AC grid, the load supply, under poor illumination, can be carried out from the grid; under excess illumination, the electric energy may be sent to the grid. Such a system is modeled in the model **Photo_m7**. Its block diagram is depicted in Figure 5.31.

PV optimization is carried out with the circuits that have been already used in the model **Photo_m4V**. The reference PV voltage is fabricated by the measuring PV cells, which operate in the no-load condition. The voltage of these PV cells recalculates in the reference for the voltage V_i^* controller, taking into account the optimizing coefficient and the number of series-connected PV cells in the measuring and working PV units. The voltage controller gives the reference I_i^* for the inverter AC current controller of the hysteresis type. The output of the PV unit is connected to the inverter DC input.

Owing to PLL, I_i^* is in antiphase with the AC grid voltage, providing, with this way, the maximum transfer of the PV energy in the grid and load.

The same process, as in the model **Photo_m6_2**, is simulated. The reference (optimal) and the real PV voltages, the reference for current controller and DC

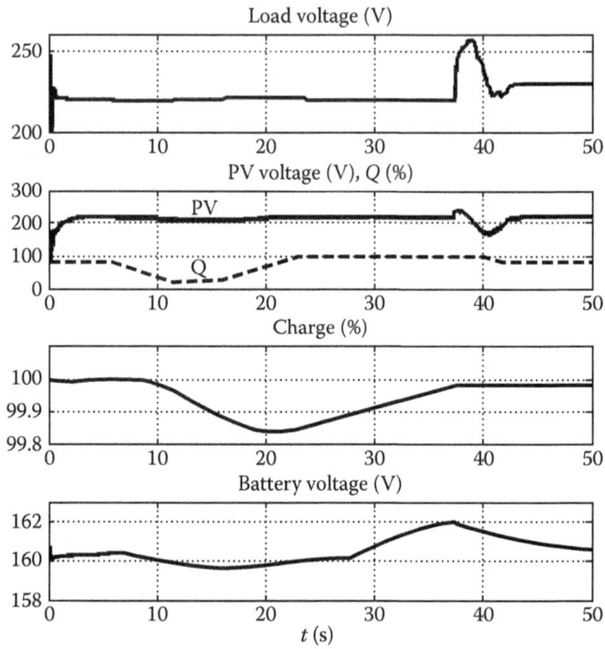

FIGURE 5.29 Processes in the model **Photo_m6_1**.

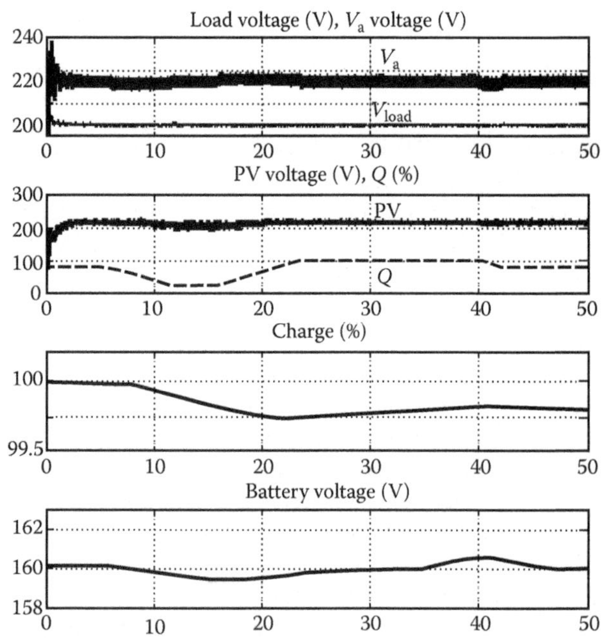

FIGURE 5.30 Processes in the model **Photo_m6_2**.

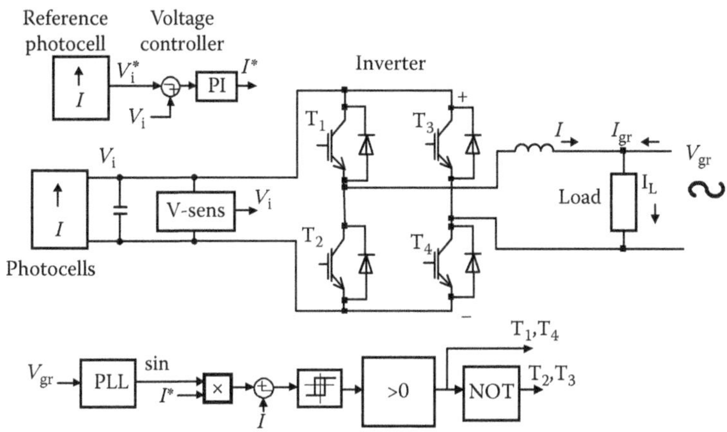

FIGURE 5.31 Block-diagram of the model **Photo_m7**.

inverter voltage, and the PV power are shown in Figure 5.32. The plots of the grid voltage together with the plots of the currents of the grid, the load, and the inverter in the steady state are shown in Figure 5.33 under the irradiance of 0.8, 0.25, and 1. It is seen that, under the irradiance of 0.8 and 1, the PV energy transfers in the grid (the voltage and the current of the grid are in antiphase); under the irradiance of 0.25, the load consumes the grid energy. The inverter power under these irradiance values is 10.5, 3.1, and 13.1 kW, respectively, and the load power is 5 kW.

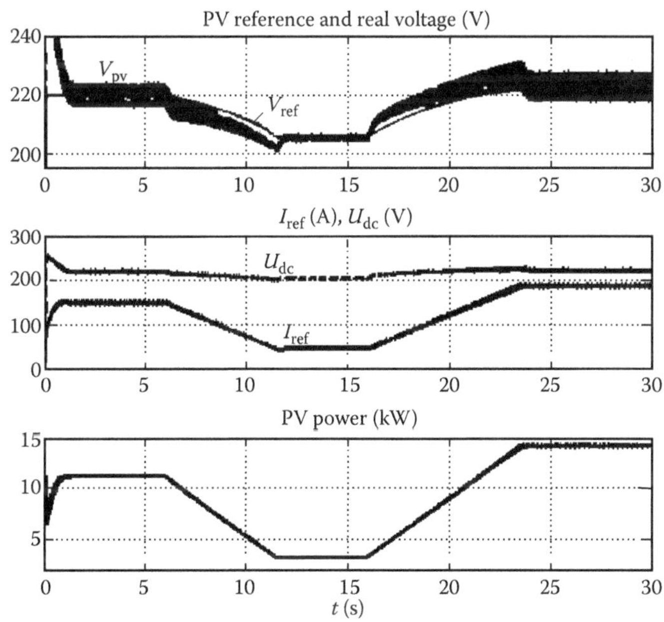

FIGURE 5.32 Processes in the model **Photo_m7**.

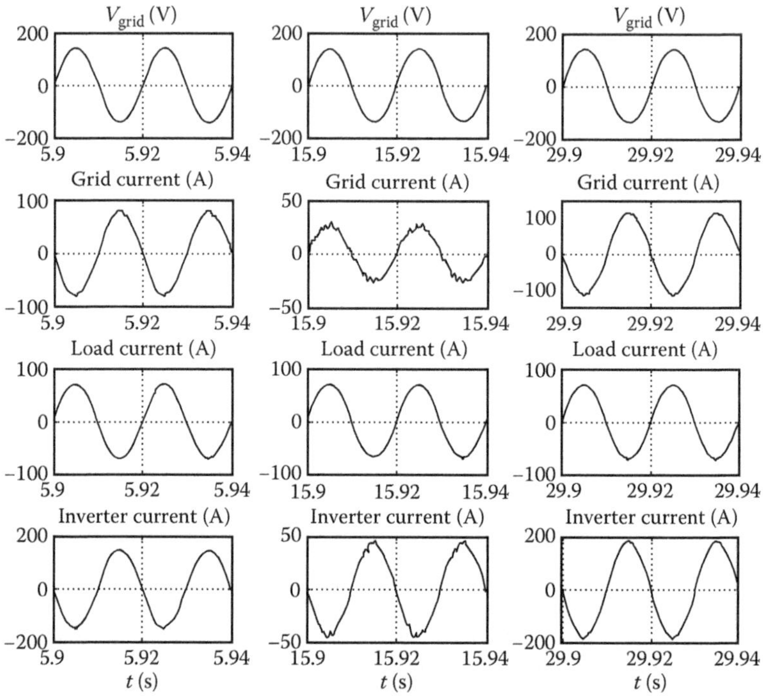

FIGURE 5.33 Voltage and currents in the model **Photo_m7**.

The model **Photo_m71** is a development of the previous one. The opportunity to power the load either separately, from the grid or from the PV unit, or together is supposed. Because the PV unit can operate individually, it must have the generator of the sinusoidal signal. But on the transfer from isolated operation to collaboration, before turning on the corresponding breakers, a synchronization of the sinusoidal signal with the grid voltage has to be fulfilled. At this, when the grid is connected while the PV unit operates, the synchronization must be carried out smoothly, in order not to distort the voltage of the load connected to the PV unit.

The power circuits of the considered model differ from the previous one by the presence of the controlled breakers **Br** and **Br1** in the circuits of the grid and the PV unit, respectively. For simplicity, these breakers switch over at the assigned instants of time, but in reality, the breakers can switch over independently of other signals: disappearance of the grid voltage or of the necessary irradiance, time of the day, and so on. On appearance of the turning-on signals, the preliminary signals PBr and PBr1 are formed that actuate synchronization, and later, the proper signals of switching on **Br** and **Br1** are fabricated.

The subsystem **Control** contains the hysteresis current controller considered earlier; the sinusoidal reference signal, having an amplitude equal to 1, is fabricated either by **PLL** block that is driven with the grid voltage, when both breakers are turned on, or by the subsystem **Br_Control**.

This subsystem contains the controlled generator of the saw-toothed oscillations that employs (utilizes) **Integrator** with reset to 0 when its output reaches the value of 2π. The frequency is defined by the constant **Freq**. The block **Sin** converts the integrator output into the sinusoidal signal for the current controller. In order to synchronize this signal with the grid voltage, in the case when the PV unit is being connected to the load that is supplied from the grid, the short pulses (with the period of 110 ms) are formed at the output of the AND element, when the **Br** is closed and the preliminary signal PBr1 comes. These pulses come, through the element OR, to the reset integrator input; the saw-toothed waveform *SyncX* that is fabricated by the PLL comes to the integrator input x_0 that defines the initial value under reset. In this way, the saw-toothed waveforms of the integrator and the PLL begin to coincide; the connection of the PV unit to the load occurs without perceptible distortion of the load voltage. After **Br1** closing, the reference for the current controller is fabricated with PLL, as mentioned earlier.

The circuits for synchronization in the case, where the grid is being connected to the load that is supplied from the PV unit, contain the device for the measurement of the phase displacement that begins to function when the element **Trigger** is set with an activation of the signal PBr. The device operates in the following way. When the saw-toothed waveform *SyncX* that is fabricated by PLL changes from 2π to 0, the block **Edge Detector** sets **Trigger1**, and the block **Integrator 2** begins to integrate the signal, whose amplitude is equal to 1. When the same happens with the saw-toothed waveform at the output of the block **Integrator**, the trigger resets and the integration stops. So the output of **Integrator 2** is equal to the phase displacement of two saw-toothed waveforms expressed in the time units. When the trigger is resetting, the short pulse comes to the input *S* of the block **Sample & Hold** that results in the storage of the **Integrator 2** output value at the output of the block **Sample & Hold**. This quantity, multiplied by the factor *Gain*, is integrated with the block **Integrator 1**, whose output signal is added to the saw-toothed waveform of the block **Integrator**, shifting it in the direction of the phase displacement decrease. This process stops when the input value of the block **Integrator 1** is getting less than the value fixed in the comparison block **Compare2** (the **Trigger** resets). The breaker **Br** may be turned on in this moment, connecting the grid to the load. The big negative signal is sent to the input of **Integrator 1** and set to 0.

The following process is simulated under the irradiance of 100%: at $t = 0.01$ s, the signal for **Br** turning on is fabricated; at $t = 1.2$ s, the signal for **Br1** turning on is fabricated; at $t = 2.5$ s, **Br** is turned off; at $t = 4.5$ s, the signal for **Br** turning on is fabricated. The phase of the grid voltage is 180°. The plots of the inverter voltage and currents are shown in Figure 5.34; in Figure 5.35, the same variables are plotted, but in the larger scale, in the instants of breaker switching over. It is seen that, under the switching, the load current is not exposed to the abrupt alterations.

The model **Photo_m8** differs from the model **Photo_m7** that the three-phase system, instead of the single-phase, is used. In Figure 5.31, the inverter, the reactor, the load, and the AC source V_{gr} are three-phase units. The V_{gr} phase-to-phase rms value is 220 V, and the load power is 100 kW. The PV unit has 750 PV cells in series and 50 PV cells in parallel. It is supposed that under the irradiance of $Q_r = 75\%$, the load

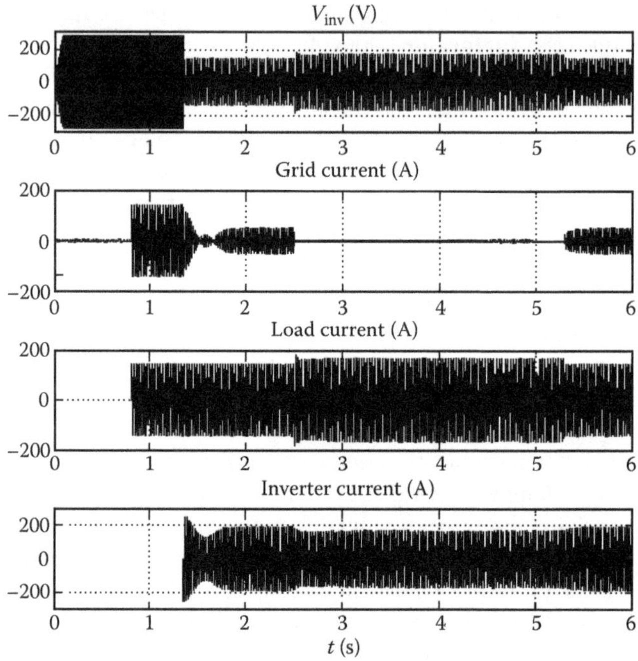

FIGURE 5.34 Inverter voltage and currents in the model **Photo_m71**.

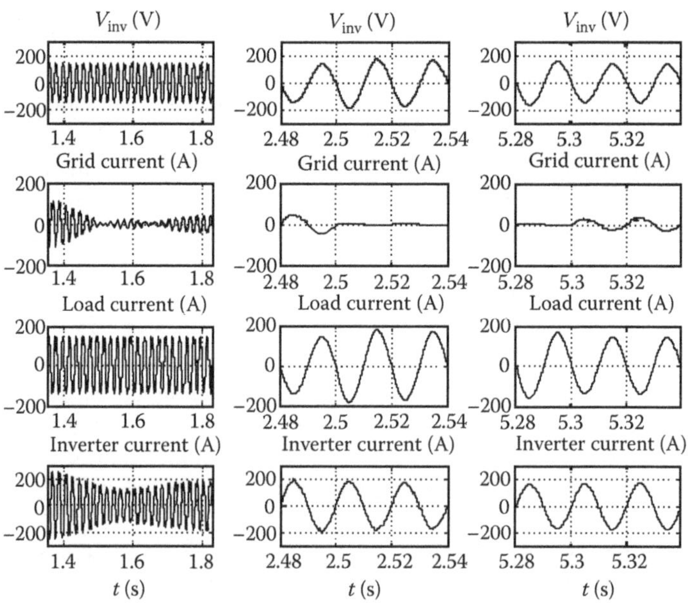

FIGURE 5.35 Inverter voltage and currents in the switching over instants in the model **Photo_m71**.

is supplied only with the PV unit; when $Q < Q_r$, the load is supplied from the PV unit and the grid simultaneously; when $Q > Q_r$, the energy surplus is sent to the grid.

The three-phase hysteresis current controller is utilized in the model that is taken from the demonstration models of SimPowerSystems. The grid voltage, instead of the PLL, is used for aligning the current reference.

The changes of the powers of the grid, the load, and the PV, together with the PV voltage, are shown in Figure 5.36. Under $Q = 0.75$, all PV power is consumed for the load supply. Under $Q = 0.25$, the PV power is only 34 kW, and the shortage of 66 kW is taken from the grid. Under $Q = 1$, PV power is 135 kW, and the surplus of 35 kW is sent to the grid. The optimal PV power quantities under these irradiance values are 107, 34, and 143 kW, respectively, which proves (taking into account the losses in the inverter and the power circuits) that the point of the maximum power is tracking satisfactory.

The model **Photo_m81** differs from the previous that the hysteresis current controller is replaced with the three-phase current controller. The number of the series-connected PV cells increases to 825. The change of the powers of the grid, the load, and the PV, together with the PV voltage, are shown in Figure 5.37. Under Q = 0.75, the PV power is 118 kW, and the surplus of 18 kW is sent to the grid. Under

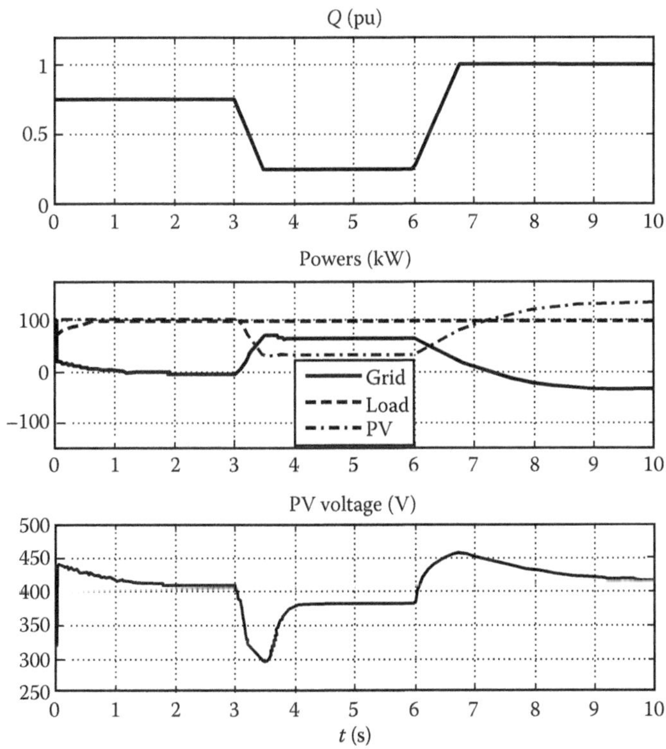

FIGURE 5.36 Powers of the grid, the load and the PV, the PV voltage for the model **Photo_m8**.

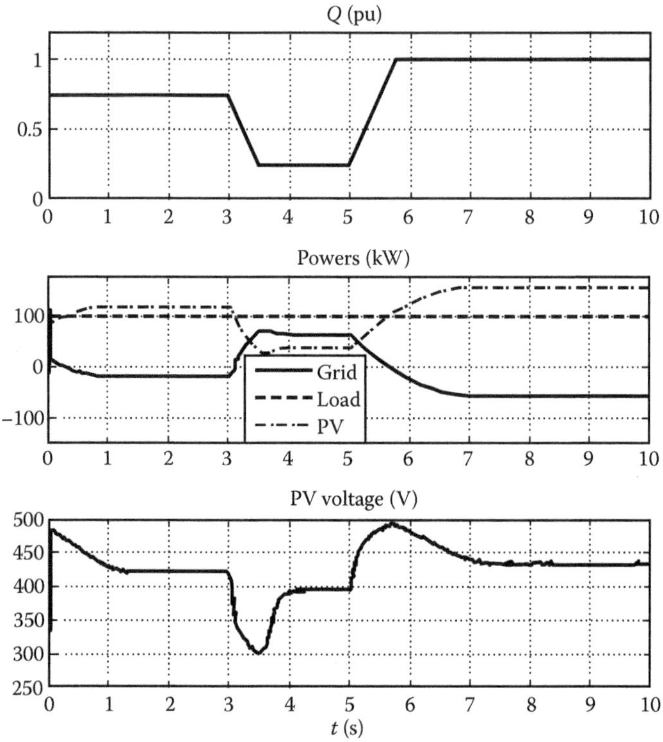

FIGURE 5.37 Powers of the grid, the load and the PV, the PV voltage for the model **Photo_m81**.

$Q = 0.25$, the PV power is only 37 kW, and the shortage of 63 kW is taken from the grid. Under $Q = 1$, PV power is 157 kW, and the surplus of 57 kW is sent to the grid. The optimal PV power quantities under these irradiance values are 118, 37, and 157 kW, respectively, that corresponds to the simulated quantities.

Today, the persistent trend exists to increase the power of PV sets—up to 100 MW and more. These sets usually consist of the same modules that are connected in parallel at the set output. The diagram of one such module that has the power of 1.76 MW is shown in Figure 5.38.

Two PV units, each of them has 1370 PV cells in series and 170 PV cells in parallel, are connected to the two-level inverters, which, in turn, are connected to the secondary windings of the three-winding transformer. The transformer has the power of 1.76 MVA and the primary voltage of 10 kV and is connected to the grid via transmission line. The secondary voltages are 360 V. Under nominal conditions, the PV unit power is 887 kW under the voltage of 700 V. **Photo_m9** models such a system.

The diagram of the control system is depicted in Figure 5.39. The current references are aligned with the corresponding secondary voltages. But these voltages are distorted essentially; therefore, they are filtered by the first-order filters with the time constant of 5.5 ms and the gain of −2, which, under the frequency of 50 Hz, give the

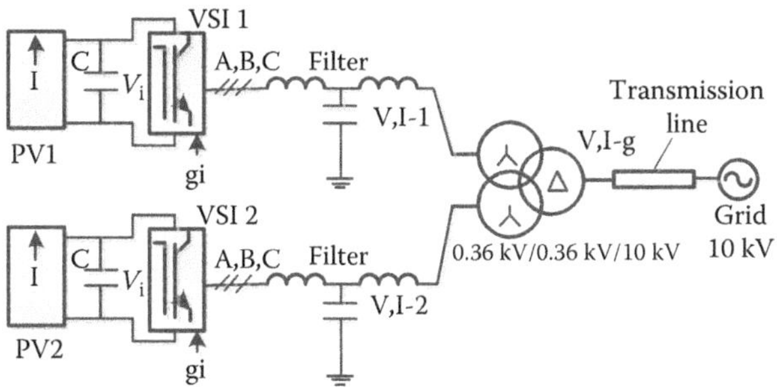

FIGURE 5.38 Diagram of the PV module with power of 1.76 MW.

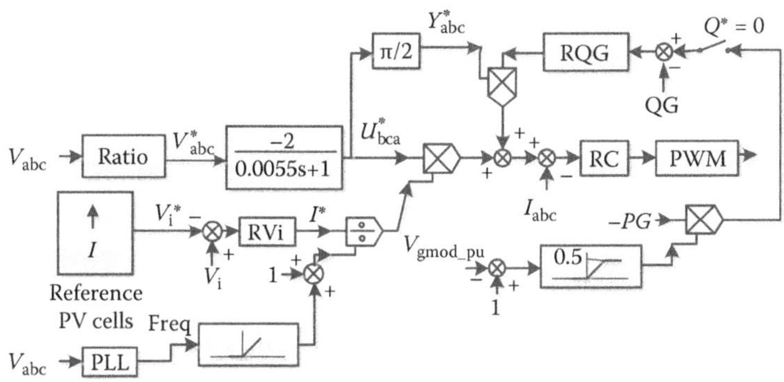

FIGURE 5.39 Diagram of the control system of the model **Photo_m9**.

lag of 60°. The filtered voltages that have the amplitude equal to 1 are designated as U_a^*, U_b^*, U_c^*. Instead of the vector V_{abc}^*, the vector U_{bca}^* that has the same phase but is free from higher harmonics is used.

The module of the current reference is formed, as in the previous models, with the reference PV cells and the voltage controller RVi. When the grid frequency is getting more than 50 Hz, the current reference decreases and is halved under $Fr = 50.2$ Hz.

The reactive power control is intended in the system. For this, the three-phase system Y^* is fabricated that leads the system U_{bca}^* by 90°. For example, for phase A,

$$Y_a^* = U_b^* - U_c^* = -\frac{1}{\sqrt{3}}(\sin\omega t + 2\sin(\omega t - 120°)) = -\sin(\omega t - 90°) = \cos\omega t.$$

The signal Y_{abc}^* is modulated by amplitude with the output of the reactive power controller RQG. The reference for this controller can either be zero or change as a function of the grid voltage V_g. In the latter case, when the grid voltage decreases, the reactive power QG that delivers to the grid increases; this reactive power is proportional to the active power PG and does not exceed 0.5 PG.

The processes in the system under irradiance change are shown in Figure 5.40 under the reactive power control with zero reference (tumbler **Switch2** is in the top

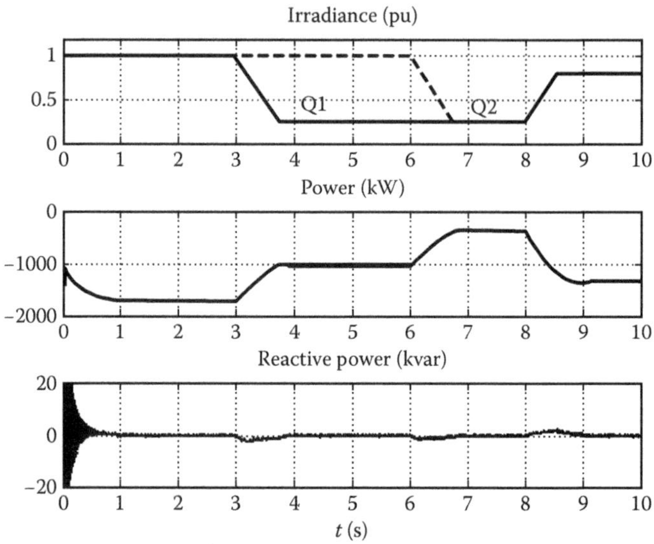

FIGURE 5.40 The processes in the model **Photo_m9** under irradiance changing.

position; tumbler **Switch** in the **Subsystem** is in the low positions). The steady-state power values are 1720, 1050, 390, and 1370 kW, respectively. The optimal values for two PV arrays in the same conditions are 1780, 1102, 424, and 1420 kW, without taking into account the losses in the inverters and the transformers. The reactive power is approximately zero. The system response for frequency increase is shown in Figure 5.41; that for the sag of the grid voltage is shown in Figure 5.42.

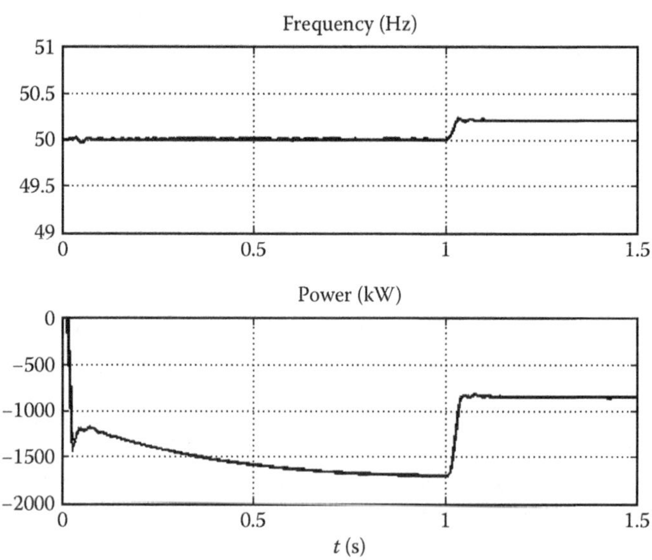

FIGURE 5.41 The processes in the model **Photo_m9** under frequency changing.

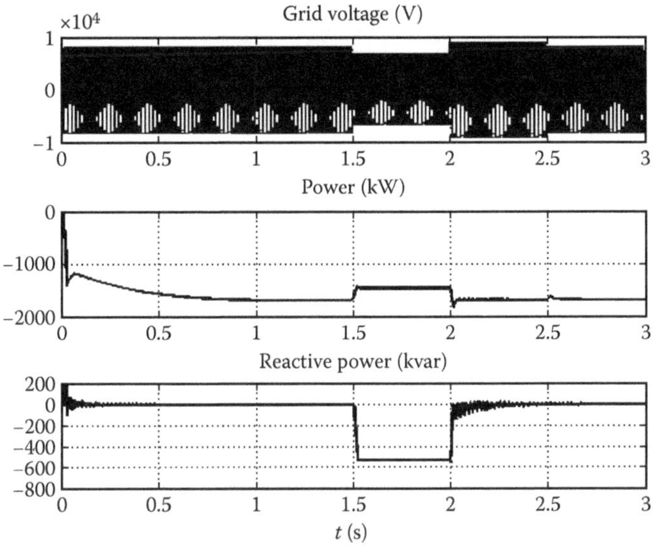

FIGURE 5.42 The processes in the model **Photo_m9** under grid voltage sag.

5.3 SIMULATION OF THE FUEL CELLS

The fuel cell (FC) is a device for the direct conversion of fuel chemical energy into electric energy. The intermediate conversion of the chemical energy into the thermal or the mechanical energy is missed out, so that the device efficiency can be very high. Unlike the battery, which must be disconnected periodically for recharge, the FC can produce electric energy without interruption, as long as it contains the fuel (such as hydrogen) and an oxidant (such as oxygen), which can be refilled during FC operation.

The FC consists of two electrodes separated with an electrolyte and, perhaps, with a membrane. The hydrogen comes to the anode where its molecules, under the action of the electrolyte and the catalyst, lose the electrons and convert into positive ions. These ions diffuse through the electrolyte to the cathode, and the electrons reach it through the external circuit that connects the electrodes; in this way, the electrical current is fabricated. The oxygen of the air comes to the cathode, where (oxygen creates on the cathode), connecting with the ions and the electrons, creates the water; the heat is produced under this reaction. The water must be removed from the FC; it has to be cooled off. The heat can be utilized. The surplus of the reacting gases has to be removed as well (Figure 5.43). The FC output voltage is about 1 V; in order to gain higher voltage, the FCs are connected in series. FC disadvantage is a slow response to the load change.

Various types of FC exist that differ, mainly, by the electrolyte composition, the catalyst, the membrane (see, for example, Refs. [7,8]).

In the solid oxide fuel cells (SOFC), the electrolyte is the dense ceramic membrane of zirconium oxide; these fuel cells operate below the temperature of 600°C –1000°C without the expensive catalyst and can utilize the natural gas. In the polymer

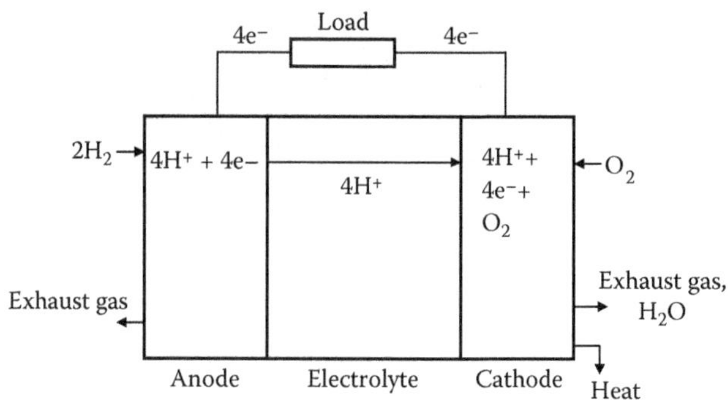

FIGURE 5.43 FC structure.

electrolyte fuel cells (PEMFC), the electrolyte is a very thin solid membrane; the operating temperature is not more than 100°C, and the fuel is pure hydrogen.

The main part of the FC voltage is determined as [9]

$$E = E^0 + \frac{JT}{2F} \ln\left(\frac{P_{H_2}\sqrt{P_{O_2}}}{P_{H_2O}}\right), \tag{5.13}$$

where E^0 is the open circuit voltage under standard pressure and temperature and with the pure reagents (equal to 1.17 V under 100°C), J is the universal gas constant that is equal to 8314 J/kmol/K, T is the absolute temperature (K), F is the Faraday's constant equal to 96.485 × 10^6 C/kmol, and P_{H_2}, P_{O_2}, and P_{H_2O} are the partial pressures of hydrogen, oxygen, water vapor (on the cathode).

When the FC conducts the current, the voltage drop appears, so it may be written as

$$V = E - IR - \Delta V_{act} - \Delta V_{con}, \tag{5.14}$$

where I is the FC current; R is the resistance of the whole electrical circuit including the membrane and various interconnections; ΔV_{act} is the activation losses, owing to the slowness of the polarization reactions, taking place on the electrodes; and ΔV_{con} is the voltage drop, owing to the change of the concentration of the active components in the reaction area.

The corresponding expressions are intricate and depend on many parameters, so that for the creation of the FC model, the empirical and semiempirical formulas are used, as, for instance,

$$\Delta V_{act} = A \ln\left(\frac{i}{i_0}\right), \tag{5.15}$$

where i is the FC current density (mA/cm^2), i_0 is the density of some reference current, and A is constant; the formula is valid under $i > i_0$.

If in Equation 5.14, instead of E, to substitute $E_c = E + A \ln(i_0)$, then

$$\Delta V_{act} = A \ln(i). \tag{5.16}$$

The following formula is suggested for ΔV_{con} in Ref. [7]:

$$\Delta V_{con} = m e^{ni}, \tag{5.17}$$

where m and n are constants.

Some data for SOFC (T = 650°C) and PEMFC (T = 70°C) are given in Ref. [7]. For the former, E_c = 1.01 V, r (resistance for the current density) = 2×10^{-3} (kΩ cm²), A = 0.002 (V), $m = 10^{-4}$ (V); for the latter, E_c = 1.031 V, $r = 2.45 \times 10^{-4}$ (kΩ cm²), A = 0.03 (V), $m = 2.11 \times 10^{-5}$ (V); $n = 8 \times 10^{-3}$ for both cases. The FC output characteristics are shown in Figure 5.44.

FC sets are rather complicated systems that have the units for fuel, oxidant, water supply, for withdrawal of the exhaust gas and the generated water, and for temperature maintenance. However, when the FC set is investigated as the additional source in the systems with WG or PV, a number of simplifications are taken [9]. It is often supposed that the controllers work ideally, and the only channel for action on the output voltage is the hydrogen consumption.

The model of SOFC is developed in Ref. [10]. This model is used in the demonstration model **Solid-Oxide Fuel Cell Connected to Three-Phase Electrical Power**

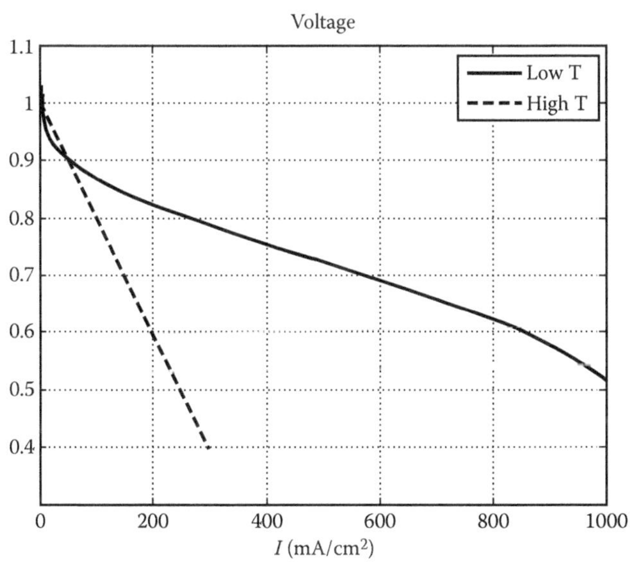

FIGURE 5.44 FC output FC characteristics.

System that is included in SimPowerSystems; SOFC has a power of 40 kW. In the model **Fuel_SOFC** developed for this book, the model is somewhat modified, in order to have the possibility to utilize the voltage controller; the model dialog box that is displayed in Figure 5.45 is slightly changed too. The meanings of the fields of this dialog box are as follows:

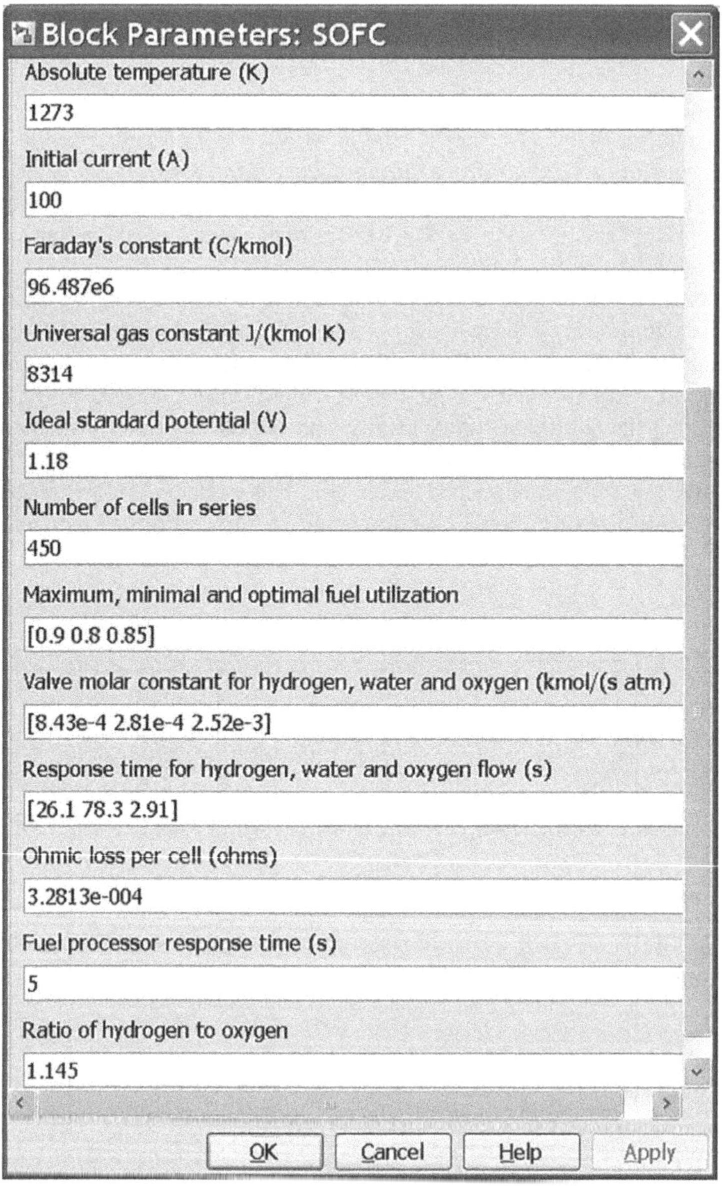

FIGURE 5.45 Dialog box of the developed model SOFC.

1—The absolute temperature T, at which FC operates; it is supposed to be constant during simulation. 2—The initial current under simulation I_0; its value determines the initial values of the integrators by formulas

$$\begin{aligned} &q\_H2\_0 = 2\ K\_r\ I_0/U\_opt; \\ &p\_H2\_0 = (1/K\_H2)\ (q\_H2\_0 - 2\ K\_r\ I_0); \\ &p\_O2\_0 = (1/K\_O2)\ (q\_H2\_0/r\_HO - K\_r\ I_0); \\ &p\_H2O\_0 = (1/K\_H2O)\ 2\ K\_r\ I_0. \end{aligned} \tag{5.18}$$

Here, $K\_r = K\_r = N/(4\ F)$; for the meanings of the parameters, see below. The value $I_0 > 0$, for example, $I_0 = 5$ A, must be assigned; otherwise, infinity can appear under the equation solution. 3—Faraday's constant, F. 4—The universal gas constant, J. 5—The E^0 quantity. 6—The number of FC connected in series, N. 7—Operation range for the fuel utilization; it is taken in the model $U\_opt = 0.85$. 8—The quantities that define the relationships between the rate of the agent amount change and the alternation of its partial pressure K_H2, K_H2O, and K_O2 for hydrogen, water, and oxygen, respectively. 9—The time constant of the reaction of the agent partial pressure to the change of its amount T_h, T_w, and T_o for hydrogen, water, and oxygen, respectively. 10—The value R in Equation 5.14. 11—The time constant of the device that controls the hydrogen supply, T_f. 12—The ratio between the supplied amounts of the hydrogen and oxygen, r_{HO} ($r_{HO} > 1$, because one molecule of the oxygen reacts with two molecules of the hydrogen).

The model block diagram is shown in Figure 5.46. It is seen that only the first subtrahend from Equation 5.14 is taken into account. The gas feed mechanism can operate either by the load current signal, boosting the gas supply under current increase

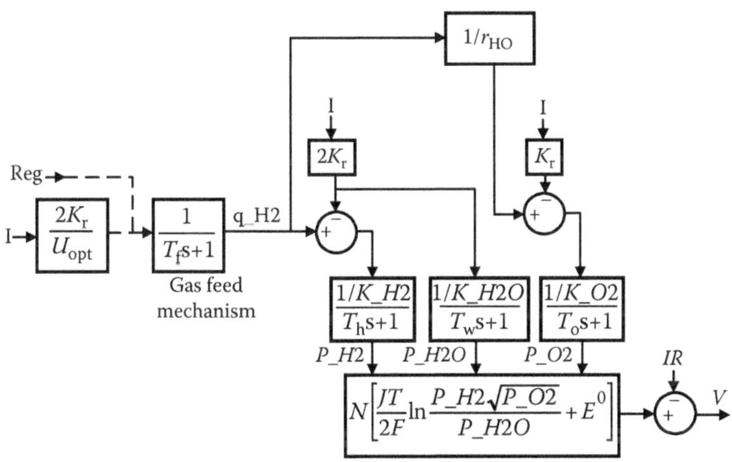

FIGURE 5.46 Block diagram of SOFC model.

(disturbance-compensating or feed-forward control), or by the signal of the external voltage controller.

Fuel_SOFC models the supply; with this FC set, the load resistance 4 Ω decreases to 2 Ω at $t = 250$ s. The processes, when the tumbler **Switch** is in position A (feed-forward control) and in position B (voltage feedback), are shown in Figure 5.47.

The model of the analogous structure can be used for PEMFC modeling. Such a model is utilized in **Fuel_PEMFC**. The FC parameters are taken mainly from Ref. [11]. The dialog box is displayed in Figure 5.48. The model output voltage is calculated by Equations 5.13 through 5.17. The model contains the reformer, which processes the natural gas into pure hydrogen. The diagram of the reformer model that replaces the gas feed mechanism shown in Figure 5.46 is depicted in Figure 5.49. It consists of the model of the device for gas processing and of the PI controller that adjusts the rate of the natural gas supply in such a way that the demanded amount of hydrogen would be obtained. The voltage and current variations, when the load resistance decreases from 6 Ω to 5 Ω at $t = 30$ s and to 4 Ω at $t = 60$ s (under power increasing from 6.7 to 10 kW), are shown in Figure 5.50; the voltage controller with the reference 200 V is actuated.

In order to moderate an unfavorable effect of the slow FC response on the load variations, and also to keep the load voltage under short FC failure, the other energy storage systems connected parallel to the FC are employed, for instance, the batteries or (and) the ultracapacitors. In the model **Fuel_PEMFCB**, the battery has about the same voltage as the FC set; therefore, their outputs are connected directly via a diode; when the FC voltage drops lower than the battery voltage, the diode begins to conduct, and the battery determines the load voltage and the discharges. When

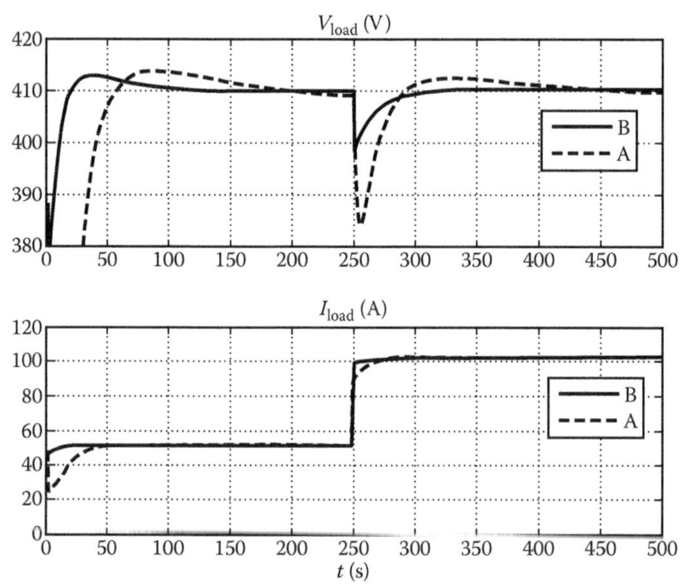

FIGURE 5.47 Processes in the model **Fuel_SOFC**.

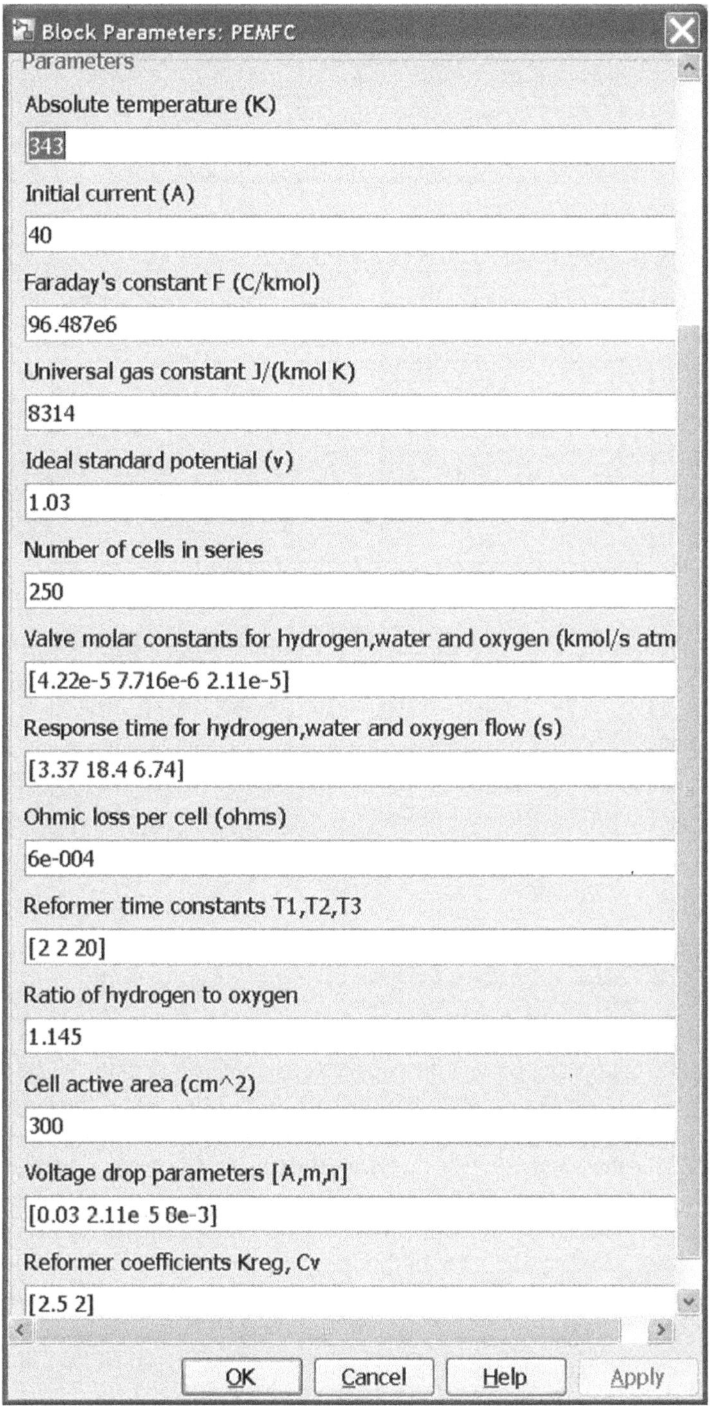

FIGURE 5.48 Dialog box of the developed model PEMFC.

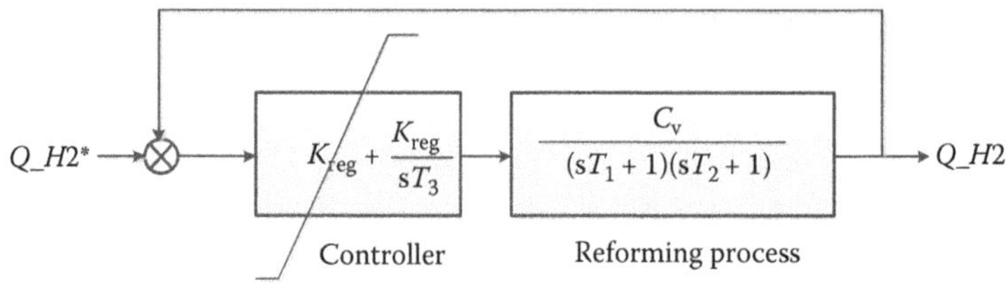

FIGURE 5.49 Block diagram of the reformer model.

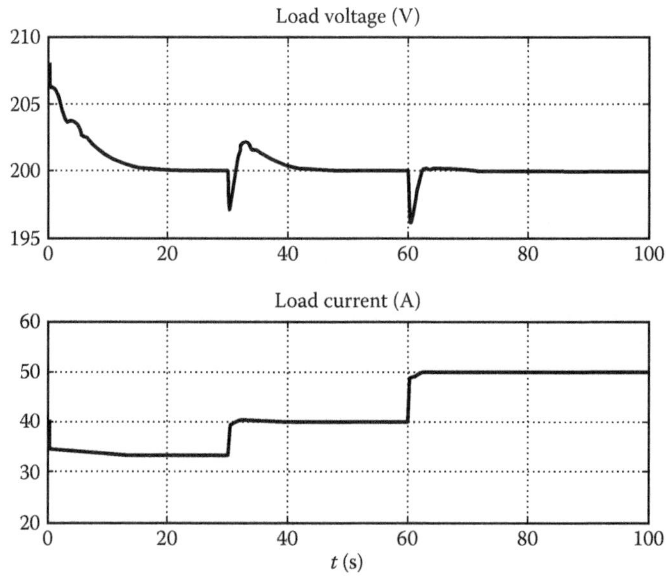

FIGURE 5.50 Processes in the model **Fuel_PEMFC**.

the FC voltage increases, the diode is cut off, and the battery is charging through the IGBT. The charge stops when the battery voltage increases to the wanted value.

The load circuits as in the previous model are used. At t = 100 s, the FC is turned off from the load; at t = 220 s, it is turned on again. The load voltage is shown in Figure 5.51, on the upper part in the large timescale. The change of the battery parameters is fixed in Figure 5.52. The voltage sags are seen to decrease in comparison with the previous model. It follows that with the selected (small) battery capacity, the load voltage decreases by 10% for 2 min. The demerits of this simple circuit are the impossibility to control the load voltage when the FC is turned off and the small rate of the battery charge.

More complicated circuits for the battery control that are taken from the model **Photo_m6** (Figure 5.26) are utilized in the model **Fuel_PEMFCB1**. The simulation sampling time is essentially decreased, which results in the simulation time increasing. It is supposed that the breakers of the battery and the load are closed at the

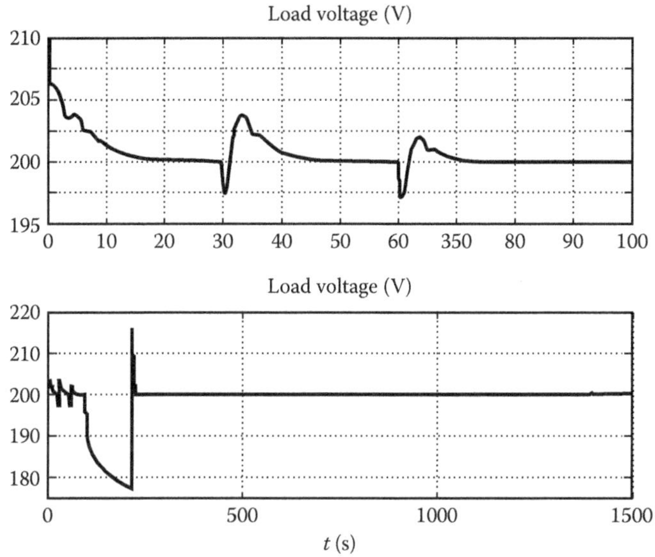

FIGURE 5.51 Load voltage variations in the model **Fuel_PEMFCB**.

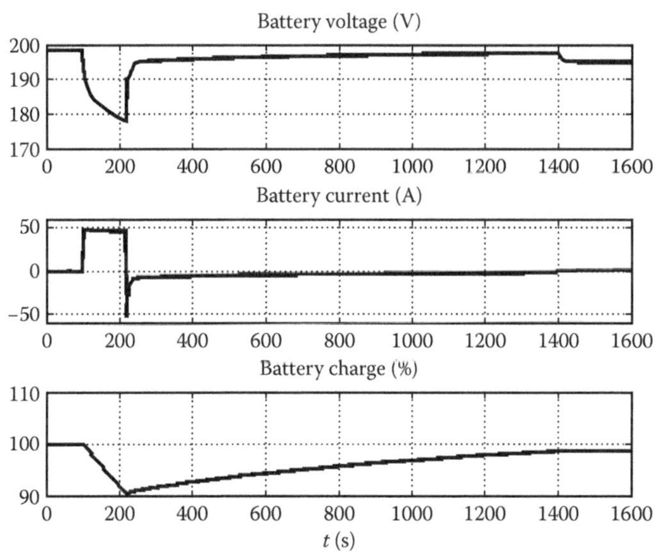

FIGURE 5.52 Battery parameters variations in the model **Fuel_PEMFCB**.

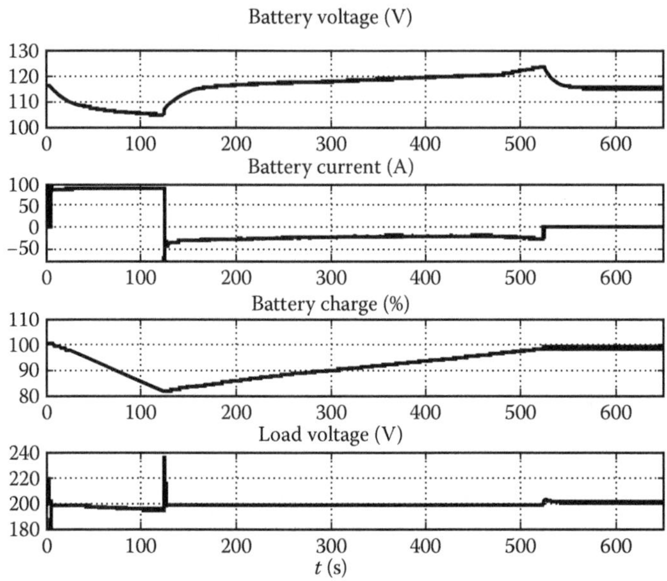

FIGURE 5.53 Processes in the model **Fuel_PEMFCB1**.

simulation start; at $t = 5$ s, the FC set is disconnected from the load; at $t = 125$ s, it is reconnected to the load. The process is displayed in Figure 5.53. During FC unit disconnection (2 min), the load voltage decreases by 6 V, which is much less than for the previous model with the same battery capacity, whose charge proceeds quicker as well.

In the considered models, the DC load is employed. However, the AC load is mainly utilized. In this case, the single-phase or the three-phase VSI is used. Such a system is utilized in the model **Fuel_PEMFC2**. The singe-phase load has AC rms voltage of 220 V. An FC set has 400 FCs in series. Three options for FC control are intended in the model: the voltage controller (feedback control; the tumbler **Switch** is in the top position); the feedforward control with utilization of the load current (the tumblers **Switch** and **Switch1** are in the low positions); and the combined control (the tumbler **Switch** is in the low position; the tumbler **Switch1** is in the top position). The single-phase VSI is connected to the FC set via an L-C filter and has the output voltage controller with the reference of $220 \times \sqrt{2} = 310\,\text{V}$. The active–inductive load, having the active power of 6.5 kW, at $t = 20$ s, increases to 10 kW. The variations of the FC set output voltage for options of the control system mentioned earlier under temperatures 343 and 373 K are shown in Figure 5.54.

To conclude this section, the model **SRGFLY** will be considered, in which the switched reluctance machine (generator—SRG) rotates the flywheel and is intended for damping the voltage deviations under load variations and under short hydrogen supply interruptions of the high-temperature FC unit of SOFC. The FC unit control system is the same as in the model **Fuel_SOFC**. The SRG control system differs from the used one in the SRG model that the current reference is not the speed controller output, but the output of the voltage proportional controller in the DC link V_{am}.

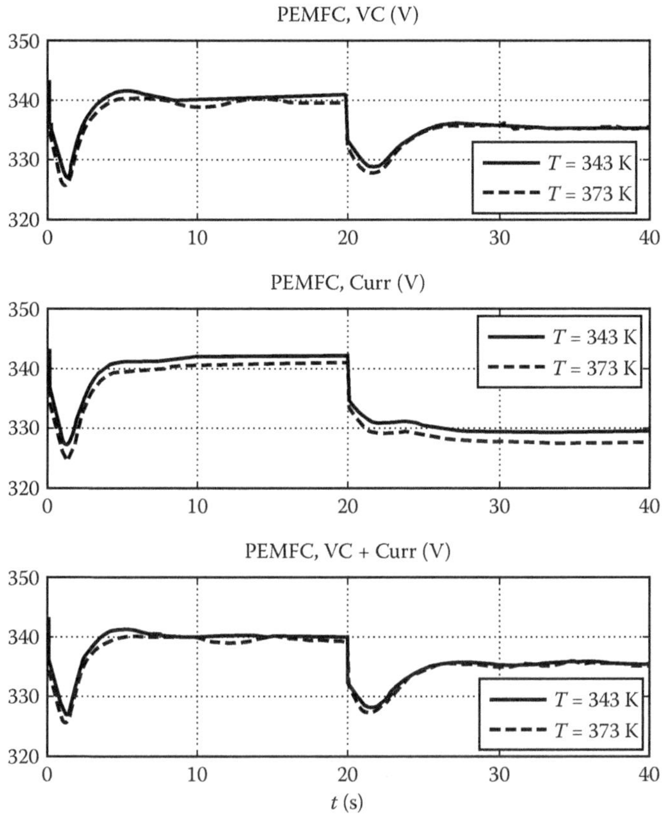

FIGURE 5.54 Variations of the FC output voltage for the model **Fuel_PEMFC2**.

Moreover, with the help of the block **Saturation Dynamic**, the rotation speed ω is limited in the motor mode by the value of ω = 600 rad/s and has to be more than ω = 200 rad/s in the generator mode.

The block diagram of the control system is depicted in Figure 5.55. It is taken that $V_1^* < V^*$, so that under $V_{am} > V_1^*$ the current $I_g > 0$ and the SRG accelerates, increasing the flywheel kinetic energy; when $V_{am} < V_1^*$, the current $I_g < 0$, SRG gives the stored energy to the DC link, preventing further voltage decrease. DC voltage is converted into a three-phase voltage by the VSI that operates in PWM mode with the modulation factor m close to 1. This voltage, via the transformer, supplies a three-phase load.

Two types of disturbances are considered in the model: load abrupt increasing, for instance, under the start of the induction motor, which is supplied from the FC unit, or the short interruption hydrogen supply. For the simulation of the first case, the powers of **Load** and **Load_1** are set as 30 and 240 kW, respectively; the breaker **Br** closes and afterward discloses at $t = 2$ and 15 s; the constant *Mode* = 0. The corresponding process is shown in Figure 5.56; to save a simulation time, the initial SRG speed was accepted as 600 rad/s. The plot of the DC voltage change without

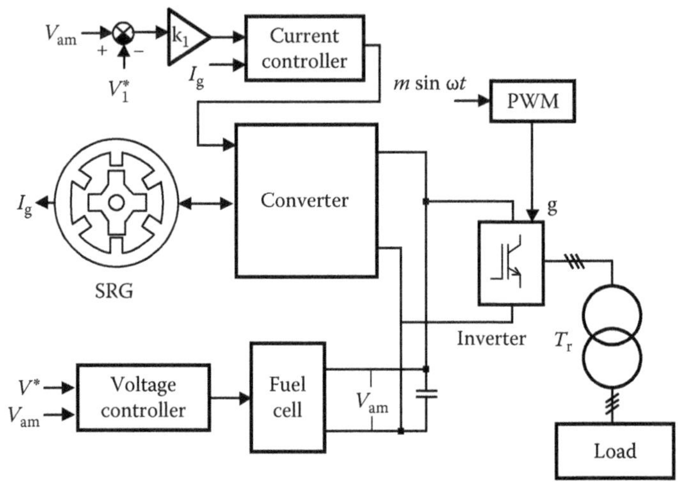

FIGURE 5.55 Block diagram of the control system in the model **SRGFLY**.

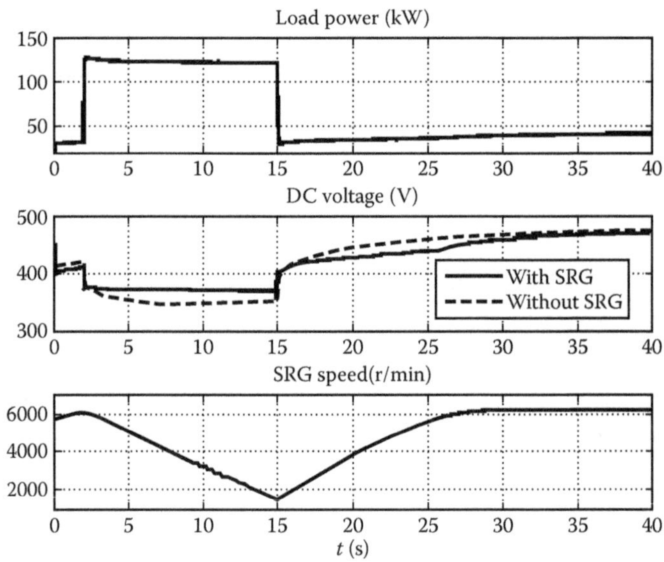

FIGURE 5.56 Processes in the model **SRGFLY** under load variations.

SRG is shown in Figure 5.56. It is seen that, owing to the flywheel, the voltage sag halves, from 50 to 25 V.

In order to simulate the process of the short hydrogen supply interruption, it is necessary to set *Mode* = 1, to increase the power of **Load** to 60 kW, and to forbid the switching over of **Br**. Therefore, the supply interruption lasts 12 s, from 2 to 14 s. The corresponding process is shown in Figure 5.57. It is seen that the system continues to operate, whereas without SRG, the DC voltage drops quickly.

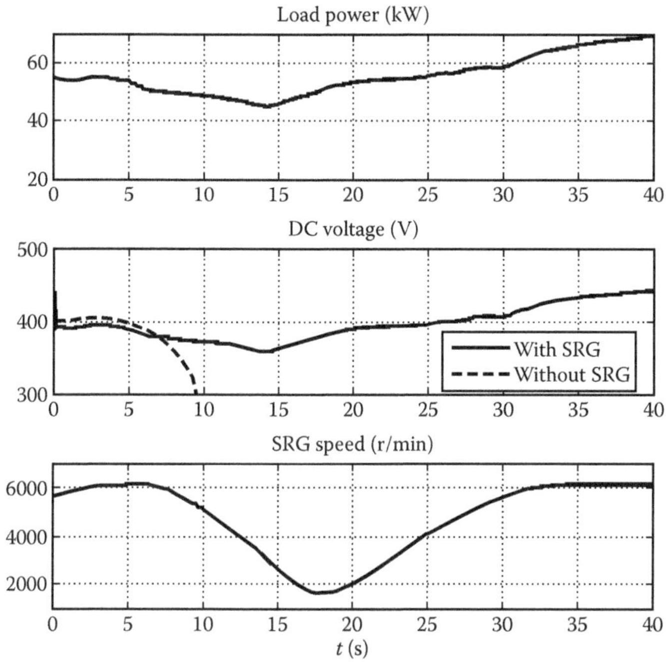

FIGURE 5.57 Processes in the model **SRGFLY** under interruption of the hydrogen supply.

5.4 MICROTURBINE SIMULATION

Microturbines that have a power from some dozens to some hundreds of kilowatts are relatively new technology for electric energy generation. They have better efficiency in comparison with the diesel generator sets. The microturbine consists of the radial compressor for the air compression that, together with the fuel, comes into the combustion chamber; the gas of the high pressure and the high temperature that is fabricated in so doing rotates the turbine, which, in turn, rotates the compressor and the generator. Moreover, there is the heat exchanger recuperator. The heat of the exhausting gas can be used for heating water or air in absorption chillers [8] or for additional heating of the air in the compressor, increasing, in this way, the set efficiency. There are the single-shaft and the split-shaft microturbines. In the former, the turbine, the compressor, and the generator, which are mounted on the same shaft, have very large rotation speed—up to 100,000 r/min. Therefore, the generator has a very high output frequency and requires additional units of the industrial electronics. In the latter, the generator is driven via a gearbox, so that the generator can have the standard frequency of the output voltage and dispense with additional devices for connection with the grid. The former becomes widespread today.

The PMSG with one pole pair is used in the modern types of the microturbines, so that the output voltage frequency can reach 1500–1600 Hz. The main components of the single-shaft microturbine are displayed in Figure 5.58.

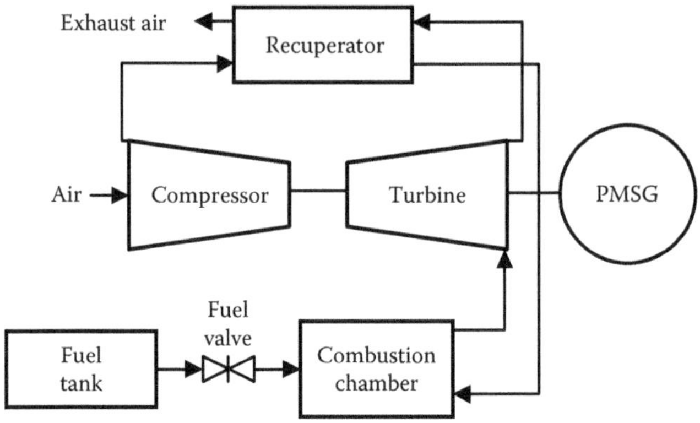

FIGURE 5.58 Main components of the single-shaft microturbine.

The structure of the electronic devices and the control system has to ensure not only that the system functions under normal operation conditions, but also the specific modes of starting and stopping. Before fuel introduction and ignition, the turbine must be speeded up to a preset rotating speed. The external source of the electric energy is necessary; as such, the grid or the battery can be used; the generator in the motor mode can be utilized for speeding up. Therefore, the electronic devices at the generator output have to fulfill bidirectional power transfer; so the electronic unit consists of two back-to-back connected VSIs as in the model **PMSMFLY**.

One distinguishes *cooldown*, or normal, and *warmdown* stop sequences. In the former, the fuel supply is first cut off; the generator begins to decelerate, producing some amount of electric energy. When the thermal energy stored in the recuperator and the other elements of the system decreases to such a low level that the compressor stops operation, the generator passes into the motor mode, supporting the airflow through the system for its cooling, until its temperature reaches the preset level. At this point, the turbine may be quickly stopped by simply shorting the generator terminals.

Under the warmdown stop, the kinetic energy of the rotating parts is absorbed by the brake resistor, as in the dynamic braking mode of the electrical drive. One can acquaint oneself with more details about the microturbine operations through Ref. [8].

Different microturbine control systems, including electronic devices, are utilized. So depending on the external conditions and the demanded generator power, the optimal turbine rotating speed exists; at that, the speed controller can affect the rate of the fuel supply via the fuel valve. In such a structure, the converter connected with the generator (VSI-Ge) keeps the DC reference voltage, and the converter connected with the grid (VSI-Gr) is responsible for the transfer of the electric energy of the given power to the grid and, under isolated operation, the fabrication of the three-phase voltage with the specified voltage and frequency. The monitoring of the exhausting gas temperature is fulfilled; the fuel supply decreases when the temperature exceeds the set limit. It also decreases when the turbine acceleration exceeds the permissible limit, for instance, more than 5%/s.

If the microturbine operates together with a grid, the system structure is possible, when VSI-Gr keeps the DC voltage and VSI-Ge controls the turbine rotating speed. The position of the fuel valve is determined by the external conditions and the demanded power, with the control of the temperature of the exhausting gases.

A number of microturbine models exist, which, with different accuracy, take into account the thermodynamic processes in them [12]. For microturbine modeling in electric power systems, more simple models are of interest, in which the start and stop processes are not considered. The block diagram of one of the most popular models in pu is depicted in Figure 5.59 [12,13].

The speed controller affects the rate of the fuel supply V_c and has the transfer function as

$$F_c = \frac{K(sT_1+1)}{sT_2+a}. \tag{5.19}$$

It means that the control characteristic has some droop, in order to relieve the generator load under the operation with the stiff grid. The rate of the fuel supply may be decreased if the turbine acceleration exceeds the permissible limit (that can have a place, for example, under abrupt load release) and if the temperature of the exhaust gases exceeds the limit. The acceleration limiter is an integrator; the temperature limiter is a PI controller. The least of the three signals determines the value V_c.

The fuel supply system is described by the following relationship:

$$V^* = (V_{c0} + K_1\omega V_c)\frac{1}{sT_3+1}\frac{1}{sT_4+1}, \tag{5.20}$$

where V_{c0} is the rate fuel supply in the no-load condition; the first transfer function of the first order defines the fuel valve lag; the second one defines the actuator lag.

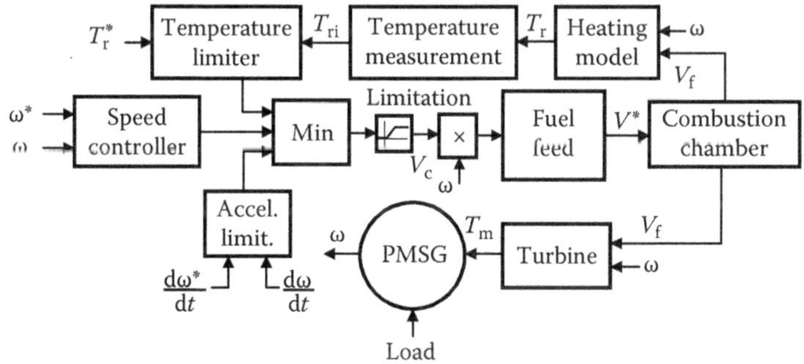

FIGURE 5.59 Block diagram of the microturbine model.

The processes on the combustion chamber are modeled with the delay θ_c. The torque developed by the turbine may be computed as

$$T_m = 1.3\left[\frac{1}{sT_5+1}V_f - 0.23\right] + 0.5(1-\omega), \tag{5.21}$$

where T_5 is determined by the gas turbine dynamic. The rotating speed may be found from the equations

$$H_1\frac{d\omega}{dt} = T_m - F\omega - T_e \tag{5.22}$$

$$H_1 = \frac{J\omega_{nom}^2}{P_{nom}}. \tag{5.23}$$

J is the moment of inertia of the rotating turbine parts; ω_{nom} and P_{nom} are the turbine nominal rotating speed and power (be reminded that the generator has one pole pair), respectively.

The exhausting gas temperature (°C) may be found from the following relationship:

$$T_R = T_R^* - 305(1 - e^{-\theta_e}V_f) + 390(1-\omega), \tag{5.24}$$

where θ_e is the exhausting gas reaction delay. The transfer function of the thermal meter is accepted as

$$F_{izm} = \frac{1}{sT_6+1}\left(K_2 + \frac{K_3}{sT_7+1}\right), \tag{5.25}$$

where T_6 is the time constant of the thermocouple, and K_2, K_3, and T_7 are the parameters of the radiation shield ($K_2 + K_3 = 1$).

The parameters in the following models are taken, mainly, from the relevant references.

The simulation of the systems with microturbine demands a very small sampling period (~1 μs), bearing in mind the high frequency of the output generator voltage, whereas the transients in the turbine pass slowly and last minutes or dozens of minutes, so that the simulation time turns out to be very large. Therefore, it is reasonable to have simplified models, which, under proper duration of simulation, give the opportunity to study and to investigate the system key features.

The model **MTURBO1** does not contain the SimPowerSystems blocks and is intended for the rapid simulation of the basic processes in the turbine. The turbine model output is its torque. The generator is modeled by Equation 5.22. At first, the turbine accelerates to the speed of 0.5 with the generator that works as a motor, when the fuel is ignited; at this point, the turbine begins to develop the torque and continues to accelerate to the working speed; afterward, the turbine operates at a steady speed with different generator loads.

When the speed is lower than 0.5, the turbine makes the load torque equal to 0.05. The generator in motor mode produces the torque equal to 0.3. When the rotating speed reaches 0.5, the turbine begins to fabricate the driving torque, and its rotating speed rises; the generator begins to send the energy into the load or in the grid. At this, the load torque is equal to the quotient of power divided by speed. The delivery power that is initially equal to 0.2, at $t = 70$ s, increases to 0.8; at $t = 150$ s, to 1; at $t = 250$ s, decreases to 0.6. The processes in the model are shown in Figure 5.60.

The part of the electronic circuits is included in the model **MTURBO2**. It is supposed that the DC voltage is constant in the steady and transient states, with the help of the VSI-Ge, so this voltage can be modeled as a DC source (700 V in the considered case). The VSI-Gr output is connected to the load with the active power of 50 kW under the voltage of 380 V and to the grid with the same voltage. The VSI-Gr control system is the same, as in the model **PMSMFLY** (the subsystem **Control**), but the current reference for I_d is not determined by the voltage controller output but is proportional to the wanted power P_{ref} (kW). Then, the power in the DC link is the power taken from the generator (neglecting losses in the rectifier). After the division by the speed, this quantity defines the generator torque, which is the load torque for turbine. The scopes **Scope_Power** and **Scope_Currents** give the opportunity to observe the values of the powers and the currents at the VSI-Gr output, in the load and in the grid. The operation close to the steady-state speed is considered. The wanted turbine power that is initially equal to 45 kW, at $t = 5$ s, increases to 60 kW. The process is fixed in

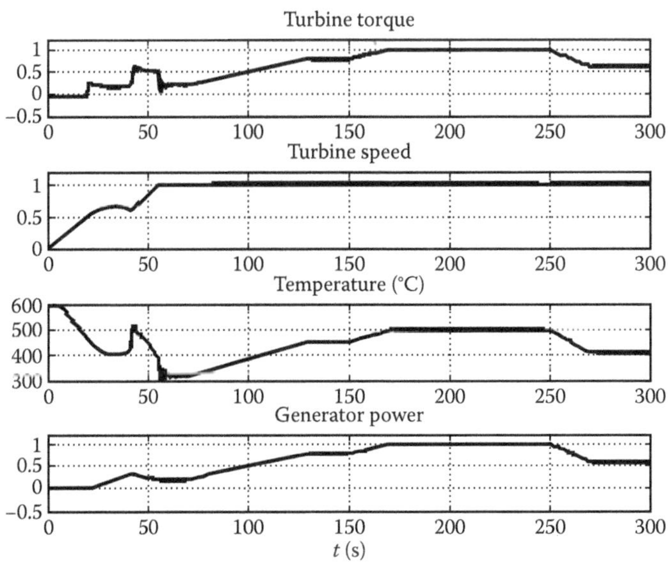

FIGURE 5.60 Processes in the model **MTURBO1**.

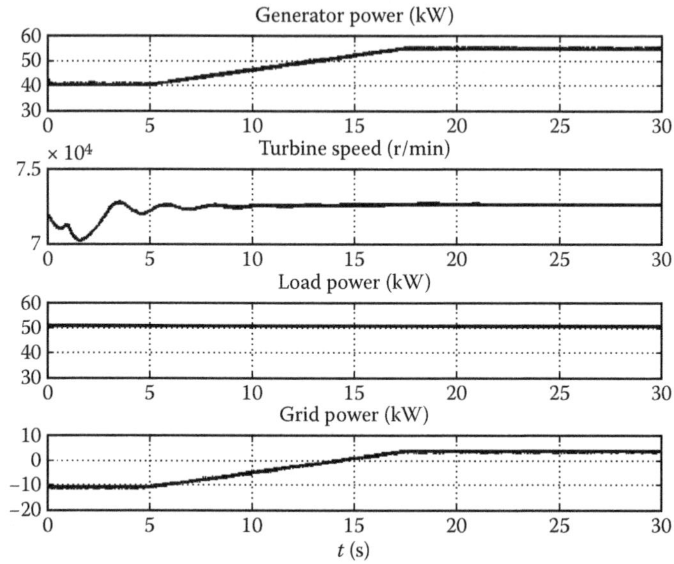

FIGURE 5.61 Processes in the model **MTURBO2**.

Figure 5.61. Firstly, the load is supplied from both the turbine and the grid, but after the turbine power increases, the power surplus of ~4 kW is delivered to the grid.

It should be noted that, with the reference power of 60 kW, taking into account the loss in the turbine that is proportional to the rotating speed, it turns out to be necessary to add to the reference speed the quantity $\delta\omega$ that is proportional to the load torque. One can check that under the power reference of 55 kW, this additional action is not in demand.

MTURBO3 models the electronic part of the system and the generator, but the turbine itself is not modeled. It is supposed that its speed is nominal and does not change. VSI-Gr is the same, as in the previous model; the VSI-Ge control system contains the controllers of the currents I_d and I_q and the capacitor voltage controller that defines the reference for the current I_d controller (as in model **HNPC_5L** considered earlier). The reference voltage across the capacitor is 800 V. The fundamental frequency is 1200 Hz, and the PWM frequency is 20,800 Hz. The generator power changes from 45 to 60 kW at $t = 0.5$ s. The plots of the processes in the system are shown in Figures 5.62 through 5.64. The currents are rather close to the sinusoidal.

In the model **MTURBO4**, the rate of the fuel supply is constant, and the rotating speed of the turbine is controlled with VSI-Ge; the DC voltage is maintained with VSI-Gr. In principle, it is the same structure, as in the model **PMSMFLY**, but in the case under consideration, the problem is complicated with the high-frequency generator. The current controllers I_d and I_q are used; these currents are aligned with the rotor position. The output of the speed controller determines the reference for I_q; the reference for I_d is zero. The quantities $k\omega L_s I_q$ and $-k\omega L_s I_d$ are added to the outputs of the controllers I_d and I_q, respectively, where L is the inductance of the stator winding, and k is the compensation factor.

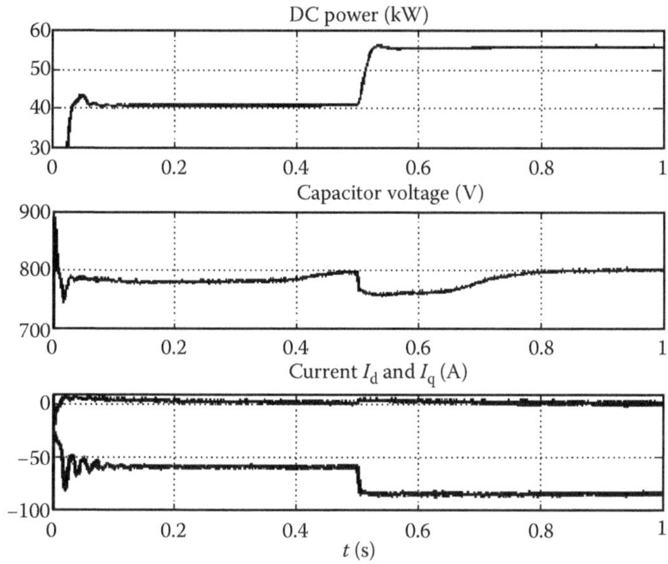

FIGURE 5.62 Processes in the model **MTURBO3**, capacitor voltage, and I_d and I_q currents.

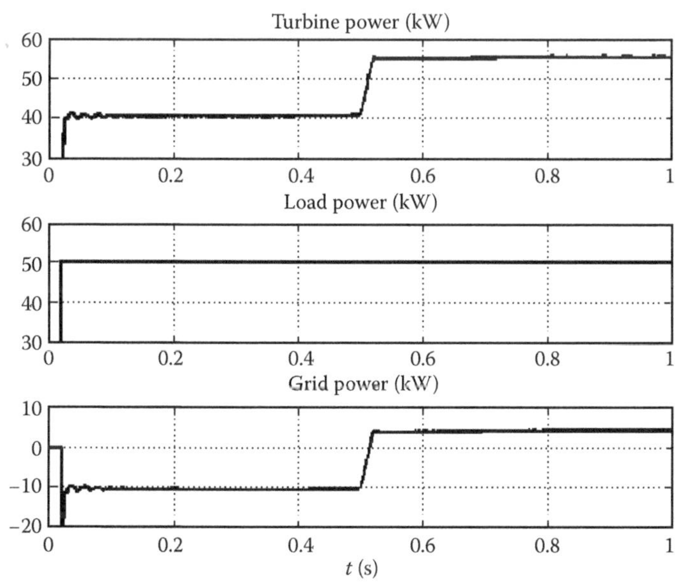

FIGURE 5.63 Processes in the model **MTURBO3**, powers.

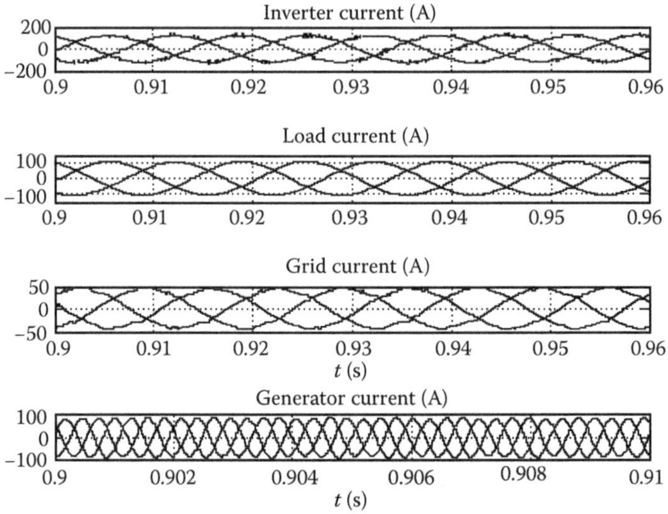

FIGURE 5.64 Processes in the model **MTURBO3**, currents.

It is supposed in the model that the DC voltage is maintained with VSI-Gr with sufficient accuracy. The process of acceleration to the speed 7500 rad/s is simulated; at the speed of 5000 rad/s, the turbine begins to produce the driving torque of 10 Nm that gives the power of 75 kW under the nominal rotating speed. The scope **Scope_Current** fixes the generator current with the small sampling time that gives the opportunity to observe the current waveform. The processes in the system are shown in Figure 5.65.

The considered model is idealized because it is supposed that the generator rotor position and the speed are measurable, which is technically difficult under the large rotating speed. Therefore, the operation without the mechanical transducers is considered, the so-called sensorless control. The block diagram of the circuits for the estimation of the rotor position and its rotating speed are shown in Figure 5.66.

The stator flux linkage components Ψ_α and Ψ_β are computed as

$$\Psi_{\alpha(\beta)} = \frac{T}{sT+1}\left[V_{s_{\alpha(\beta)}} - R_s I_{s_{\alpha(\beta)}}\right], \tag{5.26}$$

where $V_{s\alpha(\beta)}$ and $I_{s\alpha(\beta)}$ are the components of the generator voltage and the current space vectors; the condition $T\omega \gg 1$ is valid in the essential frequency range, so that the frequency characteristics of the lag elements are very close to the integrator frequency characteristics; and R_s is the stator winding resistance. The components of the rotor flux that are created with the permanent magnets are

$$F_{r_{\alpha(\beta)}} = \Psi_{\alpha(\beta)} - L_s I_{s_{\alpha(\beta)}}. \tag{5.27}$$

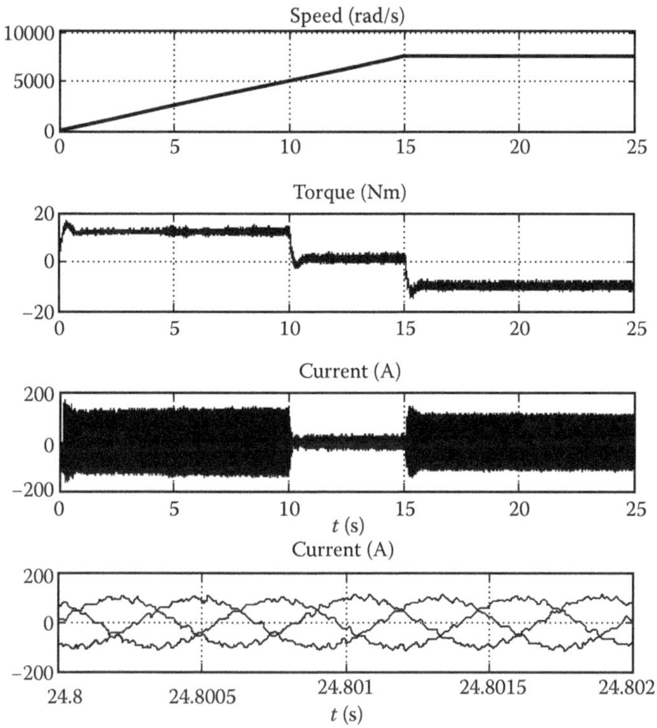

FIGURE 5.65 Processes in the model **MTURBO4**.

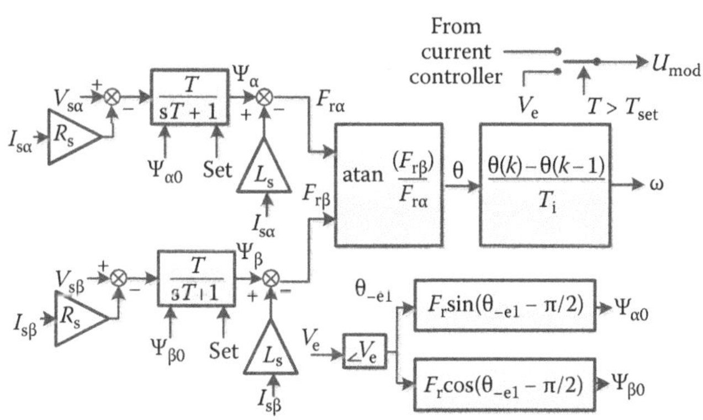

FIGURE 5.66 Block diagram of the circuits for estimation of the rotor position and its rotating speed.

The rotor position may be found as

$$\theta = arctg\left(\frac{F_{r_\beta}}{F_{r_\alpha}}\right) \tag{5.28}$$

and the rotating speed as

$$\omega(k) = \frac{\theta(k) - \theta(k-1) + 2\pi a}{T_i}, \tag{5.29}$$

where T_i is the sampling period; $a = 1$ if $\theta(k)$ and $\theta(k-1)$ are taken from different periods of rotor rotation (from the different teeth of the saw-tooth waveform $\theta = f(t)$), and $a = 0$ for the rest of the cases.

In order for the system to work properly, the initial estimation of the flux linkage has to be done under the start that corresponds to the real rotor position. It is carried out in the following way. During the start, after the DC link voltage has reached the steady state, the modulation voltage V_e of the small frequency, for instance, 1 Hz, and of the corresponding small amplitude are fabricated. The rotating field is created in the stator that leads the rotor; the axis of the rotor magnet flux lags the space vector V_e at about $\pi/2$. In this way, by calculating the position $\theta_{\_el}$ of the space vector V_e, the rotor position may be estimated. When the rotor rotates at a steady speed, the signal *Set* is formed; at that, the quantities $F_r \sin(\theta_{\_el} - \pi/2)$ and $F_r \cos(\theta_{\_el} - \pi/2)$ are set at the outputs of the lag elements; the signal from the control system that corresponds to the normal operating mode comes at the modulator input, instead of the signal V_e.

Such a system is realized in the model **MTURBO5**, in which the same process is simulated as in the previous model. The subsystem **Control** contains the current controllers I_q and I_d that are analogous to the ones in the previous model; they are activated after the command *Set*. The subsystem for the flux estimation **Flux_Est** is made according to Figure 5.66; the developed subsystems for the lag elements give the possibility to set the outputs in the state X_0 by the signal *Reset*. Subsystem **Teta_Est** computes the rotor position estimation by Equation 5.28, taking into account a suggestion about its initial position (see option *Advanced* in the generator dialog box). The rotating speed is calculated in the subsystem **Speed_Est** by Equation 5.29; moreover, two values of the sampling time are used: $T_i = 1$ ms when $\omega < 3000$ rad/s and $T_i = 0.1$ ms when $\omega > 3000$ rad/s.

The subsystem **Start** contains the source of the three-phase modulation signal with the frequency of 1 Hz and with the amplitude of 0.0055 and the circuits for the calculation of the angle $\theta_{\_el}$ and for the shift of this angle by $\pi/2$. At $t < 2.2$ s, the signal of the system **Start** comes to the modulator input; at $t = 2.2$ s, the signal *Set* is fabricated, the initial values of the lag elements are fixed, and the signal from the current control system begins to come to the modulator. The variations of the speed and torque are shown in Figure 5.67.

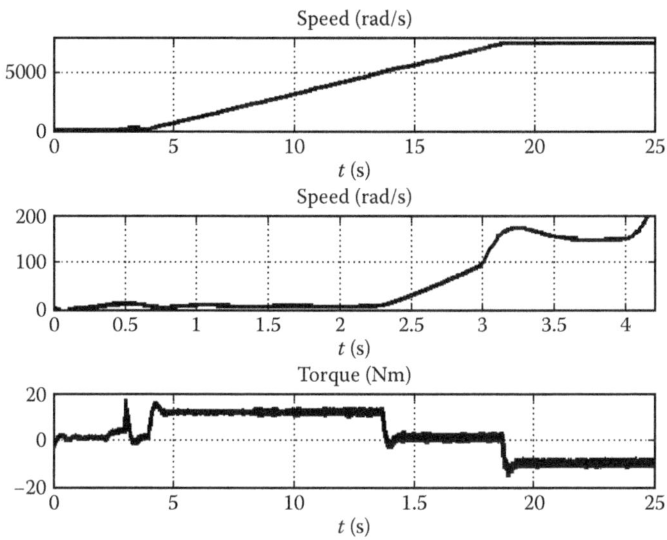

FIGURE 5.67 Variations of the speed and torque in the model **MTURBO5**.

5.5 HYDROELECTRIC POWER SYSTEM SIMULATION

By conversion of the water flow energy into electric energy, 20% of all electric energy is produced in the whole world. One distinguishes between the hydroelectric stations of the large power and the small hydroelectric systems. The former are very big constructions, which to build; the river stream is blocked with a dam that forms a large water reservoir, which is intended for both the increase in the water potential energy and water storage during drought periods. The great area is gone out from the land utilization during the building of these electric stations. The power of such electric stations can reach some gigawatts. The small hydroelectric stations have the power from some kilowatts to some megawatts. When they are designed, one tries to use the natural water level difference; the special reservoirs, when they are built, serve only to increase the pressure of the water column, not to accumulate the water storage; the loss of the ground is minimal.

The simplified diagram of the set of the hydraulic turbine SG is depicted in Figure 5.68. The water stream moves through the water conduit that has length L under the action of the hydraulic pressure that is caused by the difference H of the water level surface and the turbine position (the so-called hydraulic head); this water stream falls on the turbine blades with the velocity U. The amount of the coming water is controlled with the position of the gate that is driven with the servomotor. The SG is coupled with the turbine shaft.

The mechanical turbine power (W) may be found as

$$P_m = \eta \rho a_g Q H, \tag{5.30}$$

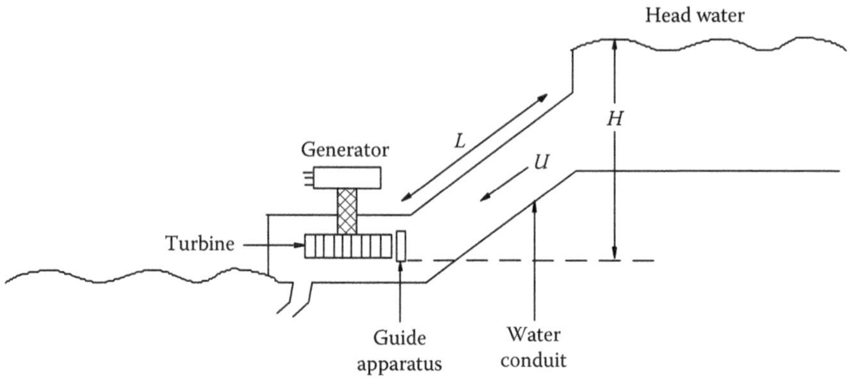

FIGURE 5.68 Simplified diagram of the unit the hydraulic turbine SG.

where Q (m^3/s) is the rate of the water discharge ($Q = AU$), A is the cross-section area of the water conduit, ρ is the water mass density (1000 kg/m^3), a_g is the gravitation acceleration (9.81 m/s^2), and η is the efficiency factor.

The rate of the water discharge is controlled with the gate. It is proved in Ref. [14] that for the small deviations around U_0 and H_0, the transfer function, which connects the power change with the gate relative position change, may be written as

$$\frac{\Delta P_m^*}{\Delta G^*} = c\frac{(1 - bT_w)}{(1 + aT_w)} \tag{5.31}$$

$$\Delta G^* = \Delta G/G_0,$$

where for the ideal turbine, $c = 1$, $b = 1$, and $a = 0.5$;

$$T_w = LU_0/a_g H_0 \tag{5.32}$$

is the time constant that changes with the load change and is in the range 0.5–4 s. For the real turbine, the coefficients a, b, and c can essentially differ from values mentioned earlier. For instance, for the turbine with power of 40 MW, under the nominal load, $c = 1.5$, $a = 0.58$, and $b = 0.45$ [14].

A more complicated nonlinear model is accepted in SimPowerSystems, whose block diagram is shown in Figure 5.69. Its ground is given in Ref. [14].

The real gate opening g^* and the ideal gate opening G^* are related as

$$G^* = A_t g^* \tag{5.33}$$

$$A_t = 1/\left(g_{max}^* - g_{min}^*\right). \tag{5.34}$$

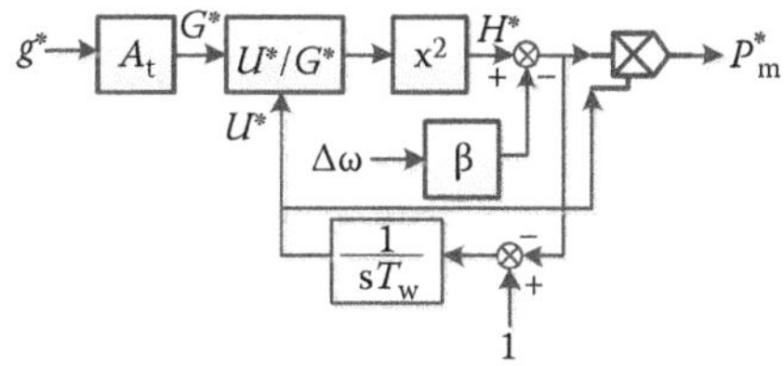

FIGURE 5.69 Hydraulic turbine model in SimPowerSystems.

Here g^*_{min} is the initial gate opening that corresponds to the losses in the no-load conditions, when $G^* = 0$; g^*_{max} is the final gate opening, when $G^* = 1$ [14]. The factor β determines the system damping; its value is about 1–2 pu.

The model **Hydraulic Turbine** depicted in Figure 5.69 is inside the block **Hydraulic Turbine and Governor** in the folder *Machines*; the parameters g_{min}, g_{max}, T_w, and β are specified in the block dialog box.

The turbine has a governor device, whose function is to control the turbine rotating speed by gate opening with the help of the servomotor. It ensures the integral (stiff) relationship speed power. Such dependence is permissible for the isolated block turbinegenerator, but usually, some generators are connected in parallel; in order to have even generator load, the characteristic droop is used; for that, the servomotor position feedback is employed, with the factor R_p. For example, if $R_p = 0.05$, the speed change by 5% leads to the full gate opening change.

The controller block diagram is depicted in Figure 5.70a. The servomotor controls the gate opening by means of a proportional–integral–derivative controller (PID controller) regulator as a function of the difference of the reference ω_{ref} and the accrual turbine rotating speed ω_e. Two options for the turbine characteristic droop are intended, depending on the binary signal d_{ref}: In the top switch position, the gate

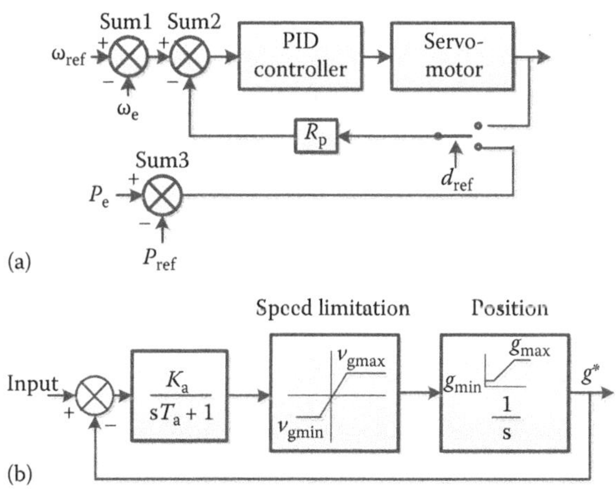

FIGURE 5.70 Control of turbine power: (a) controller block diagram and (b) servomotor block diagram.

position negative feedback is used; in the low position, the droop signal is determined by the difference between the reference P_{ref} and the measured P_e quantities of the generator power.

The block diagram of the servomotor model is depicted in Figure 5.70b. The limitations of the rate of both gate opening and gate boundaries are intended.

A number of examples of the considered model utilization are given in Refs. [1,15]. Three instances demonstrate the employment of the block **Hydraulic Turbine and Governor**.

The model **HydroP1** shows the separate unit turbine–generator supplying the isolated load. The initial load power is 50 MW; at $t = 5$ s, it rises to 75 MW, and at $t = 35$ s, it rises to 100 MW. The SG has the power of 200 MVA under the voltage of 13.8 kV. The SG inertia coefficient takes into account the inertia of the turbine rotating parts too. SG is driven with the hydraulic turbine with $T_w = 2.5$ s; the rest of the parameters of the block **Hydraulic Turbine and Governor** can be seen in the block dialog box. Since the unit operates isolated, $R_p = 0$. SG provides with standard system for voltage control **Excitation**, described earlier (see model **Excitation_Examples**).

The step-up transformer delta/star with power of 210 MVA and voltage of 13.8 kV/230 kV is set at the SG output. The opportunity to use the block **Generic Power System Stabilizer** that is located in the folder *SimPowerSystems/Machines* is intended; this block consists of the low-pass filter of the first order; the gain; the high-pass filter of the filter order; two lead/lag elements, depending on the selected parameters; and the limiter of the output signal. The diagram of the block is depicted in Figure 5.71.

Powergui enables the start of simulation from the steady state that corresponds to the load of 50 MW. For this aim, the option *Machine Initialization* has to be activated. At that, the vector of the SG initial conditions and the constant P_{ref} change. The **Scope** fixes the load current of phase *A*, the SG output voltage, and the excitation voltage; the **Scope_g,Pm,w** shows the turbine output power, the gate opening, and the turbine rotating speed.

In Figure 5.72, the process is shown under load change with operating **Generic Power System Stabilizer**. The process without this block is shown for the speed as well; it is seen that the stabilizer essentially decreases the system oscillations. At the initial instant of the load increasing, the power P_m drops a little, although the water amount rises (*g* increases). It can be explained by the fact that after gate opening, the water stream does not increase instantly, because of water inertia, and the pressure inside the turbine drops [14].

The next model **HydroP2** shows two nearly similar units, supplying the common load of 250 MW. The generators have the powers of 250 and 200 MW, respectively; they connect to the load via the 13.8/230 kV transformers, whose secondary windings are connected in parallel with the help of the transmission line of 100 km in

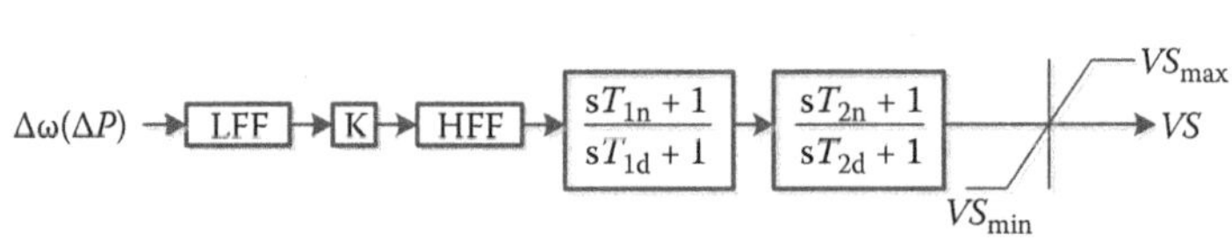

FIGURE 5.71 Block diagram of the block **Generic Power System Stabilizer**.

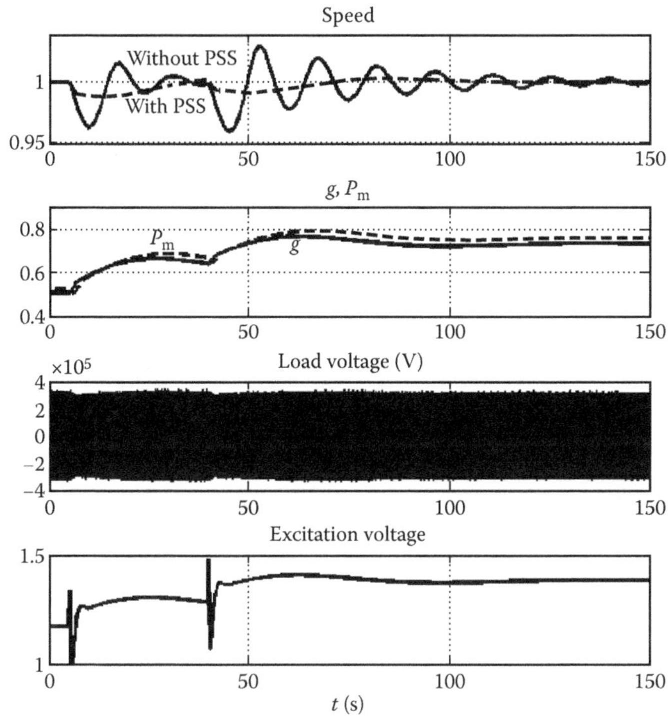

FIGURE 5.72 Processes in the model **HydroP1**.

length. Moreover, direct to the generator terminals, the loads of 5 and 15 MW are connected, respectively; the additional load of 100 MW is connected to the secondary winding of the second transformer at $t = 25$ s.

Because the demanded load power can be obtained with different combinations of the generator powers, some suggestions have to be done. Suppose that before the connection of the additional load, the generators carry the same load: (250 + 5 + 15)/2 = 135 MW, the additional load of 100 MW is powered from the first SG (with power of 250 MW). In the model of the first turbine, $R_p = 0$; that of the second turbine, $R_p = 0.05$; moreover, the droop depends on the power (*Droop Reference* = 0).

If, before the simulation start, one opens the window of the item *Powergui/Machine Initialisation*, it can be seen that for SG 250 MW, *Bus Type* as *Swing Bus* is selected under voltage of 13,800 V; for SG 200 MW, *P &V Generator* with the power of 135 MW. After the execution of the command *Compute and Apply*, the power references and the initial conditions in the model change in a proper way. The results of the simulation with **Generic Power System Stabilizer** and without it are shown in Figures 5.73 and 5.74, respectively. It is seen that, in the first case, the frequency dynamic deviation is about half as much as in the second case, but the load power dynamic deviation can be observed, whereas the power is nearly constant in the second case. In both cases, after rather long transients, the power of the second SG returns to the initial value, and that of the first SG increases by 100 MW.

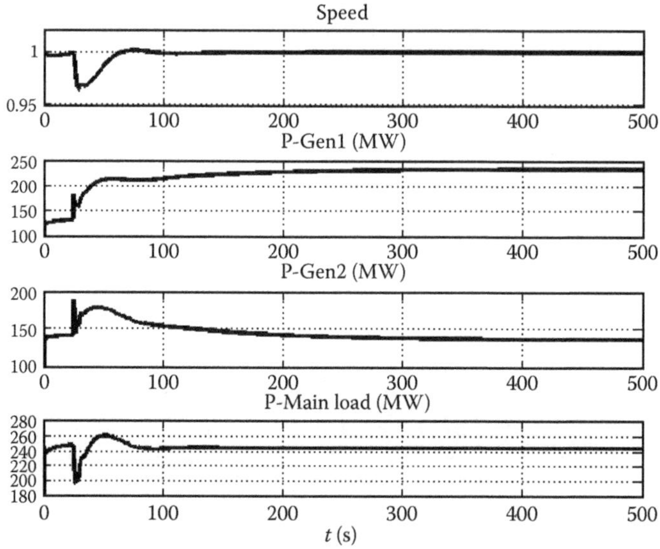

FIGURE 5.73 Processes in the model **HydroP2** with **Generic Power System Stabilizer**.

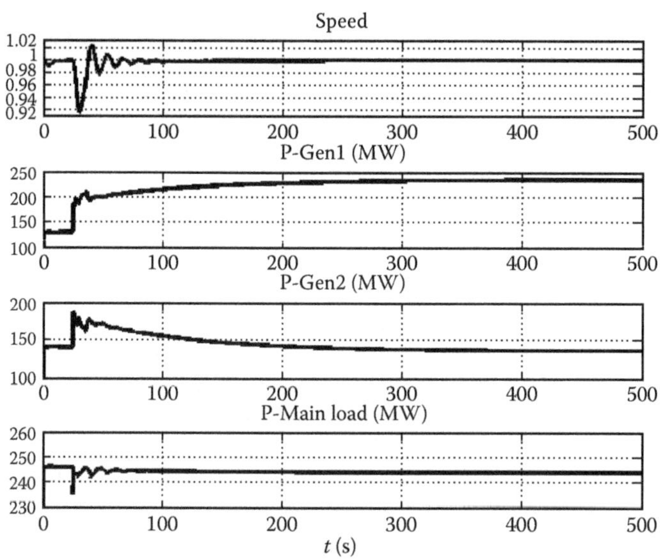

FIGURE 5.74 Processes in the model **HydroP2** without **Generic Power System Stabilizer**.

In the model **HydroP3**, the main load of the previous model is replaced with the source of 230 kV with a short-circuit level of 5 GVA; it means that the parallel operation of two SGs and the power network are simulated. The SG rotating speed is defined now only by the network frequency, and the action to the references of SG rotating speeds results in generating power change. The initial SG power quantities are the same, as in the previous model; at $t = 5$ s, the power of the first SG increases by 30%; for that, the reference of its rotating speed increases by $\Delta\omega = R_p\, \Delta P = 0.05 \times 0.3$. The load 100 MW is connected at $t = 250$ s, so that the power delivered to the network decreases by the same value.

Before simulation start, the option *Powergui/Load Flow* is activated. The mode *P &V Generator* with power of 135 MW is chosen for SGs; the mode *Swing Bus* is set for the source. The processes in the model are displayed in Figure 5.75.

The small hydroelectric sets subdivide into the full sense as small (with the power more than 100 kW), mini (with the power 30–100 kW), and micro (with the power less than 30 kW), but such a classification is not generally accepted. Most of such electric stations are run-of-the-river type. For such electric stations, the normal river stream does not change essentially; the water reservoir is small or is completely absent. The schematic drawing of such a set is depicted in Figure 5.76. The essential part of the water stream comes to the turbine via the canal or the conduit, and then the water returns to the river.

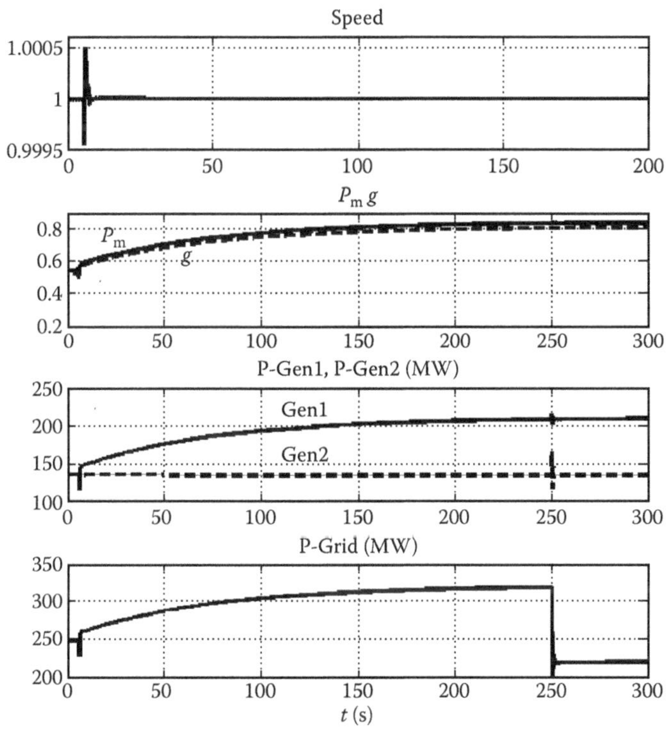

FIGURE 5.75 Processes in the model **HydroP3**.

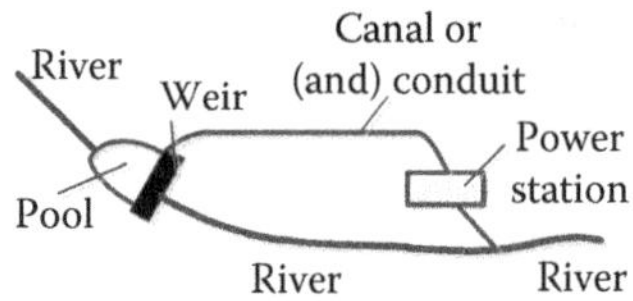

FIGURE 5.76 Design of the run-of-the-river electric station.

The different types of the turbines are used in the small hydroelectric sets, including the propeller turbines and the Francis and Pelton turbines. In the first ones, the water stream comes parallel to the axis, and in the Francis turbines, the water stream comes perpendicular to the axis, in the radial direction. The turbine axes can be vertical or horizontal and have a slope or a variable blade position (Kaplan turbines).

The propeller turbines operate under small H values: 3–35 m and with Q change in the range of 50–115% of the design value. The Francis turbines operate under H values of 12–240 m and with Q change in the range of 40–110% of the design value [16]. Both turbines belong to the reactive-type turbines, in which the water pressure applies directly to the turbine blades. The Pelton turbines belong to the action, or impulse, turbine, in which the nozzles form the water streams with high kinetic energy; these streams hit the blades of the special form, so that the overwhelming part of the kinetic energy is converted into mechanical energy that rotates the turbine impeller. The turbines operate under H values of 60–1500 m and with Q change in the range of 10–115% of the design value.

Power control of the small turbines is carried out, as for the big ones, by means of the gate; for turbines with power less than 100 kW, the dummy load can be used as well. The SGs or the IGs with the squirrel cage rotor are employed in such turbines; when the turbine operates together with the grid, the turbine rotating speed is constant for the former and nearly constant for the latter. The dependence of the turbine torque on its rotating speed, under different water discharge Q, is a straight line, as displayed in Figure 5.77. The torque is maximum under zero speed (it is designated as T_{m0}) and is equal to zero under the runaway speed n_m, so that it may be written as

$$T_m = T_{m0}\left(1 - \frac{n}{n_m}\right). \tag{5.35}$$

Therefore, the dependence of the turbine power on its rotating speed is a parabola with an explicit maximum (Figure 5.77), and the control of the rotating speed, when the water discharge changes, can give the perceptible effect. The dependence of the optimal rotating speed (that gives the maximum efficiency) of the Francis turbine on the water discharge is shown in Figure 5.78; this dependence is computed using the data given in Ref. [17].

One of the first hydraulic turbines that have the power of 10 MW with variable rotating speed is described in Ref. [18]. The IG with wounded rotor that is supplied

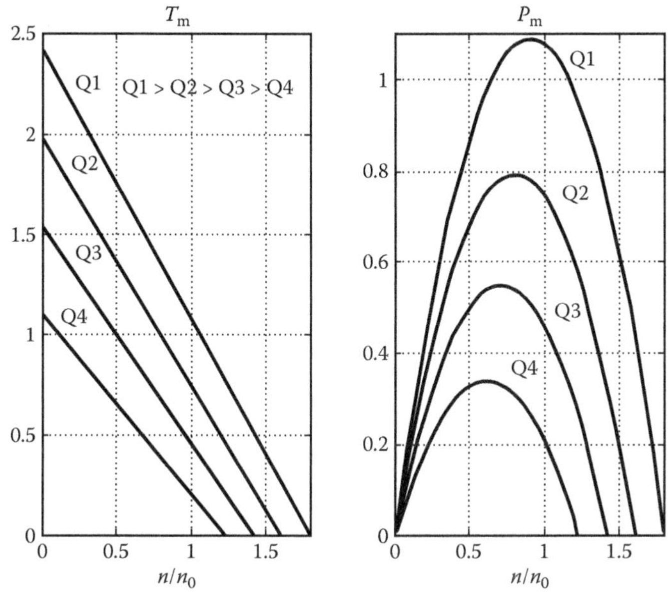

FIGURE 5.77 Dependences of the torque and power on the water discharge.

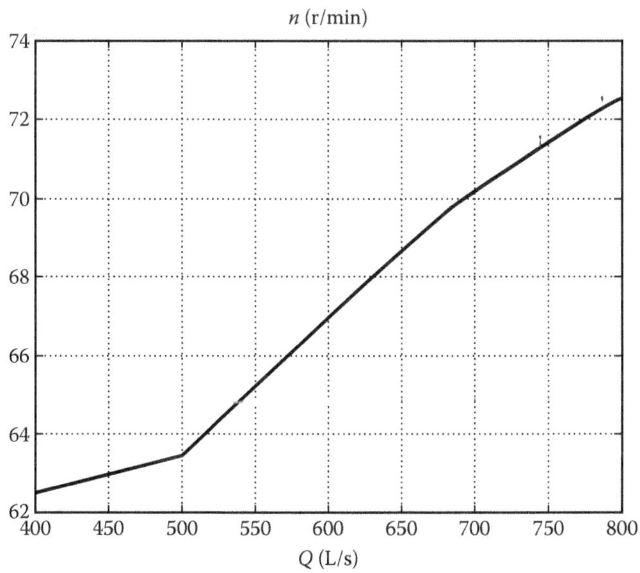

FIGURE 5.78 Dependence of the optimal rotating speed on the water discharge.

from cycloconverters is utilized. The PMSG are in widespread use today; the units that consist of two generators on the same shaft, IG plus PMSG or two IGs with wounded rotor, are also used [19], which give the possibility of dispensing with slip rings in IG. The voltage source inverters are used for connection with the grid or with isolated load in these cases.

If the devices such as the water gate are used for power control of a small turbine, the models of big hydraulic sets considered earlier can be utilized. In the model **HydroM1**, the dummy (ballast) load is employed for this aim. According to Equation 5.35, the dependence of the turbine power on the rotating speed is taken as

$$P_m = 2.25\omega(g - 0.555\omega), \tag{5.36}$$

where g is the actual water discharge. It is accepted that with the fully opened gate (G = 1) g = 1.2, then the nominal water discharge g = 1 is provided with G = 0.833. The nominal SG power is 60 kVA; the useful load power is 51 kW and 0.85 pu. From Equation 5.36, it follows for P_m = 0.85 that under g = 1.2, 1.1, and 1, the rotating speeds are ω = 1.78, 1.54, and 1.26, respectively; therefore, the dummy load is necessary. By accepting $P_m = 0.85 + \gamma P_b$, where P_b is the relative power of the dummy load, and γ is its utilization factor, with g = 1.2, ω = 1, and γ = 0.95, we find

$$P_b = 0.63 \times 60 = 38 \text{ kW}$$

The dummy load diagram is depicted in Figure 5.79. Two thyristors are connected antiparallel in each phase. The instant when the firing pulse comes to the thyristor

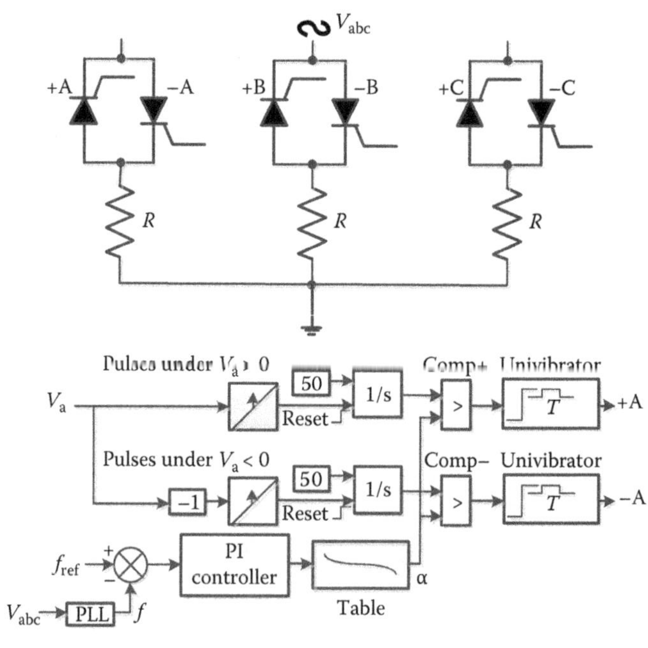

FIGURE 5.79 Dummy load diagram.

delays relative to the phase voltage crossing the zero level for $t_\alpha = \alpha/\omega_0$ and $\omega_0 =$ 314 rad/s. Then the power that is absorbed by the dummy load for half of the supplying voltage period is

$$P_b = \frac{3}{T}\int_{t_\alpha}^{T} \frac{V_{ph-m}^2}{R}\sin^2(\omega_0 t)\,dt = \frac{3V_{ph-m}^2}{2\pi R}(\pi - \alpha + 0.5\sin 2\alpha) = \frac{3V_{ph-m}^2}{2R}\varphi(\alpha), \quad (5.37)$$

where V_{ph-m} is the amplitude of the phase voltage.

The reference and the actual frequencies of the SG output voltage are compared at the input of the PI controller; the controller output defines the demanded dummy power. For the linearization of the control system characteristic, angle α is computed by the relationship that is inverse to $\varphi(\alpha)$; this dependence is given in the option *File/Model Properties/Callbacks/InitFcn* calculated under model start and is realized as the table.

The given angle α is reproduced in the following way. When the phase voltages cross the zero level, the short pulses are fabricated, individually for voltage increasing and decreasing that reset the proper integrator, which, afterward, begins to integrate the constant quantity. When the integrator output equals the value that is proportional to angle α, the firing pulse is fabricated; the thyristor keeps the conducting state, until the phase voltage crosses the zero level the next time.

The SG excitation winding is supplied from the exciter ST2A. Since the simulation time is very big, the system variables are fixed with the large sampling time, but the scope **Scope Voltage, Currents** records the variables during the small time interval with *Decimation* = 1 that gives the opportunity to observe the current and voltage curves in detail.

The simulation begins with $Q = 1$ ($g = 1.2$); beginning from $t = 60$ s, the value Q decreases smoothly to $Q = 0.833$. The processes in the model are shown in Figure 5.80; the steady-state rotating speed is equal to the wanted one.

The model **HydroM2** is a modification of the previous one, in which the SG is replaced with the IG with squirrel cage rotor. The model of such IG with the self-excitation is taken from the model **Diesel_IG**. The program `IM_MODEL_01.m` has to be executed before the simulation start. The process is simulated when, under constant water discharge $G = 0.9$ ($g = 1.09$), the load is equal to 50 kW and halves at $t =$ 50 s. The processes in the model are shown in Figure 5.81. It is seen that increasing the dummy (ballast) load limits the rotating speed and maintains the load voltage.

In the next model **HydroM3**, the IG with power of 160 kVA delivers power to the grid with a relatively small short-circuit power of 1.6 MVA. The standard IG model with the squirrel cage rotor is used. The capacitors are employed to decrease the consumption of the reactive power from the grid. The main problem in this system is carrying out the smooth connection of the IG to the grid. For that, the thyristors are used, according to the diagram in Figure 5.79, where, instead of the resistors, the generator windings are placed (see Figure 6.3a as well). The firing pulses are fabricated with the block **Synchronized 6-Pulse Generator** considered in Chapter 3; the firing angle is determined either with the generator current controller output or

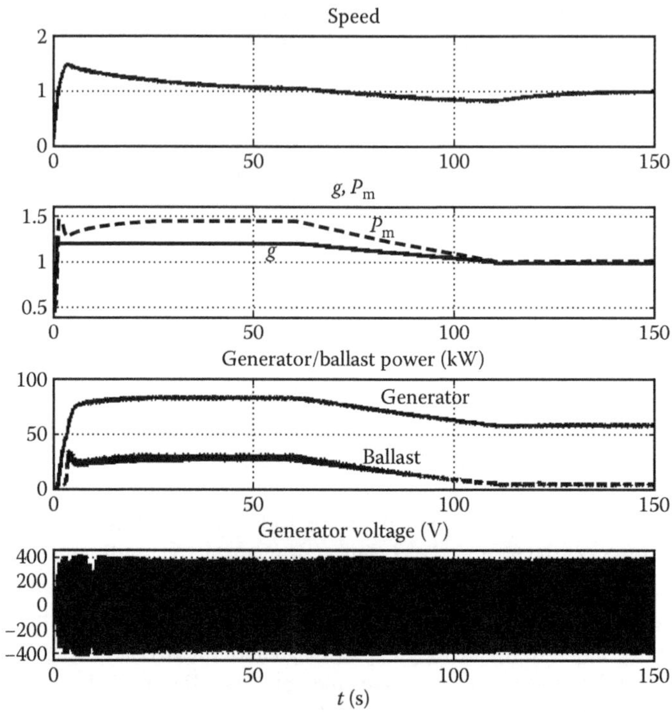

FIGURE 5.80 Processes in the model **HydroM1**.

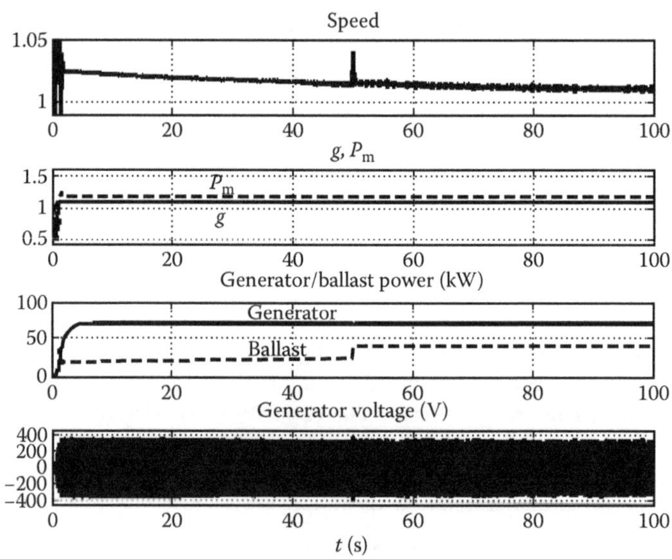

FIGURE 5.81 Processes in the model **HydroM2**.

as a decreasing function of time. After selected start time elapses, the thyristors are shunted with the breaker. The sync voltages are filtered with the filters of the first order with time constants of 5.5 ms; therefore, their connections change, as was described under consideration of the model **Photo_m9** (Figure 5.39). For example, instead of U_{ca}, the filtered voltage $-U_{ab}$ is used and so on.

The process is investigated when the water discharge $Q = 0.9$ ($g = 1.08$), at $t = 15$ s, decreases to $Q = 0.7$, and at $t = 30$ s, to $Q = 0.5$. The processes in the system are shown in Figure 5.82 for the case when the current control is chosen during system start. The turbine rotating speed is nearly constant (it changes from 1.01 to 1 in a steady state); the reactive power, consumed from the grid, is nearly zero.

As mentioned earlier, the turbine rotating speed control under the alternation of the operating conditions may give a notable effect. For the speed control, two VSIs connected back-to-back, as depicted in Figure 5.83, are utilized in the model **HydroM4**. The task of the generator inverter (VSI-Ge) is to control the generator rotating speed; the task of the grid inverter (VSI-Gr) is to transfer the produced electric energy into the grid. The latter is carried out by keeping the voltage across the capacitor C in the DC link constant, when the amount of the fabricated power changes. The circuits of the control system are depicted in Figure 5.83 too. The output of the voltage controller *VR* is the reference for the current component I^*_{sd} that is aligned with the grid voltage space vector, the reference for the current component $I^*_{sq} = 0$. The block θ **Estimation** computes the position θ of the grid voltage space vector U_s. With the utilization of this angle, the quantities I^*_{sd} and I^*_{sq} are

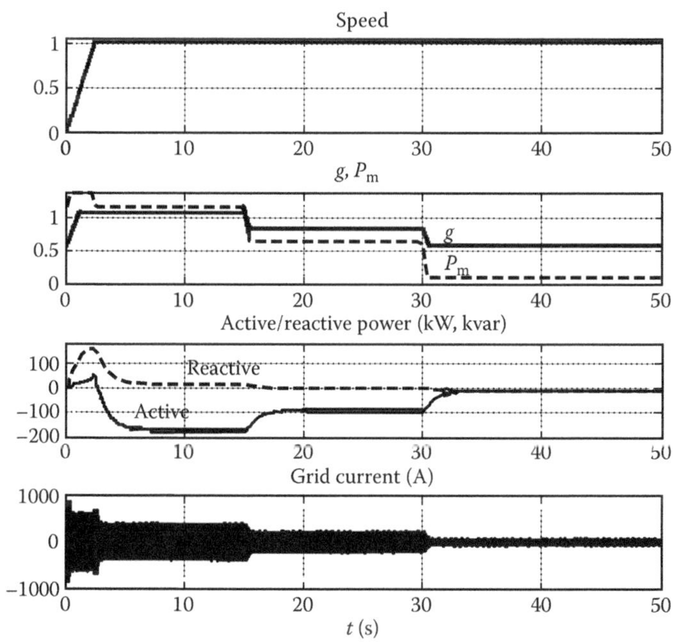

FIGURE 5.82 Processes in the model **HydroM3**.

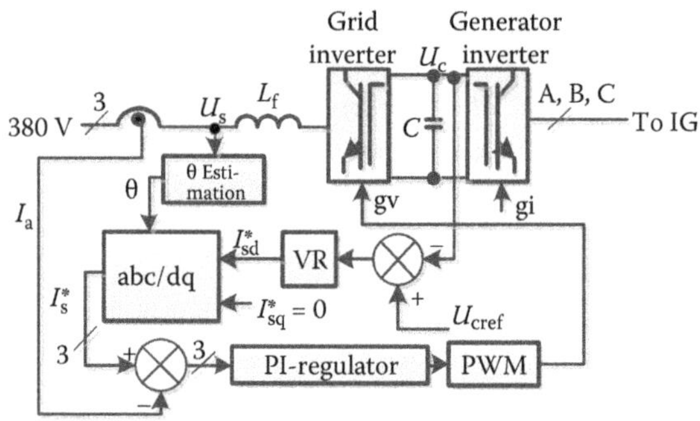

FIGURE 5.83 Diagram of the VSI connection and of the VSI-Gr control system for the model **HydroM4**.

converted into the three-phase quantity I^*_s of the reference for the grid current; this reference, together with the grid current I_s, comes to the input of the three-phase current controller, whose output, with the help of PWM, fabricates the gate pulses gv for VSI_Gr.

The indirect vector control is utilized for the IG speed control that was already used in the model **Matr_Conv2**. Unlike the latter, the phase current controllers are employed in the considered model, not I_d and I_q controllers. The diagram of the control system is depicted in Figure 5.84. The relationships in Equations 3.23 through 3.25 are valid; relationship 3.26 is written as

$$I^*_q = \frac{2}{3}\left(\frac{T^*_e L_r}{Z_p L_m \Psi_r}\right), \tag{5.38}$$

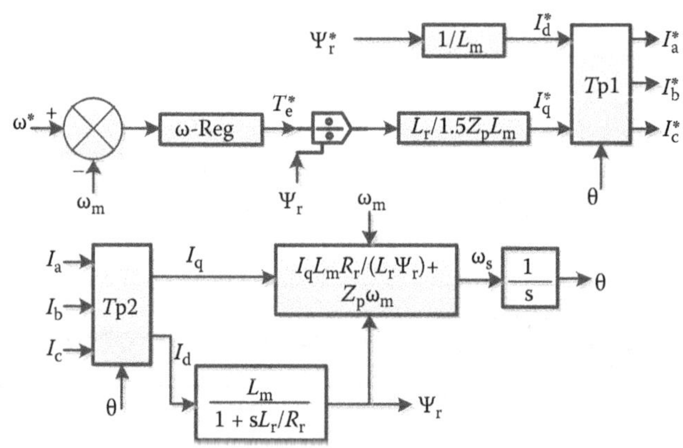

FIGURE 5.84 Diagram of the indirect vector control.

where T_e^* is the output of the speed controller **ω-Reg**, because the IG parameters are specified in SI units; Equation 3.22 is written as

$$\theta = \int \omega_s \, dt = \int (Z_p \omega_m + \Delta\omega_s) \, dt. \tag{5.39}$$

In the transformation block **TP1**, the reference quantities I_d^* and I_q^* are transformed into the phase current reference I_{abc}^* for the three-phase current controller. The parameters of the control system can differ from the generator parameters of the same names (or of the same designations), in order to have the possibility to investigate the influence of the errors in their estimation on the control performance. These parameters are specified in the option *File/Model Properties/Callbacks/InitFcn*. The current controller is of the hysteresis type.

The 37 kVA and 400 V IG is used in the considered model. The R-L circuit at the terminals of the ideal AC source takes into account the smoothing reactor and the grid reactance. It is accepted in the turbine model $G = g$; the block **Timer1** defines this quantity. The block **Timer** assigns the rotation speed fall from 157 to 20 rad/s with a deceleration of 5 rad/s². The plots of the power that delivers to the grid, depending on the turbine rotating speed, under a different water discharge, are shown in Figure 5.85. It is evident that the change of the rotating speed in conformity with the straight line, when the water discharge changes, leads to the increase in the hydraulic unit efficiency. In practice, the speed control may be carried out, depending on the present weather conditions or the season, with the direct measurement of

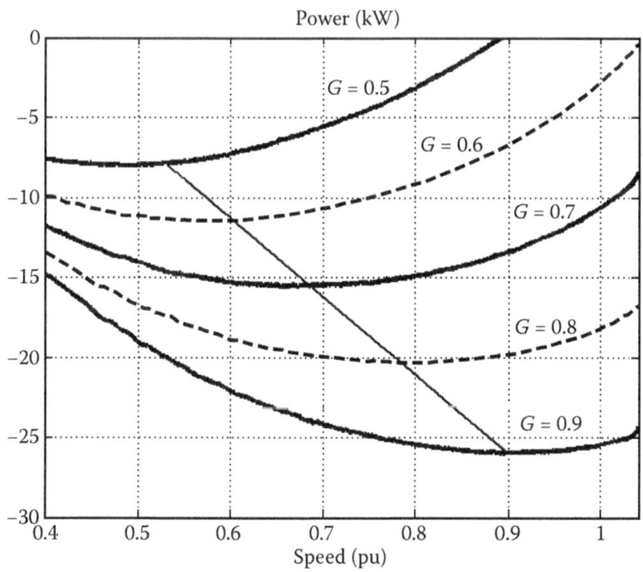

FIGURE 5.85 Delivered to the grid power depending on the turbine rotating speed under different water discharge.

the water discharge by means of one of the methods considered in Ref. [8], with the utilization of the optimum search method, as was made for the model **Photo_m4S**. For the latter, although the necessary speed changes demand additional energy consumption, the change of the water discharge takes place very slowly, so that the search process can be fulfilled very rarely.

The employment of the IG with wounded rotor and with two VSIs in the rotor circuit, as displayed in Figure 5.86, gives an opportunity to decrease the inverter installed power by three to four times. Such systems are widespread in WG units, as will be shown in the next chapter. The principle of vector control, with the utilization of the rotor position transducer, is used in WG. This transducer makes the unit more expensive and lowers the unit reliability, but the power of such a WG is usually more than some hundred kilowatts, so that the relative increase in the cost is low. A number of WG units usually form a WG farm, so that the temporary stop of one WG, owing to transducer failure, slightly affects the system performance; access for service is usually not difficult. But the hydraulic turbine generators can have much less power, are located in a hard-to-reach place, and can be the sole source of the electric energy, so that hydraulic turbine generator failure can have grave consequences. Therefore, the control systems without the position transducer are reasonably employed. In the model **HydroM5**, the direct torque control (DTC) is used, which was described in the Section 4.3.1 under control by the stator.

The VSI-Ge controls the generator rotating speed; the VSI-Gr controls the power flow, maintaining the voltage across the capacitor C. Subsystem **Gen_Contr**, in which DTC is realized, controls the VSI_Ge. The DTC is analogous to the system utilized in the model **PMSMFLY**. The principal difference is the following. In **PMSMFLY**, the stator flux linkage is controlled. Therefore, when it is necessary to increase the torque, the vector Ψ_s rotation has to be speeded up. For this aim, with the vector Ψ_s in the first sector, the state $Vk = 6$ or $Vk = 2$ has to be chosen. In the system considered in the model **HydroM5**, the rotor flux linkage Ψ_r is controlled; the stator flux linkage vector rotates with constant speed; therefore, to increase the torque, the rotation of Ψ_r has to be slowed down; when Ψ_r places in the first sector, the state $Vk = 1$ or $Vk = 5$ has to be selected. Such an algorithm change is achieved by adjusting of the block **Relay** in the subsystem DTC_REG: the output *On* is equal to 0, not 1; the output *Off* = 1.

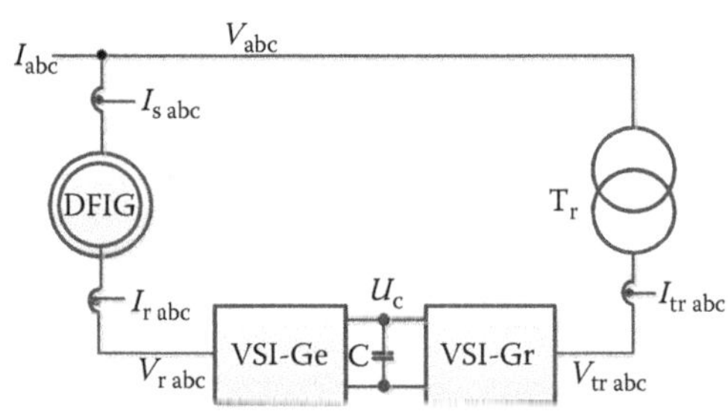

FIGURE 5.86 Diagram of the IG with wounded rotor and two VSIs.

For VSI-Gr control, the subsystem **Contr_Inv** is used, which is the same as the subsystem **Control** in the model **PMSMFLY**. The reference of the capacitor voltage is 1500 V. The currents in the secondary transformer winding are controlled. The generator power is 1600 kVA under a voltage of 1200 V.

At the unit start, the stator must be connected to the grid. In order to smooth the voltage surge and the current inrush during this process, the stator voltage and the grid voltage have to be synchronized beforehand by the amplitude, the frequency, and the phase. The stator voltage control is carried out by action on the rotor current. The diagram of the device is depicted in Figure 5.87. It contains the three-phase generator, whose output signal may be written as

$$I^* = A\sin\left(\theta + \varphi - \frac{2\pi}{3}ik\right). \tag{5.40}$$

$i = 0, 1, 2$; $k = \pm 1$; φ is the phase shift; and I^* is the current reference. The amplitude A is defined by the output of the controller that compares the voltages of the grid and of the generator stator. The quantities

$$|V| = \max\left(|V_a|, |V_b|, |V_c|\right) \tag{5.41}$$

that imitate the full wave rectification of the three-phase voltage are used for comparison. The quantity

$$\theta = \int \pm\omega \, dt, \tag{5.42}$$

where ω is defined by the output of the controller that compares the frequencies of the grid and of the generator voltages. The frequency signals are fabricated with

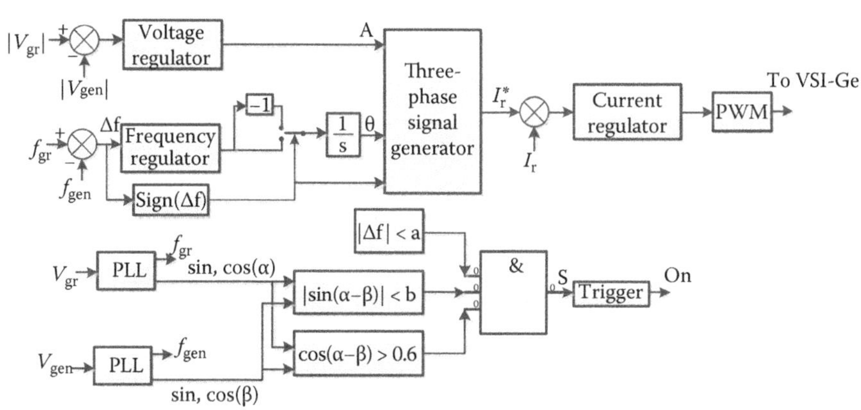

FIGURE 5.87 Diagram of the device for stator and grid voltage synchronization.

PLL. The sign depends on the Sign (Δf); this signal changes the sign of parameter k in expression 5.40; owing to that, the rotor flux rotates in the direction that is opposite to the direction of the generator rotation, and the stator voltage frequency decreases. The operations of both controllers are separated in time.

The synchronization is carried out by the observation of the difference between the angles of the grid voltage space vector α and of the generator voltage space vector β. The quantity

$$\sin(\alpha - \beta) = \sin\alpha\cos\beta - \cos\alpha\sin\beta$$

is used for this aim. Since $\sin(\alpha - \beta) = 0$ not only under $\alpha = \beta$ but also under $\alpha - \beta = \pi$, the quantity $\cos(\alpha - \beta)$ is calculated, which is equal to 1, when $\alpha = \beta$.

When, during the synchronization, the frequency difference decreases to the small value a, and $\sin(\alpha - \beta)$ to the small value b, under the condition $\cos(\alpha - \beta) > 0$,

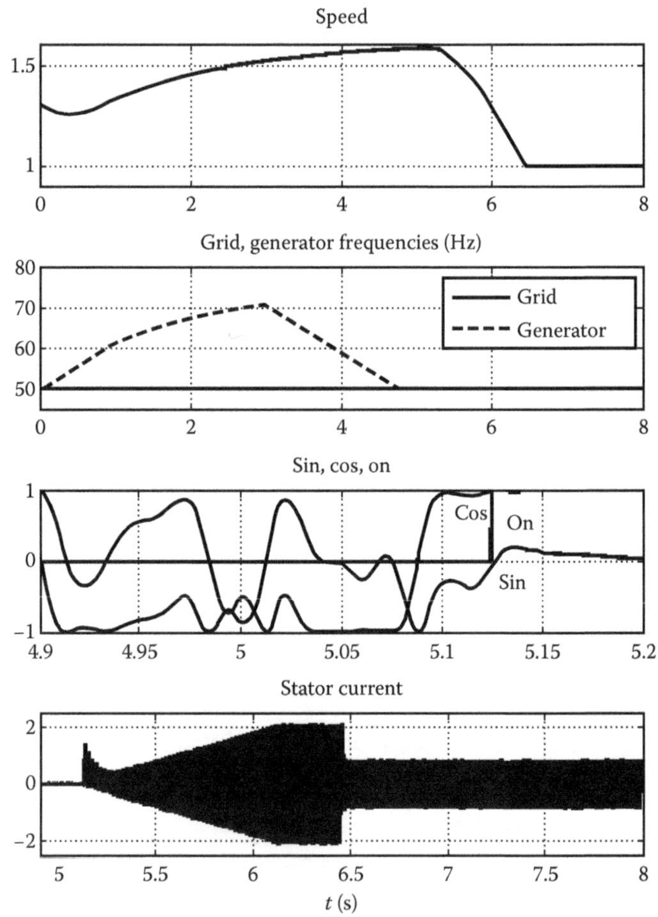

FIGURE 5.88 Process of the synchronization under water discharge $Q = 0.9$.

the turning-on signal *On* is fabricated that turns on the breaker **Br** and connects the stator windings to the grid. Simultaneously, the gate pulses for IGBT begin to come from the DTC system.

The process of the synchronization under a large water discharge ($Q = 0.9$), when, before synchronization, the turbine rotates at a speed close to the maximum runaway, is shown in Figure 5.88; the same process under $Q = 0.7$ is shown in Figure 5.89. In order to make the process of synchronization easier, it is desirable to have the possibility to limit the water flow in the turbine. Really, in order to obtain the stator nominal voltage (for the generator in the considered model), the rotor current amplitude in pu must be $I_r = 1/L_m = 1/2.416 = 0.41$. Under $Q = 1$, the runaway speed is 1.8; it means that the rotor current space vector must rotate for synchronization in the direction that is opposite to the generator direction of the rotation with the frequency of 40 Hz. The rotor voltage $V_r = (40/50) \times I_r(L_m + L_{1r}) = 0.8 \times 0.41 \times (2.416 + 0.0955) = 0.82 = 800$ V, because the voltage base value is $1200 \times 1.41/1.73 = 978$ B. So the DC

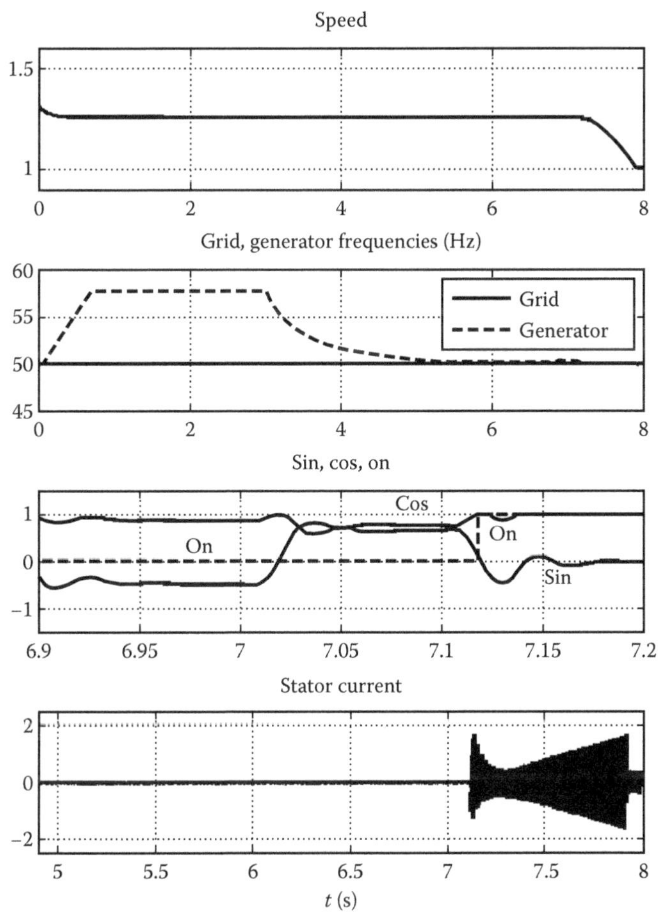

FIGURE 5.89 Process of the synchronization under water discharge $Q = 0.7$.

VSI voltage must be more than 978/0.612 ≈ 1600 V, which is more than the value accepted in the model. After turning on the breaker, the speed control system sets the nominal rotating speed.

In the model **HydroM6**, for its simplification, the circuits of synchronization are excluded. The process is simulated when the generator rotational speed decreases with the deceleration of 0.05 pu/s, beginning from 1.3, under different values of water discharge Q; the factor g is taken as 1.3. The values of the power delivered to the grid versus the referred speed are displayed in Figure 5.90.

The operation of the IG with wounded rotor and the isolated load is investigated in the model **HydroM7**. The generator power is 160 kBA, and the load power is 120 kW. In order to limit the rotating speed under a large water discharge, the dummy (ballast) load is used with the power of 40 kW, taking from the model **HydroM1**. The control system of the VSI connected to the load is the same as the control system of VSI-Gr in the previous model; the control system of the VSI-Ge is the simplified version of the synchronization control system in the model **HydroM5**: the three-phase signal is fabricated that defines the amplitude and the frequency of the rotor current. The amplitude of the space vector of this signal is defined by the output of the load voltage controller; the angle is computed by Equation 5.42, where ω is defined by the output of the controller, whose input is the difference of the wanted and actual frequencies of the load voltage. The process is recorded in Figure 5.91, when the initial water discharge equal to 0.95, at $t = 7$ s, increases to 1.1 with the rate of 0.5 1/c, and at $t = 13$ s, decreases to 1. It is seen that the load voltage and frequency are maintained with sufficient accuracy; under a large water discharge, the rotating speed is limited with the aid of the dummy load. The rotor current in the time, when the rotating speed crosses the synchronous speed, is shown in Figure 5.92.

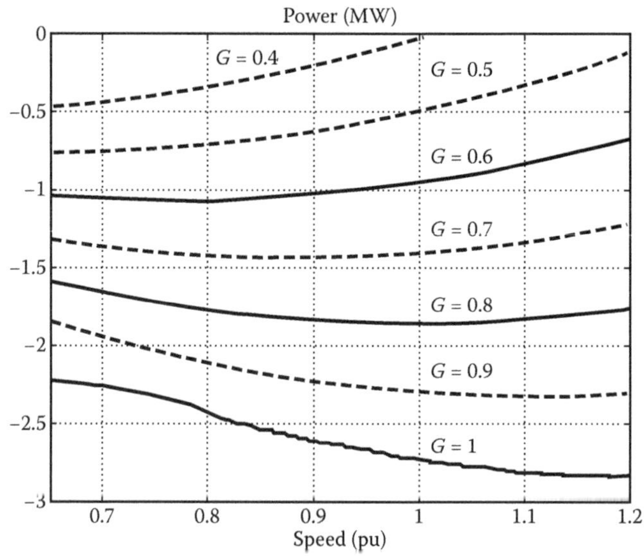

FIGURE 5.90 Values of the power delivered to the grid versus the referred speed.

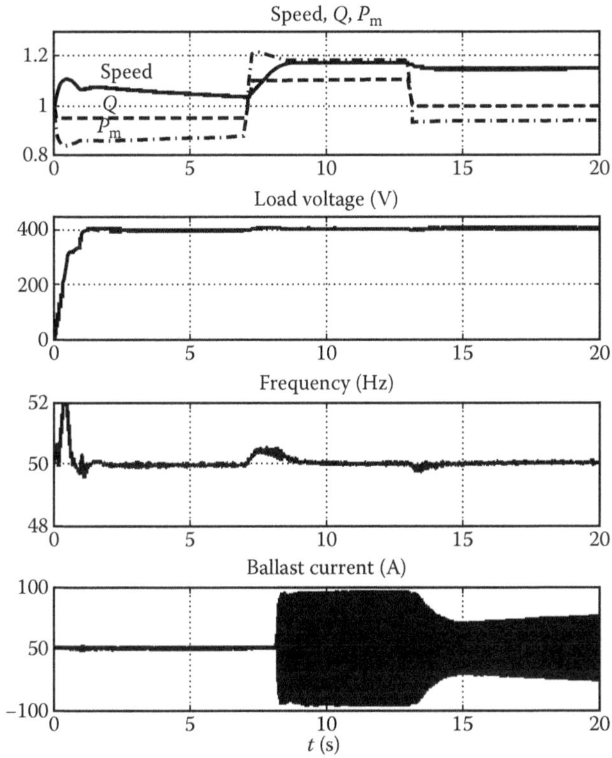

FIGURE 5.91 Processes in the model **HydroM7**.

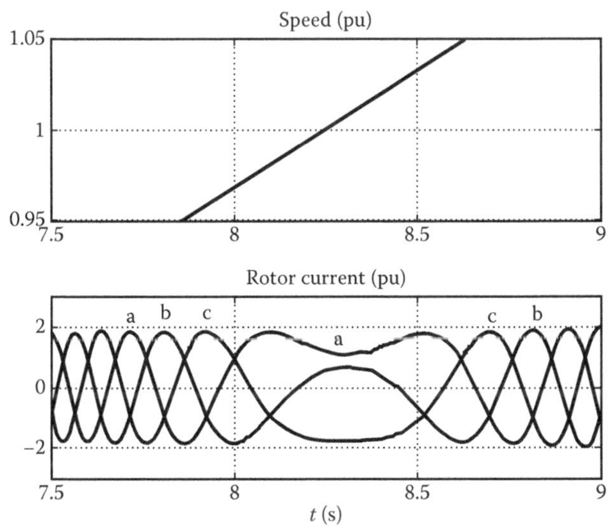

FIGURE 5.92 Rotor current in the neighborhood of the synchronous speed.

REFERENCES

1. MathWorks. *SimPowerSystems™: User's Guide*. Natick, MA, 2004–2016.
2. Mohan, N., Undeland, T. M., and Robbins, W. P. *Power Electronics*. John Wiley & Sons, Inc, NY, 2003.
3. Inoue, S., and Akagi, H. A bidirectional isolated DC–DC converter as a core circuit of the next-generation medium-voltage power conversion system. *IEEE Transactions on Power Electronics*, 22(2), 2007, March, 535–542.
4. Femia, N., Petrone, G., Spagnuolu, G., and Vitteli, M. *Power Electronics and Control Techniques for Maximum Energy Harvesting*. CRC Press, Boca Raton, FL, 2012.
5. Jiang, Y., Abu Qahouq, J. A., and Batarseh, I. Improved solar PV cell Matlab simulation model and comparison. *Proceedings of 2010 IEEE International Symposium on Circuits and Systems, (ISCAS)*. 2770–2773.
6. Jiang, Y., Abu Qahouq, J. A., and Orabi, M. Matlab/PSpice hybrid simulation modeling of solar PV cell/module. *Applied Power Electronics Conference and Exposition (APEC)*. 201, March, 1244–1250.
7. Larminie, J., and Dicks, A. *Fuel Cell Systems Explained*. John Wiley & Sons Ltd, West Sussex, 2003.
8. Chakraborty, S., Simões, M. G., and Kramer W. E. *Power Electronics for Renewable and Distributed Energy Systems*. Springer-Verlag, London, 2013.
9. Gou, B., Na, W. K., and Diong, B. *Fuel Cells, Modeling, Control and Applications*. CRC Press, Boca Raton, FL, 2010.
10. Zhu, Y., and Tomsovic, K. Development of models for analyzing the load-following performance of microturbines and fuel cells. *Electric Power Systems Research*, 62, 2002, 1–11.
11. Uzunoglu, M., and Alam, M. S. Dynamic modeling, design and simulation of a combined PEM fuel cell and ultracapacitor system for stand-alone residential applications. *IEEE Transaction on Energy Conversion*, 21(3), 2006, 767–775.
12. Yee, S. K., Milanovic, J. V., and Hughes, F. M. Overview and comparative analysis of gas turbine models for system stability studies. *IEEE Transactions on Power Systems*, 23(1), 2008, February, 108–118.
13. Gaokar, D. N., and Patel, R. N. Modeling and simulation of micro turbine based distributed generation system. *IEEE Power India Conference*, 2006, April, 256–260.
14. Kundur, P. *Power System Stability and Control*. McGraw-Hill, New York, 1994.
15. Perelmuter, V. M. *Electrotechnical Systems, Simulation with Simulink and SimPowerSystems*. CRC Press, Boca Raton, FL, 2013.
16. Institute of Electrical and Electronics Engineers (IEEE). IEEE Std.1020: Guide for control of small (100 kVA to 5 MVA) hydroelectric power plants. IEEE, Piscataway, NJ, 2011.
17. Michailov, L. P. (ed.). *Small Hydraulic Power engineering*. Energoatomizdat, Moscow, 1989.
18. Merino, J. M., and Lopez, A. ABB varspeed generator boosts efficiency and operating flexibility of hydropower plant. *ABB Review*, N3, 1996, 33–38.
19. Kato, S., Hoshi, N., and Oguchi, K. Small-scale hydropower. *IEEE Industry Applications Magazine*, 9(4), 2003, August, 32–38.

6 Wind Generator System Simulation

6.1 FUNDAMENTALS

Today, the fabrication of electric energy with the help of the wind force is getting more and more widespread. In 2014, the global power of installed WG sets in the world was 370 GW [1]; about 4% of all electric energy this year had been produced by WGs [2]. According to the forecast [1], the global WG power will reach about 666 GW in 2019.

The power of the modern WG reaches 5 MW; WGs with power up to 10 MW are underway. But the WGs of much less power (dozens of kW) are utilized very extensively that are intended for supplying to remote and isolated installations.

The WG unit consists of a tower with a nacelle above (Figure 6.1), in which the main equipment is placed. The output step-up transformer is installed usually within the tower, at its base. The WG can have a horizontal or vertical axis of rotation; only the former is utilized at present. The electric or the hydraulic drive is used, in order to rotate the nacelle in the wind direction. The arrangement of the main WG equipment in the nacelle is schematically displayed in Figure 6.2.

The rotor can have two to four blades, but the modern WG has three blades; in WGs with large power, their length can reach 60 m; the tower height can be 70–80 m. The blades have a rather complicated design, because the WGs with big and moderate power are equipped with devices for rotating the blades about their long axes, with the aim to change the angle of attack β and, in this way, to change the power extracted from the wind and to limit the WG power under the strong wind, exceeding the accepted value.

The rotor rotating speed is not more than 15–30 r/min, whereas the standard electric generators have the rated rotating speed of 1.5–2 orders greater. Therefore, the speed-up gearbox with the big gear ratio is set between the rotor and the generator, at triple stage often. Such gearboxes are bulky, have a heavy weight, and need regular maintenance. Therefore, WGs without the gearbox, with direct connection of the rotor and the generator, are developed. The special multipole generators, synchronous with the wounded rotor or with the permanent magnets, are used for this aim. Because such generators have a large diameter, the compromised design can be utilized, at which the generator has more poles than in the usual design, but their number is not sufficient for gearless connection, so that the single-stage gearbox can be used.

IG with the squirrel cage (SCIG), IG the wounded rotor (WRIG), and SGs, with the excitation winding and with the permanent magnets (PMSG) are employed in WG sets. They are connected to the power system with the help of many interfacing devices. In the simplest case, for the SCIG, it can be the step-up transformer and the soft starter. Such WGs are the units with a fixed rotating speed that are defined by the

FIGURE 6.1 Nacelle appearance.

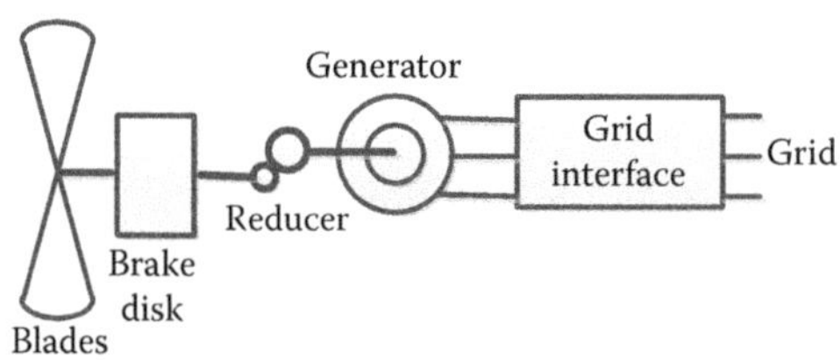

FIGURE 6.2 Arrangement of the main WG equipment in the nacelle.

grid frequency, because the slip changes not more than 1–1.5% under power alteration. As will be seen further, in order to harvest more wind energy, the WG rotating speed must change with variation of the wind speed that is not provided in this WG; nevertheless, owing to their simplicity and less cost, they are very popular, particularly in the units that had been installed in the end of the last century. In some cases, for harvesting more wind energy, the SCIG with the changing number of stator poles or the WRIG with the resistors in the rotor circuits are utilized.

But in up-to-date WGs, for optimizing the mode of operation, the semiconductor converters are employed. Two converters are usually used: the first is connected to the generator, and the second, to the grid, with a coupling capacitor between them. The function of the first converter is to control, in an optimal way, the generator rotating speed; the function of the second one is to transfer the fabricated energy to the power system with the control of the reactive power. The main configurations of the WG are schematically shown in Figure 6.3.

WGs with moderate and large power located not far from one another are united in so-called farms, or parks, which consist of some or dozens of WGs connected in parallel and connected with the power system at one point (PCC), so that the total WG power in the farm can reach hundreds of megawatts. A WG farm in the state of Texas, United States, consists of 627 WGs with a total power of 782 MW and occupies the area of 405 km^2 [3]. The power system very often has an infinitely large power in comparison with the individual WG in the farm, so that the operation of the electric part of every WG may be considered independent; but this does not apply to the power produced by the WG, because of the mutual influence of the WG rotors

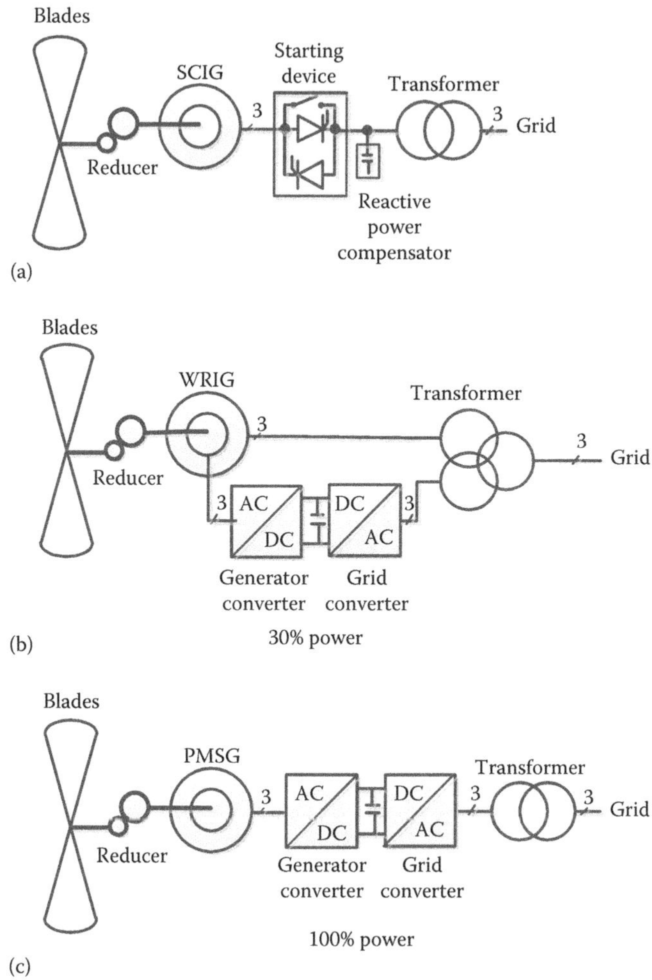

FIGURE 6.3 Main configurations of the WG. (a) WG with SCIG; (b) WG with WRIG; and (c) WG with PMSG.

via airflow (so-called Wake effect) [4]. Even the distance between individual WGs is taken to be big enough (some rotor diameters); the wind speed for WGs in the first rows of the farm that directly face the main wind direction can be perceptibly more than for the WG in the last row [4].

When WGs are connected to the grid of relatively little power, both the WG mutual influence and their common influence on the grid in PCC appear under load and wind speed change [4].

Since the WG farms demand a large area and, besides, exercise negative influence on the environment, offshore wind farms have been developed and built. There is a sufficient space offshore and the wind speed is most of the time higher. The delivery of the produced electric energy to the shore is carried out with an underwater cable; in this case, the DC power transmission is seen as preferable. At present, the London offshore wind park is the greatest in the world. It consists of 175 WGs; each of them has a power of 3.6 MW, so that the total WG park power is 630 MW. The park takes the area of 245 km^2 and is located at an average distance of 20 km from the coast. Despite the advantages of this technique, the investment costs for offshore wind power are much higher than for onshore installations.

Although most WGs operate in parallel to the power network, WG is rather often employed for supplying to the isolated load (island mode). Unlike the former, when the output voltage and frequency are determined by the power network, for the latter, these parameters have to be defined by the WG control system. At this, the load power has to conform to the power produced by WG. Since the latter is defined, mainly, by the wind speed, it is necessary, under its change, in the way that depends on the load type, to change the load power: to switch over the load parts, to change the load and the speed of the connected motors, and so on. Since the pointed measures change the load power by the steps, but the WG power can change continuously and exceed the load power, measures have to be taken, to limit the WG rotating speed. If the WG is equipped with a device for blade position change, it begins to operate, limiting the WG power. Such a device is usually absent in a low power WG; in such a case, the dummy (ballast) load must be used.

Since it is necessary to power at least the part of the load under a weak wind, together with WG, other sources of electric energy are utilized: photovoltaic, chemical (battery, FCs), electrical (ultracapacitors), electromechanical (flywheels), and diesel generators, which are united in the common system and are controlled in a proper way.

The wind turbine power may be written as

$$P_m = 0.5\rho A V_w^3 C_p(\lambda,\beta) = K_p V_w^3 C_p(\lambda,\beta), \tag{6.1}$$

where ρ is the air density that is equal to the average of 1.2 kg/m^3, A is the sweep area of the turbine rotor blades and equal to πR^2, R is the blade length, V_w is the wind speed, C_p is the power coefficient that depends on the blade angle position β and on the tip speed ratio λ, $\lambda = R\omega_r/V_w$, and ω_r is the rotor rotating speed ($2 < \lambda < 13$ usually).

The dependence C_p on λ has a maximum under λ_m; thus, to harvest maximum wind power, the WG rotation speed has to vary with the wind speed. The value $C_{pmax} = 0.593$ theoretically and reaches the values 0.3–0.5 in the modern WG. The different analytical expressions for $C_p(\lambda)$ exist. In SimPowerSystems, the following relationship is accepted:

$$C_p(\lambda) = c_1\left(\frac{c_2}{z} - c_3\beta - c_4\right)e^{-c_5/z} + c_6\lambda \tag{6.2}$$

$$\frac{1}{z} = \frac{1}{\lambda + 0.08\beta} - \frac{0.035}{1+\beta^3}, \tag{6.3}$$

where $c_1 = 0.5176$, $c_2 = 116$, $c_3 = 0.4$, $c_4 = 5$, $c_5 = 21$, and $c_6 = 0.0068$.

The dependencies C_p on λ calculated by Equations 6.2 and 6.3 under various β values are given in Figure 6.4.

The value $C_{pmax} = 0.48$ under $\lambda_m = 8.1$.

The other formulas for C_p can be found in publications. For example, in Ref. [4], the following expression is used, which is obtained with the approximation of the manufacturer's data:

$$C_p(\lambda) = c_1\left(\frac{c_2}{z} - c_3\beta - c_4 - c_5\beta^{2.14}\right)e^{-c_6/z} \tag{6.4}$$

$$\frac{1}{z} = \frac{1}{\lambda - 0.02\beta} - \frac{0.003}{1+\beta^3}, \tag{6.5}$$

where $c_1 = 0.73$, $c_2 = 151$, $c_3 = 0.58$, $c_4 = 13.2$, $c_5 = 0.002$, and $c_6 = 18.4$.

The C_p curve is shown in Figure 6.5. The value $C_{pmax} = 0.44$ and corresponds to $\lambda_m = 7$.

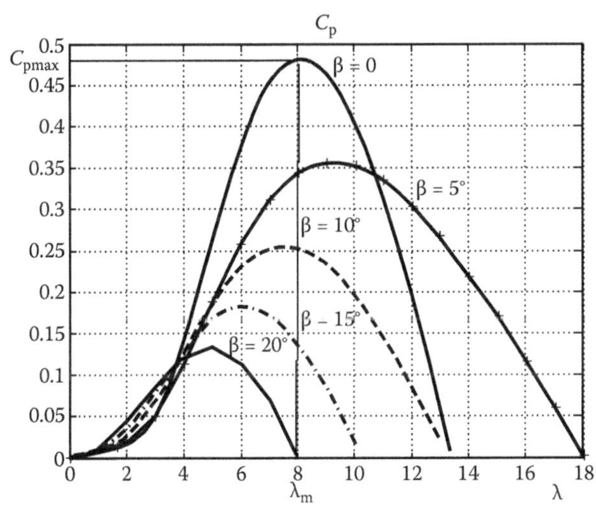

FIGURE 6.4 Dependencies C_p on λ and β.

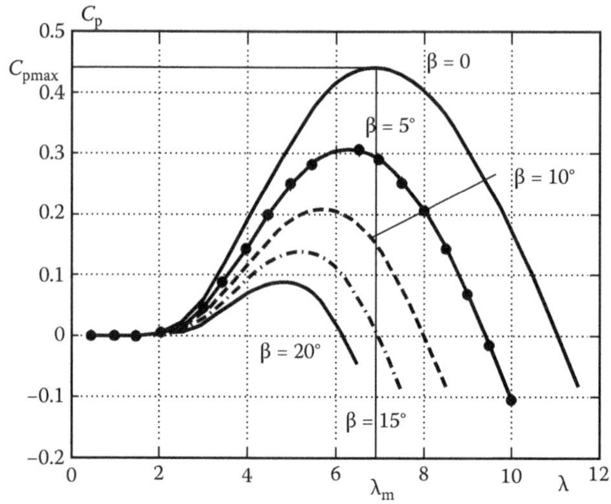

FIGURE 6.5 Dependencies C_p on λ and β for Equation 6.4.

The turbine torque is defined by the following expression:

$$T_m = K_p R V_w^2 C_p(\lambda, \beta)/\lambda = K_m V_w^2 C_m(\lambda, \beta), \tag{6.6}$$

where $K_m = K_p R$.

The plot $C_m(\lambda, \beta) = C_p(\lambda, \beta)/\lambda$ for C_p by Equation 6.2 is depicted in Figure 6.6.

The dependence of WG power on the wind speed and the turbine rotation speed N_r (r/min) for WG with the power of 3.6 MW is displayed in Figure 6.7. The curve of

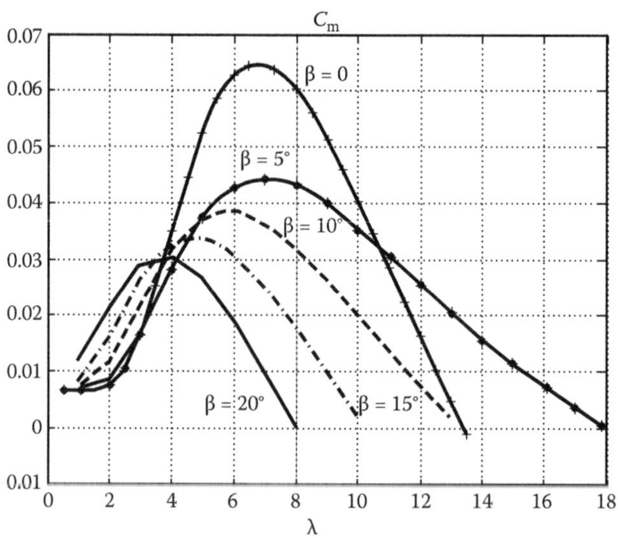

FIGURE 6.6 Dependencies C_m on λ and β for Equation 6.6.

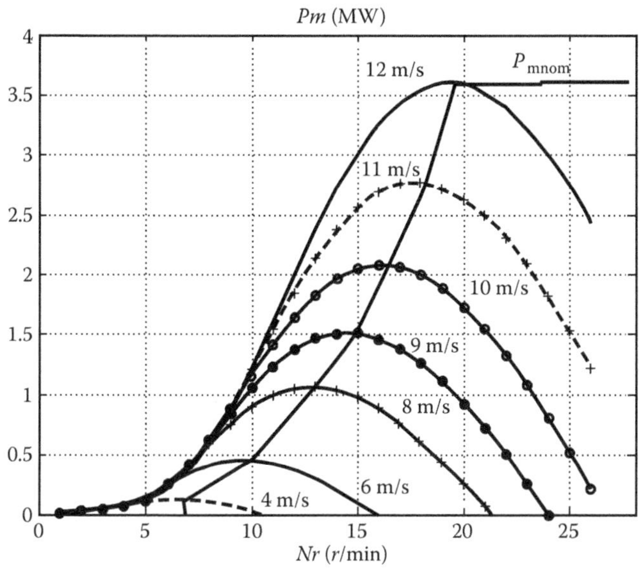

FIGURE 6.7 Dependence of WG power on the wind speed and the turbine rotation speed for WG (3.6 MW).

the maximum power connects the power curve maximums under every wind speed. It is preferable that the WG power conforms to this curve. The curve of the maximum power is limited from above with the nominal power and from the bottom by the vertical under the low wind speed, when the WG is not in operation. Sometimes, in order to have the smooth transition from the optimal power mode to the constant power mode, the short segment is formed between them (for instance, on the length of 0.9–1 of the nominal power), when the WG rotates with the constant speed, until, with the wind speed increasing, the nominal power is be reached.

When the wind speed increases, the maintenance of the nominal power is obtained with the control of the blade position (pitch control). When the wind speed reaches some critical value (cut-out speed; 20–25 m/s), the WG stops, turns away from the wind, and is braked. When the wind speed drops by 3–5 m/s from the cut-out speed, the WG begins to work again.

There are different ways to adhere to the maximum power characteristic (so-called maximum power point tracking—MPPT), including the following:

1. To control the WG rotating speed so that it corresponds to the value λ_m with the present wind speed
2. To load WG with the power P_{mmax} under present wind speed
3. To set such WG power (or torque) that is equal to the maximum possible WG power under present WG rotating speed

$$P_{mmax} = \left(K_p C_{pmax} R^3/\lambda_m^3\right)\omega_r^3 \tag{6.7}$$

Considering Figure 6.7, let WG rotate, for instance, with the speed of 16 r/min under the wind speed of 11 m/s. Under this wind speed and with the present WG rotating speed, the WG power is 2.7 MW; at this, the optimal rotational speed and power are 18 r/min and 2.8 MW, respectively. Therefore, $K_p C_{pmax} = 2.8/11^3 = 2.1 \times 10^{-3}$. Since $\lambda_m = 8.1$, the equivalent R value is $R_e = 8.1 \times 11/18 = 4.95$. Then it follows from Equation 6.7 that the value P_{mmax} that corresponds to this rotating speed is $2.1 \times 10^{-3} \times (4.95 \times 16/8.1)^3 = 1.96$ MW. Therefore, the WG will accelerate until it reaches the speed that corresponds to the maximum power under this wind speed, which is equal to 18 r/min.

Now let WG rotate with the same speed of 16 r/min, but the wind speed is 8 m/s; the maximum WG power is 1.07 MW. The WG will decelerate, until it reaches the speed of 13 r/min that corresponds to the maximum power under this wind speed.

4. The search methods, unlike those considered earlier, do not demand to know the turbine characteristics, but for the correct determination of the gradient $dP/d\omega_r$, the wind speed must be constant during calculation, which is difficult to guarantee.

The model **Wind Turbine** is included in SimPowerSystems; their inputs are the wind speed (m/s), the WG rotating speed in pu, the angle of the blade deviation from the optimal position (pitch, grad), and the output is the torque in pu, referring to the base quantities of the driven generator (Figure 6.8). The moment of inertia is not taken into account and has to be added to the generator moment of inertia. The wind turbine dialog box is shown in Figure 6.9. It specifies consecutively the turbine nominal power, W (P_{mnom}); the nominal generator power, VA (P_{gnom}); the wind base speed V_{wb} (m/s) (as such the mean value speed of the expected wind speed can be taken or the prevailing wind speed in this region); the maximum power under V_{wb} in pu (P^*_{mb}); and the WG rotating speed, at which the maximum power for V_{wb} is obtained, in pu relative to the generator base speed (ω_{gm}). These parameters are used in the following way. Because

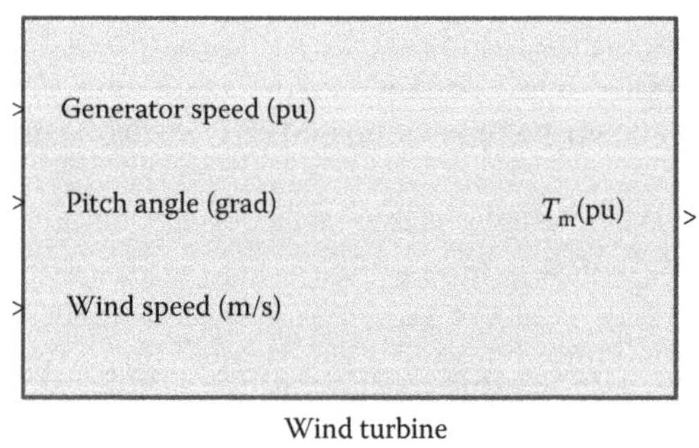

FIGURE 6.8 Picture of the WG model.

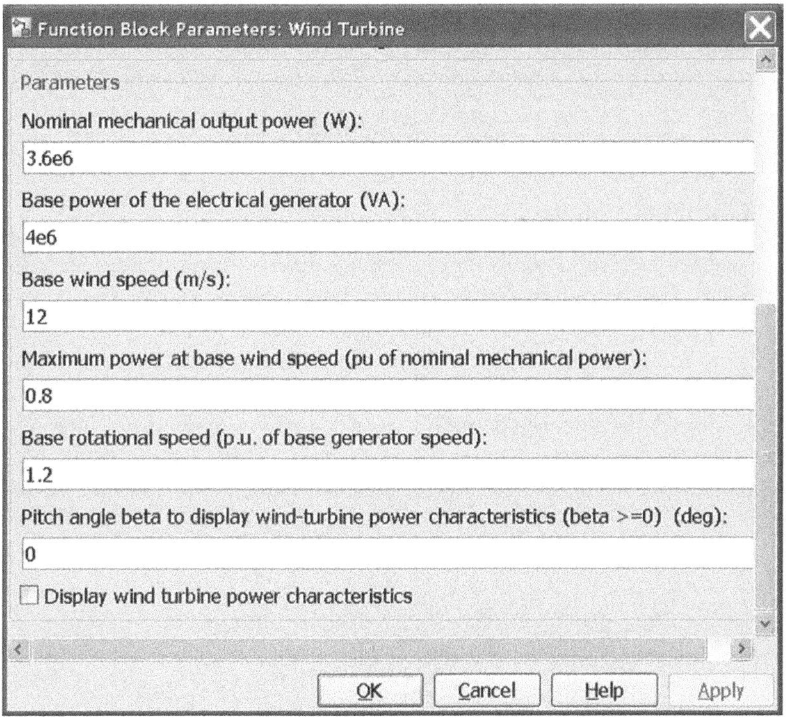

FIGURE 6.9 Dialog box of the wind turbine model.

$$P_{\mathrm{mb}} = P^*_{\mathrm{mb}} P_{\mathrm{mnom}} = K_{\mathrm{p}} V^3_{\mathrm{wb}} C_{\mathrm{pmax}}, \tag{6.8}$$

the relative turbine power P^*_m for some wind speed V_{w} may be written as

$$P^*_{\mathrm{m}} = P^*_{\mathrm{mb}} C_{\mathrm{p}} V^{*3}_{\mathrm{wb}} / C_{\mathrm{pmax}}, \tag{6.9}$$

where $V^*_{\mathrm{wb}} = V_{\mathrm{w}} / V_{\mathrm{wb}}$.

The quantity λ in the formula for C_{p} is

$$\lambda = \frac{\lambda_{\mathrm{m}} \omega^*_{\mathrm{r}}}{V^*_{\mathrm{wb}}} \tag{6.10}$$

$$\omega^*_{\mathrm{r}} = \frac{\omega^*_{\mathrm{g}}}{\omega_{\mathrm{gm}}}, \omega^*_{\mathrm{g}} = \omega_{\mathrm{r}} / \omega_{\mathrm{gnom}}.$$

The output quantity is the mechanical torque in pu of the nominal generator torque:

$$T_g^* = -P_m^* /(P_g^* \omega_g^*), \tag{6.11}$$

where $P_g^* = P_{gnom} / P_{mnom}$.

The diagram of calculations is shown in Figure 6.10. The characteristics of the turbine, whose data are given in Figure 6.9, are displayed in Figure 6.11. It is seen that the sign of the torque is negative, which is suitable when the IG is employed. When SG is utilized, the power, not the torque, comes to its input *Pm*; the quantity $T_g^*(-\omega_g^*)$ has to be sent at this input. When PMSG is used, the torque in SI unit has to come to its input $Tm{:}T_g^* P_{gnom}/\omega_{gnom}$.

The wind turbine model is defined in SimPowerSystems with its static characteristics; the processes caused, for example, by elasticity are not taken into account. It can be made according to the diagram depicted in Figure 6.12. Here, the turbine

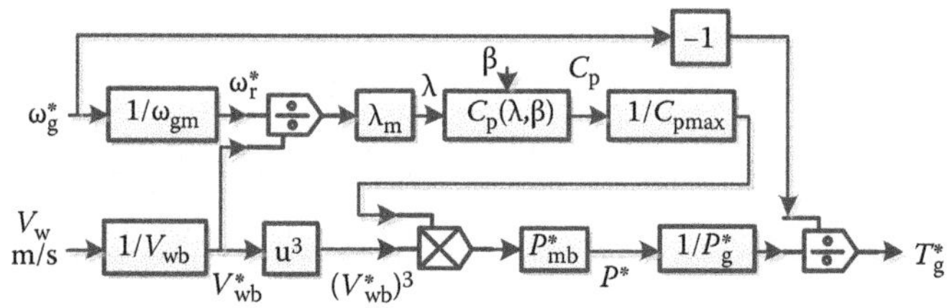

FIGURE 6.10 Block diagram of the turbine model.

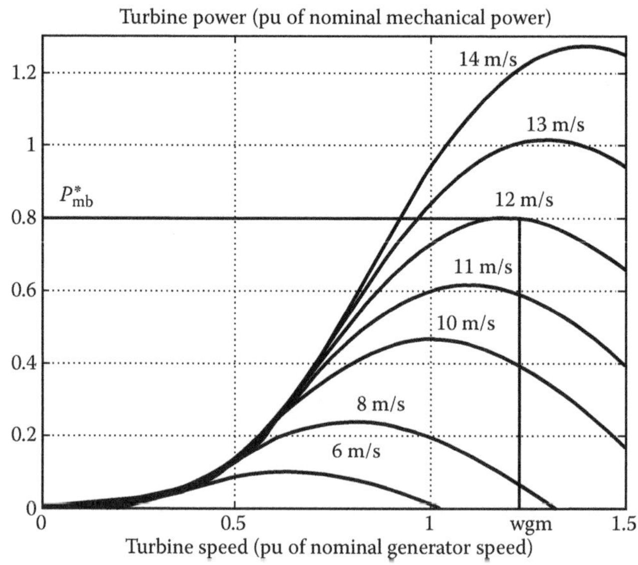

FIGURE 6.11 Characteristics of the turbine Figure 6.9.

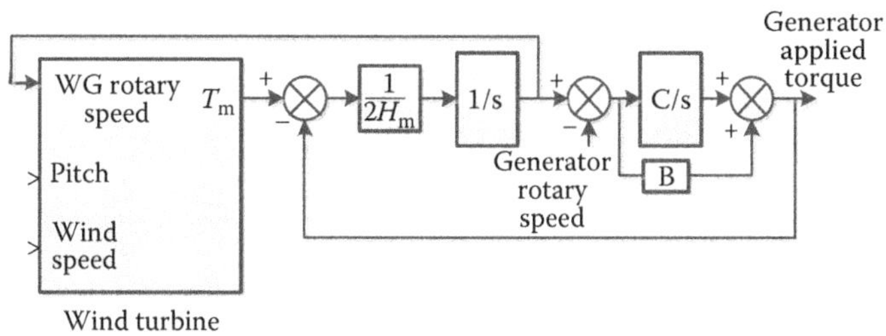

FIGURE 6.12 Scheme for taking the elasticity into account.

inertia constant H_m is considered separately from the generator inertia, C is the stiffness factor of the turbine shaft, and B is the damping coefficient. These data are obtained under turbine design. For instance, it is indicated in Ref. [4] that it may be taken as $H_m = 2–6$ s and $C = 0.3–0.6$ pu of torque/rad. In one of the demonstration models in SimPowerSystems, for the turbine with power of 2 MW with two-pole SG, it is accepted that $H_m = 4.32$ s, $C = 1.6$ kN/rad, and $B = 21$ N/rad/s. These circuits may be realized with the help of the blocks **Reduction Device** and **Mechanical Shaft** described in Chapter 4.

In some cases, it can be reasonable to investigate the influence of the tower, because, when the position of the blades coincides with the tower vertical, the wind force changes (the so-called tower shadow). This phenomenon causes the power oscillations with the amplitude of some percentage and with the frequency of $1.5 \times \omega_r/\pi$ Hz (for a three-blade turbine) [4].

6.2 WG WITH SQUIRREL CAGE IG

6.2.1 WG with Constant Rotation Speed

The WGs with SCIG that is connected directly to the grid are the simplest type of WGs and are widely utilized. Under load change, the rotating speed changes in the range of the slip, which is not more than 1–1.5% of the nominal speed. These WGs contain the devices for decreasing the reactive power consumed from the grid and the starting unit that limits the stator current during the connection to the grid. SCIG with power of 1.45 MW (the apparent power is 1.72 MVA) under the voltage of 575 V is utilized in the model **Wind_IG_1**; the SCIG parameters are taken, mainly, from Ref. [3]; its nominal slip is $s_n = -0.0072$. The SCIG is connected to the grid via the 6.3 kV/575 V step-up transformer. The turbine characteristics are displayed in Figure 6.13 under $\beta = 0$.

Two stages of the reactive power compensation with the help of the capacitors are intended in the model. The SCIG reactive power at the slip s is

$$Q = \frac{U_{1eff}^2}{X_s'} \frac{\sigma s_k^2 + s^2}{s_k^2 + s^2}, \tag{6.12}$$

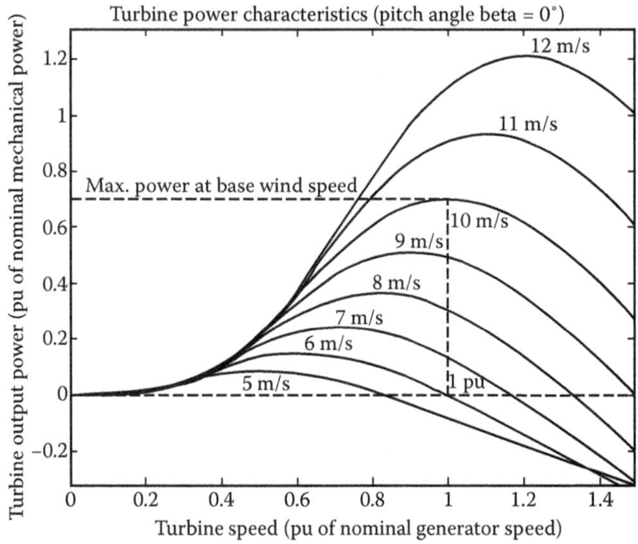

FIGURE 6.13 Wind turbine characteristics in the model **Wind_IG_1**.

where U_{leff} is the rms of the phase-to-phase voltage

$$X_s' = \omega_s L_s', X_r' = \omega_s L_r'$$

$$L_s' = L_{ls} + \frac{L_{lr} L_m}{L_r}, L_r' = L_{lr} + \frac{L_{ls} L_m}{L_s} \tag{6.13}$$

$$s_k = R_r / X_r', \sigma = 1 - L_m^2 / L_s L_r.$$

Since $X_s'/\sigma = X_s$, we have the following for the utilized SCIG under the nominal voltage, in the no-load condition, in pu

$$Q_0 = \frac{1}{2.89 + 0.1706} = 0.327,$$

that is, $Q_0 = 0.327 \times 1.72 = 0.562$ Mvar.

Under the nominal slip, bearing in mind that

$$\sigma = 1 - \frac{2.89^2}{(2.89 + 0.1706) \times (2.89 + 0.0814)} = 0.0816,$$

$$L'_r = 0.0814 + \frac{0.1706 \times 2.89}{2.89 + 0.1706} = 0.242, \quad s_k = \frac{0.0072}{0.242} = 0.0297,$$

$$L'_s = 0.1706 + \frac{0.0814 \times 2.89}{2.89 + 0.0814} = 0.25,$$

$$Q_n = \frac{1}{0.25} \times \frac{0.0816 \times 0.0297^2 + 0.0072^2}{0.0297^2 + 0.0072^2} = 0.53 = 0.911 \text{ Mvar}.$$

The following capacitances are demanded for the reactive power compensation in the no-load condition and under nominal load, respectively:

$$C_0 = \frac{0.562 \times 10^6}{575^2 \times 314} = 5.4 \text{ mF}, \ C_n = \frac{0.911 \times 10^6}{575^2 \times 314} = 8.8 \text{ mF}$$

The capacitances are reduced to a third with the capacitors connected in delta. The capacitance of the permanently connected stage is accepted as 1.8 mF; when the power reaches 1.3 MW, the second stage with the capacitance of 2.4 mF is connected.

The block for the blade position control (angle β) is included in the considered model (the subsystem **beta_control**). It is a proportional controller with the limitation of the rate of the blade position change. The block is becoming active when the SCIG power begins to exceed the nominal active power by 10%. During a start, the constant angle β can be set.

The subsystems for the soft start **Start_Power** and **Control** are the same, as in the earlier considered model **HydroM3**. Firstly, the turbine accelerates without the load to the speed of 0.96 of the nominal; then the breaker **Br** is closed and the firing pulses for the thyristors are fabricated under the initial firing angle $\alpha = 120°$, which afterward decreases by the current controller or linearly in time (depending on the tumbler **Manual Switch**). When this angle decreases to 25°, the thyristors are shunted by the breakers; the subsystem **beta_control** can control angle β.

The process is simulated when the turbine is speeding up under the wind speed of 9 m/s; the wind speed increases to 11 m/s at $t = 10$ s; increases to 14 m/s, at 20 s; and drops to 6 m/s at $t = 27$ s. To do the simulation time shorter, the SCIG initial speed is taken as 0.9.

The processes in the model are shown in Figures 6.14 and 6.15. The stator current does not exceed a two-fold value. The reactive power is not fully compensated, so that under necessity, the additional steps have to be made: to increase the capacitor stage number or to use SVC or STATCOM. Under the wind speed of 14 m/s the power limiter that acts on angle β operates.

The employment of WG that has two fixed speeds can improve WG efficiency. Such sets can be made in different ways: to use two standard SCIGs with the same pole numbers but with a different gear ratio; to use two standard SCIGs with the different pole numbers mounted on the same shaft (in such a case, the power of the

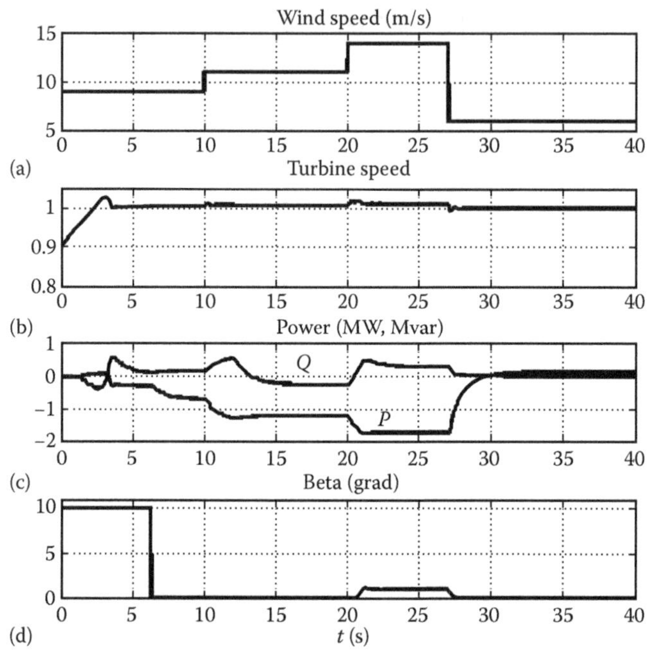

FIGURE 6.14 (a) Wind and (b) turbine speed, (c) power, and (d) angle β in the model **Wind_IG_1**.

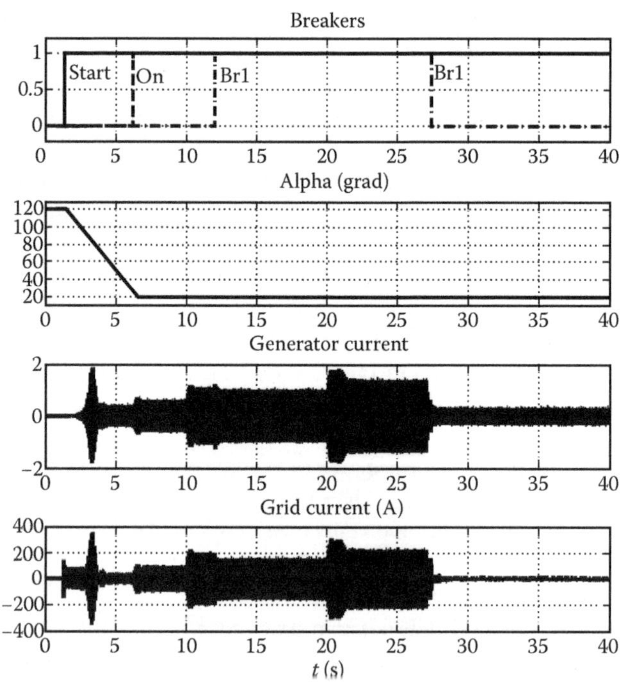

FIGURE 6.15 Angle α and currents in the model **Wind_IG_1**.

SCIG having more of the poles can be essentially less than the power of the SCIG having fewer poles); and to use the special generator with pole number change. The model of such a SCIG has already been considered in the model **IG_switchpole** and is utilized here in the model **Wind_IG_2**. The SCIG has the power of 1.72 MVA with $Z_p = 2$ and 0.86 MVA with $Z_p = 3$; the voltage is 575 V. The generator data are specified in the program `IM_MODEL_1a` that has to be executed before the simulation start.

The structure of the previous model **Wind_IG_1** remains (some parameters are different). The switching over of the pole pairs occurs by the signal *P* at the wind speed of 8 m/s. The results of the simulation are shown in Figure 6.16. The reactive power is not fully compensated, because two compensating stages are not sufficient. At the wind speed of 7 m/s, the WG power is 0.55 MW, whereas for the single-speed SCIG, it would be about 0.3 MW.

The considered model simulates only the system steady state, without regard for the electromechanical processes that occur at the transfer from one synchronous speed to another (under pole pair switching over). At changing over from SCIG G_2 ($Z_p = 3$) to G_1 ($Z_p = 2$), the initial slip of the latter is (1500 – 1000)/1500 = 33.3%, and the SCIG speeds up to the operating speed of about 1500 r/min. The big current inrush can happen under turning on; the soft start unit can be activated for it to diminish. Under changing over from G_1 to G_2, the initial slip of the latter is (1000 – 1500)/1000 = –50%; the generator does not produce the torque and cannot slow

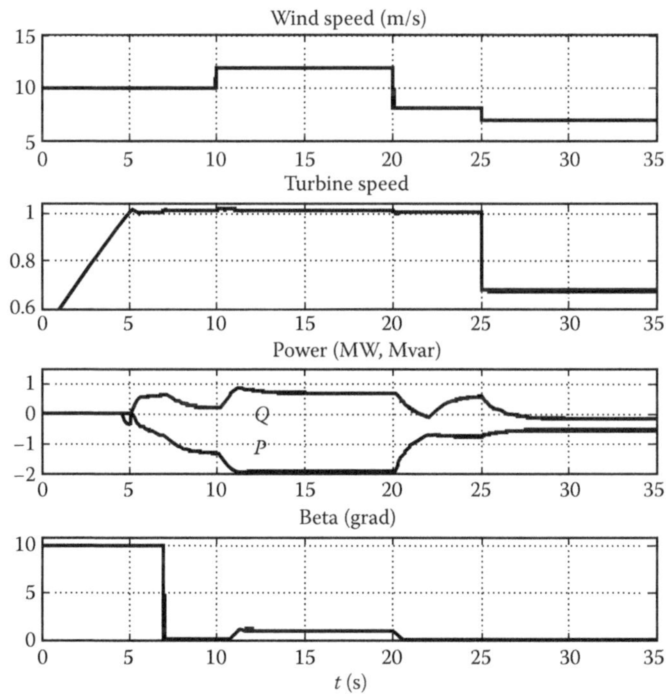

FIGURE 6.16 Processes in the model **Wind_IG_2**.

down to its synchronous speed. For this aim, the brake systems (aerodynamic, electromechanical, hydraulic) existing in the WG set must be used.

The model **Wind_IG_2a** is intended for investigation of the speed changing transients. The devices of the soft start are excluded in order to make the simulation process faster. In the SCIG model, the subsystem **Mex** is modified: the relative rotating speed ω_r for the active in the present-time generator variant can be different from the signal *Speed* that comes to the turbine and is formed in pu of the variant G_1. The transfer from G_2 to G_1 occurs at t = 10 s under the wind speed of 7 m/s; at this, the vectors [0 10] and [0 1] are assigned in the block **Timer1**; both tumblers in the subsystem **Mex** are set in the top positions. The process is displayed in Figure 6.17. The large current inrushes are seen, so that the units for the soft start must be used.

Under the simulation of the transfer from G_1 to G_2 at t = 10 s, the vectors [0 10] and [1 0] are assigned in the block **Timer1**; both tumblers in the subsystem **Mex** are set in the low positions. It is supposed that the braking torque that is equal to 25% of the nominal one is created in this mode (this quantity is chosen large enough to cut down the simulation time). When the rotating speed is getting close to the synchronous one, the braking torque is turned off. The plots of the obtained process are depicted in Figure 6.18.

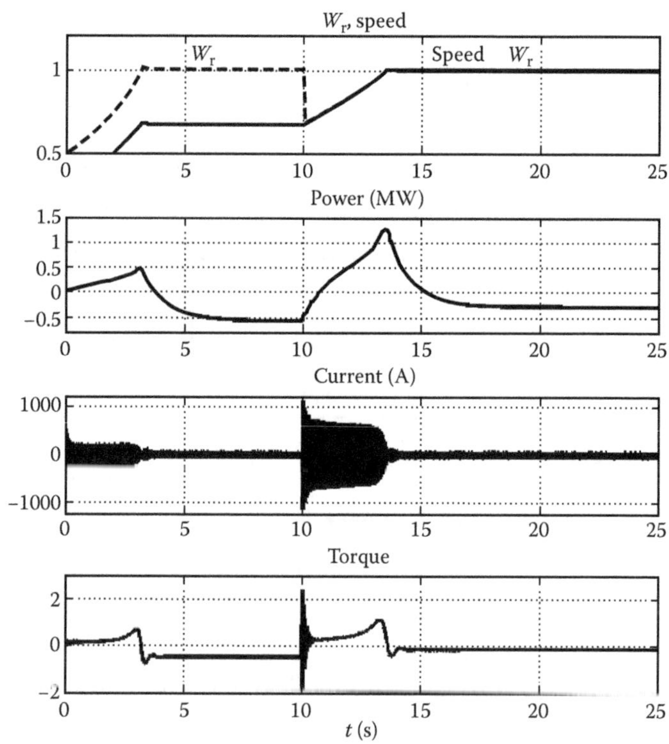

FIGURE 6.17 Changing over from G_2 (Z_p = 3) to G_1 (Z_p = 2) for the model **Wind_IG_2a**.

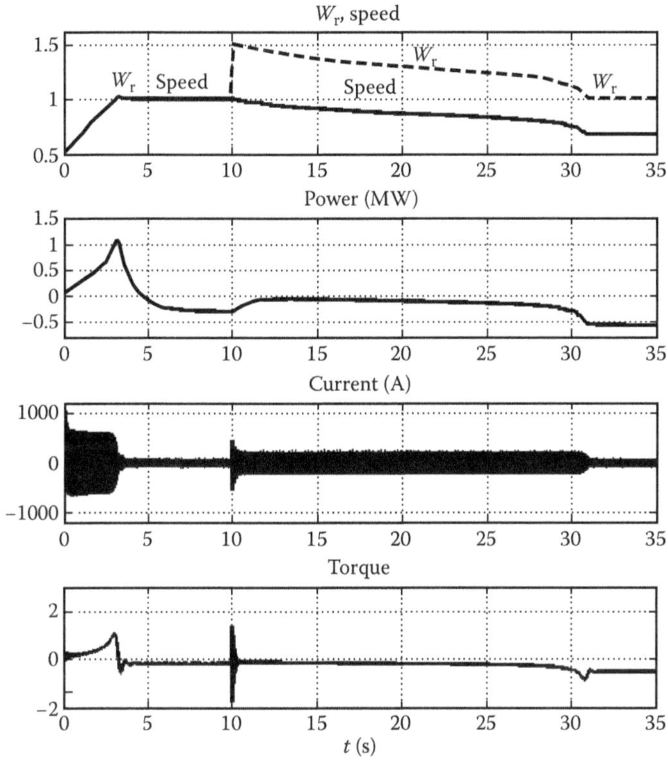

FIGURE 6.18 Changing over from G_1 ($Z_p = 2$) to G_2 ($Z_p = 3$) for the model **Wind_IG_2a**.

As has been already said, the individual WGs are united in the wind parks, or wind farms, and the modeling of their influence on the power system is of great interest. At the same time, modeling every separate WG is very time consuming; therefore, the methods of the wind farm-aggregated simulation are developed. It has been proved in Ref. [5] that for SCIG, it is sufficient to investigate the united system having the total power value. Such an approach was accepted in the model **SRC1**.

It is supposed in the model **Wind_IG_3** that each WG is made as in the model **Wind_IG_1**, with the difference that the output voltage is 35 kV (not 6.3 kV). The powers of all components of the model are increased *N* times, where *N* is the number of the WGs in the wind farm (it is specified in the option *File/Model Properties/Callbacks/InitFcn*). The farm is connected to the power network that has the voltage of 230 kV by means of the transmission line that is 30 km in length and the step-up transformer. The transformer power and the short-circuit power of the line are multiplied by *N* too; the line parameters must be selected depending on the specific conditions. In order to fully compensate for the reactive power that is consumed by the generators of the farm, the SVC is utilized that is made according to the diagram in Figure 3.56. The subsystem for thyristor control contains the **Synchronized 6-pulse Generator** and the reactive power controller with the zero reference. The controller output is converted in the firing angle value with the help of a nonlinear converter, as in Figure 3.58 and in the model **SVC1**. The capacitance and the installed inductive powers are taken constant: 15 and 40 Mvar, respectively. The filter of the 5th harmonic is employed. The processes in the model are shown in Figure 6.19. The reactive power is equal to zero in all ranges of the wind speed. For comparison, the plot of the reactive power without SVC is shown.

In the model **Wind_IG_4**, the offshore wind farm is considered. Because of that, the parameters of the line that connects the farm to a shore correspond to the underwater cable (it has much more charging currents, i.e., the larger parallel capacitance). SVC is placed at the opposite end of the line; otherwise, the additional sea platform will be necessary. The processes in the line are shown in Figure 6.20. The SVC power change is more perceptible than in the previous model, because the line reactive power has to be compensated in addition.

So far, it was supposed that the system turbine-generator is a single mass one (Chapter 4). But in reality, it is a rather complex kinematic system. It is taken as two

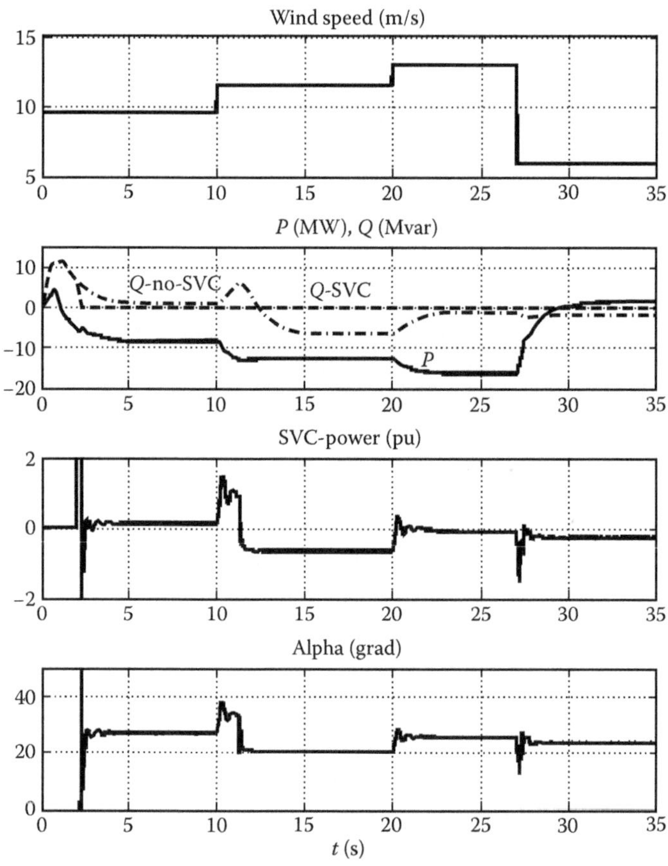

FIGURE 6.19 Processes in the model **Wind_IG_3**.

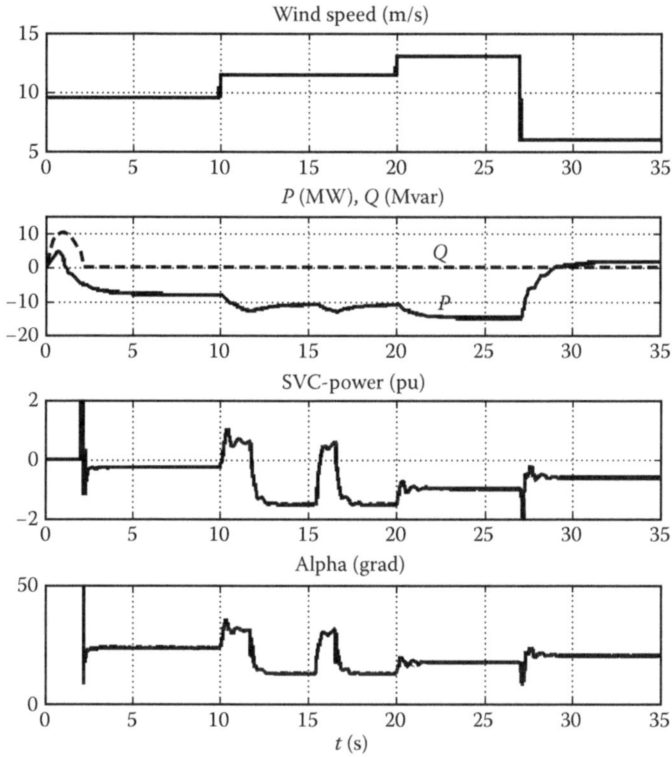

FIGURE 6.20 Processes in the model **Wind_IG_4**.

masses in the model **Wind_IG_5**; the relationships for such a system are Equations 4.168–4.170. Since the standard blocks described in Chapter 4 operate in SI units, and the WG model in pu, the model of the mechanical coupling in pu is utilized in the considered model. The same model is used in the demonstration model **Wind Farm—Synchronous Generator and Full Scale Converter (Type 4)**. On the whole, the considered system is the same, as in the model **Wind_IG_1**, but without devices for the soft start. It is accepted that the total inertia constant that is equal to 3 s in the model **Wind_IG_1** is divided in the turbine inertia constant referring to the generator shaft and is equal to 2.7 s, and the generator inertia constant is equal to 0.3 s. The stiffness factor $C = 3$, and the damping coefficient $B = 1.2$. The processes in the model are displayed in Figure 6.21. The oscillations appear with the big loads.

6.2.2 WG with Variable Rotation Speed

It was said already that the change of the WG rotating speed with wind speed change leads to the increase in the power produced with the WG. The speed change is obtained with the help of two VSIs connected back-to-back with the common capacitor in the DC link. For the WG with large power, the parallel connection of semiconductors or VSI can be necessary. At this, the circulating currents can flow between

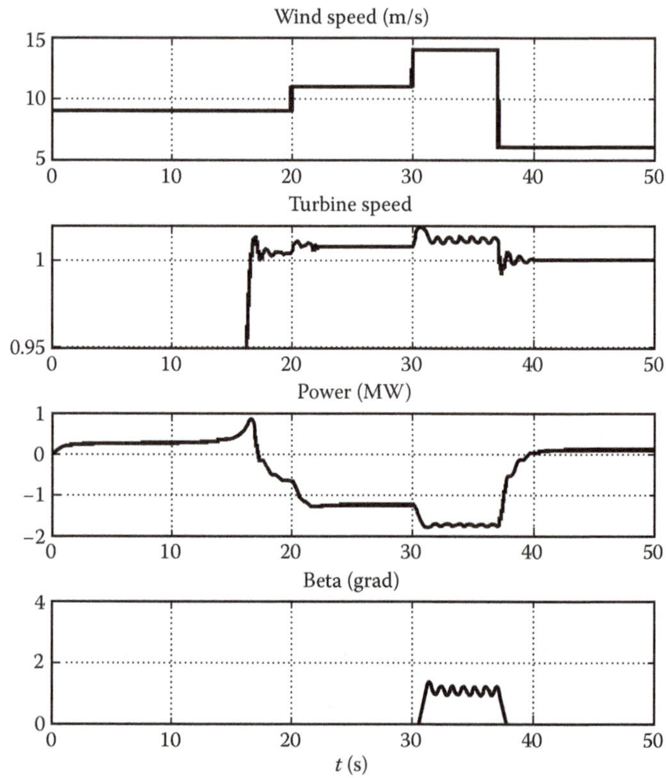

FIGURE 6.21 Processes in the model **Wind_IG_5**.

separate VSIs. This phenomenon is excluded with the employment of a generator with two three-phase windings (six-phase generator), whose model was considered in Chapter 4. Such a system is developed in the model **Wind_IG_6**; the main circuits are depicted in Figure 6.22. Each generator winding has the nominal voltage of 690 V; the generator nominal power is 1200 kVA. The generator parameters are specified in the program `IM_MODEL_6Ph_1200.m`. The converters C1 and C2 control the WG rotating speed; the converters C3 abd C4 maintain the voltages across the capacitors; the reference value is 2 kV. The converters C3 and C4 are connected in parallel via small reactors and are connected with the 35 kV/1200 V step-up transformer, which delivers the fabricated electric energy into the network.

In the subsystem **Control**, the indirect vector control is realized, whose block diagram is depicted in Figure 6.23. Here **RL** is the rate limiter; **SR** is the speed controller; **CR** are the current controllers, whose output quantities are converted into the three-phase inputs for PWM by means of the blocks for reverse Park transformations **Tp3** and **Tp4**; and their angle inputs are shifted by 30°. The relationships that describe the system function are shown in Figure 6.23. In principle, it is the same control system that was used in the models **Matr_Conv2** (Figure 3.47) and **Hydro_M4** (Figure 5.84) where its substantiations are given.

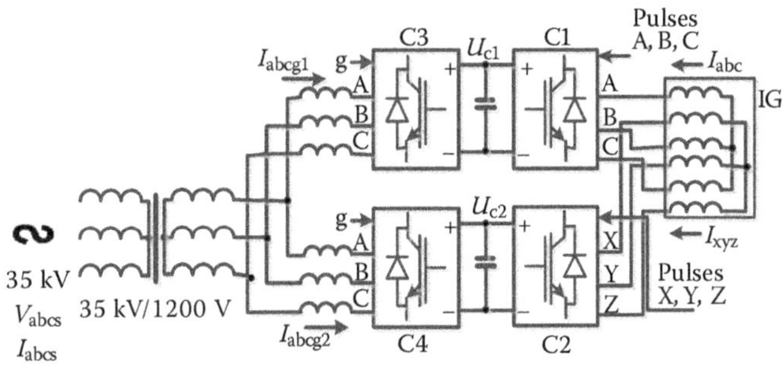

FIGURE 6.22 Main circuits of the model **Wind_IG_6**.

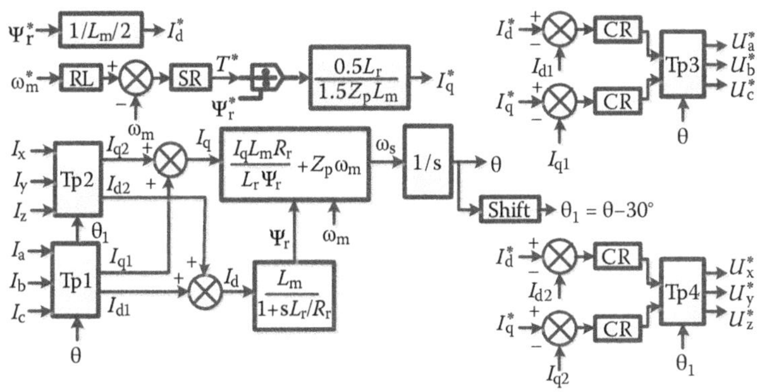

FIGURE 6.23 Diagram of the speed control in the model **Wind_IG_6**.

The diagram of the control system of converters **C3** and **C4** is analogous to that given in Figure 5.83. The voltage across each capacitor is controlled separately, but the block PWM with the triangle waveform with the frequency F_c carries out the shift of the waveforms for both controllers by $0.5/F_c$; because of that the content of the harmonics in the network decreases.

In the described model, the WG accelerates in the motor mode with the maximum angle $\beta = 25°$. At $t = 2.5$ s, $\beta = 0°$ is set (to save simulation time, this operation is supposed to be instantaneous), and WG begins to produce the electric energy. In the mode of the variable rotating speed, the speed reference is proportional to the wind speed, which is supposed to be measured. The processes under the constant speed (the tumbler **Switch** is in the top position) and under the variable speed (the tumbler **Switch** is in the low position) are shown in Figure 6.24. It is seen that in the second case, the power delivered in the network is larger.

The SCIG with power of 6 MVA under the voltage of 5 kV is utilized in the model **Wind_IG_7**. The mechanical turbine power is 5 MW. The use of three-level inverters gives the possibility to connect to the network with the voltage of 120 kV with only 120 kV/5 kV transformation stage. The DC link voltage is 9 kV; each of

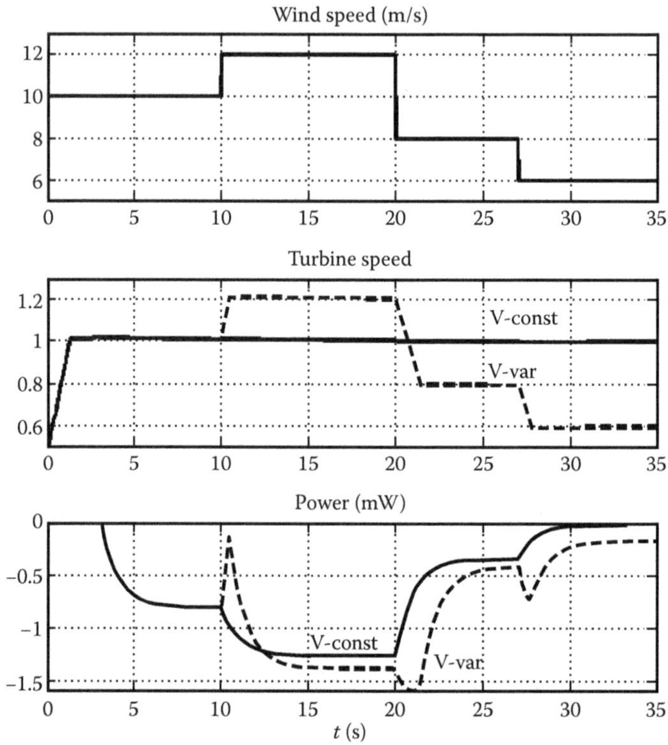

FIGURE 6.24 Processes in the model **Wind_IG_6** under constant and variable WG rotating speed.

the capacitors connected in series have the capacitance of 10 mF. For SCIG speed control, the direct vector control is used, in which the module Ψ_r and the position θ of the rotor flux linkage space vector are calculated.

From Equations 4.14, 4.15, and 6.13, it follows that the components of rotor and stator flux linkages are connected with the relationship

$$\Psi_{\alpha(\beta)r} = \frac{L_r}{L_m}\Psi_{\alpha(\beta)s} - \sigma L_s i_{\alpha(\beta)s}. \tag{6.14}$$

In turn, the components of the stator flux linkage can be found using Equation 4.151. Ψ_r and θ are computed as

$$\Psi_r = \sqrt{\Psi_{\alpha r}^2 + \Psi_{\beta r}^2}, \tag{6.15}$$

$$\theta = \text{arctg}\frac{\Psi_{\beta r}}{\Psi_{\alpha r}}. \tag{6.16}$$

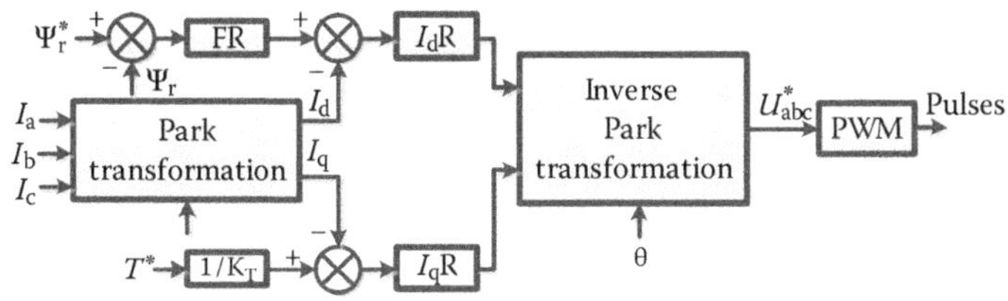

FIGURE 6.25 Block diagram of the direct vector control in the model **Wind_IG_7**.

The block diagram of the control system is depicted in Figure 6.25. It contains the rotor flux linkage controller, whose output is the reference for the inner controller **IdR** that controls the SCIG current component I_d. The reference for the controller **IqR** that controls the SCIG current component I_q is proportional to the SCIG given torque. In this model, for tracking MPPT, relationship 6.7 is used, from which it follows that the WG optimal torque, under a given rotational speed, is proportional to the square of this speed. Therefore, the SCIG reference torque is defined as

$$T_e = K_m \omega_r^2 - D - F\omega_r, \tag{6.17}$$

where K_m is defined by the parameters that are specified in the turbine dialog box (0.8 × 5/6 in the case under consideration), and D and F are the coefficients of the losses.

At the start, the given torque is positive; WG accelerates in the motor mode of SCIG with the big angle β. When the speed is getting close to the synchronous one, or as a function of time, as it is taken in the considered case, angle β is set to zero; SCIG is turned into the generator mode.

As for the control of the converter connected to the network, it is the same as in the previous model.

The processes in WG are shown in Figure 6.26. When the three-level VSIs are used, the important problem is to have equal capacitor voltages. By observing the data of the **Uc_Scope**, it can be seen that even with the capacitance difference of 5%, the difference of the voltages across them is not more than 5%.

6.3 WG WITH DOUBLY FED IG

The utilization of the wounded rotor IG (DFIG) for WG with variable rotating speed gives the possibility to decrease the power of the converters essentially. The scheme depicted in Figures 5.86 and 6.3 is used. As usual, the VSI-Ge controls the rotating speed or the torque of the DFIG, and the grid converter (VSI-Gr) is responsible for the delivery of the produced energy into the grid and for the reactive power control.

The DFIG with the power of 1.6 MVA under the voltage of 690 V is utilized in the model **Wind_DFIG_1**. The WG is connected to the grid via the transformer with the primary voltage of 6.3 kV; the 690/250 V transformer is employed for matching the stator voltage and the rotor operating voltage. The DFIG parameters are taken, mainly, from Ref. [3]. The WG, at the wind speed of 10 m/s, fabricates the power that is equal to 70% of the nominal one, which is equal to 1.5 MW. The filters are set to put down

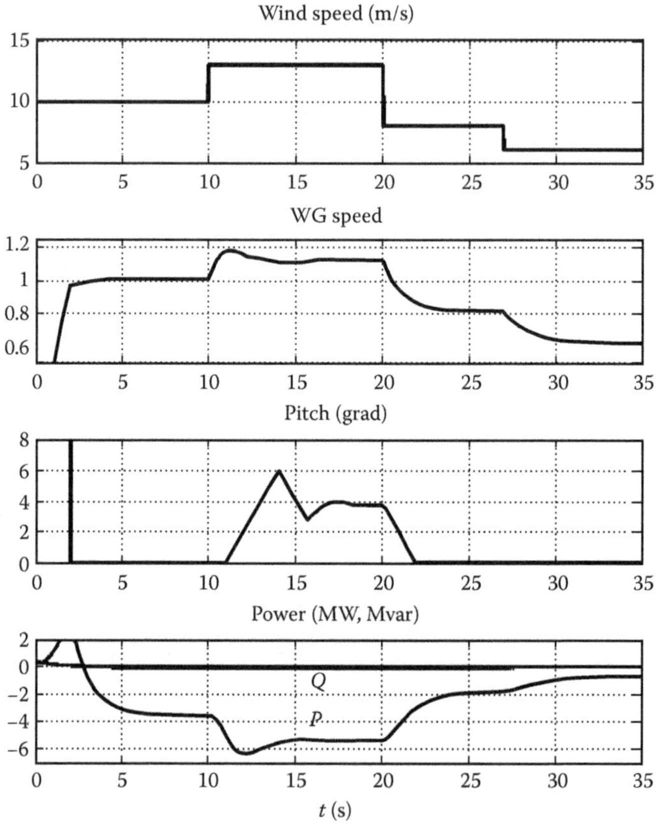

FIGURE 6.26 The processes in the model **Wind_IG_7**.

the voltage and the current higher harmonics. DTC is used for the generator converter control; this system has been described in Chapter 5 (the model **HydroM5**). The scope **Count** in the subsystem **Control_VSI-Ge/Control_PGen** gives the possibility to count the number of gate pulses sent to IGBT A1 for the given time.

The DTC system needs the information about the components of the rotor flux linkage space vector in the stationary reference frame and about the DFIG torque. These data can be obtained with the help of both the voltage model and the current model. In the first case, the stator flux linkage components are calculated by Equation 4.151, and afterward the rotor flux linkage components in the stator reference frame Ψ_r^s are computed by Equation 6.14. The following relationships are used to refer them to the rotor stationary frame:

$$\begin{aligned} \Psi_{r\alpha}^{r} &= \Psi_{r\alpha}^{s}\cos\theta + \Psi_{r\beta}^{s}\sin\theta \\ \Psi_{r\beta}^{r} &= -\Psi_{r\alpha}^{s}\sin\theta + \Psi_{r\beta}^{s}\cos\theta, \end{aligned} \tag{6.18}$$

where θ is the rotor angle.

When the current model is used, the relationship in Equation 4.15 is used; the rotor currents are used directly, but the measured stator currents are transformed in the stationary rotor frame with the formulas that are analogous to Equation 6.18. This model is much simpler, but the stator currents are usually distorted with the higher harmonics; moreover, the computation errors can appear, because the flux linkage components are found as the differences of the quantities closed by the values. In the considered model, with the tumbler in the subsystem **Control_VSI-Ge** in the top position, the voltage model is used; in the low position, the current model is used.

As for the DFIG torque computation, the expression may be used that contains the already available quantities, for instance, Equation 4.12 for the voltage model or the relationship

$$T_e = L_m\left(I^r_{r\alpha}I^r_{s\beta} - I^r_{r\beta}I^r_{s\alpha}\right) \tag{6.19}$$

for the current model.

The same system as for the previous model is employed for VSI-Gr control. Equation 6.17 is used for MPPT.

The given model is intended for the investigation of the operations when the wind speed changes, that is, the start process is not considered; the inertia constant H is taken as 1.2 s (usually H = 4–6 s). The processes in the model are shown in Figures

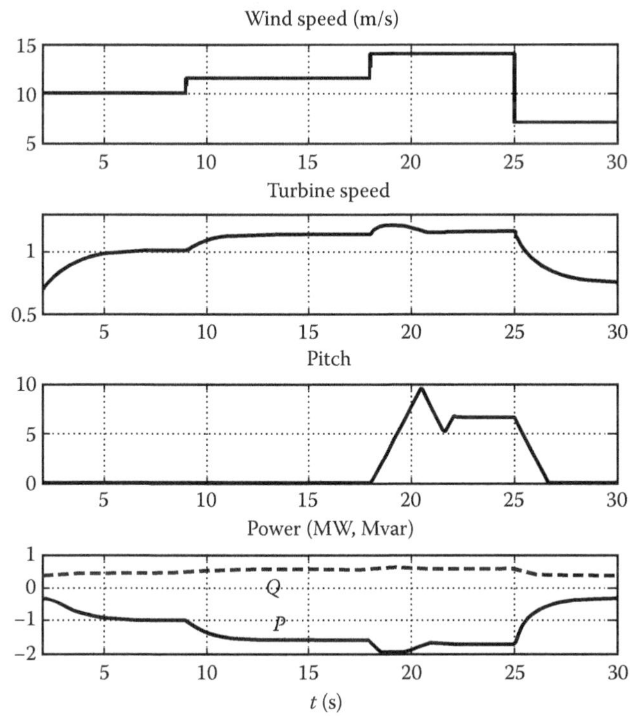

FIGURE 6.27 Processes in the model **Wind_DFIG_1**.

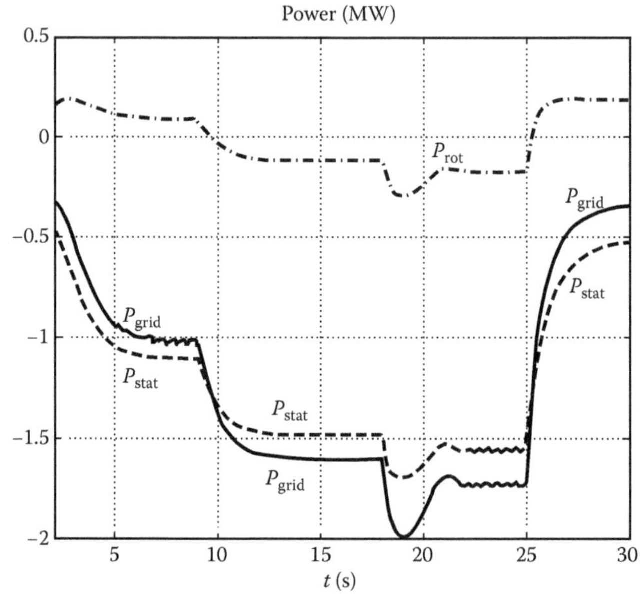

FIGURE 6.28 Power distribution under wind speed changing in the model **Wind_DFIG_1**.

6.27 and 6.28. The tracking by the optimal point is carried out; for example, at the wind speed of 11.5 m/s, the power delivered into the grid is 1.6 MW. When the wind speed increases, angle β begins to increase, limiting the rotating speed with the value of 1.15. Figure 6.28 demonstrates how the relations between the powers of stator, rotor, and grid change with the wind speed alteration. It may be seen by the scope **Uc** that the DC voltage is kept at the assigned level of 400 V. The assigned value of the rotor flux linkage influences the value of the reactive power. So at *Flux_ref* = 0.96, the average value is $Q = 0.59$ Mvar; under *Flux_ref* = 1, $Q = 0.43$ Mvar; and under *Flux_ref* = 0.92, $Q = 0.75$ Mvar.

The model **Wind_DFIG_1a** is intended for start process simulation. In comparison with **Wind_DFIG_1**, the breaker **Br** for the DFIG stator connection and the subsystem **Start** are added; the diagram of the latter is depicted in Figure 6.29; the inertia constant is increased to 3.6 s. It is supposed that the WG accelerates with turned-off stator and with $\beta = 0$. During the start, the rotor current frequency is about 20 Hz, so that the stator voltage frequency becomes equal to synchronous under $\omega_r \approx 0.6$. **Br** is turned on when the grid voltage and the induced stator voltage will be close by the amplitude, the frequency, and the phase. It is accepted that the voltage difference after filtration and rectification must be not more than 10 V. The blocks PLL are used for the frequency measurement; it is accepted that the frequency difference must be not more than 3 Hz. The saw-tooth waveforms that are formed at the PLL outputs ωt are used for the measurement of the phase difference: when such a waveform at the PLL output that is connected to the stator voltage drops from 2π to 0, the flip-flop is set and the integrator begins to integrate

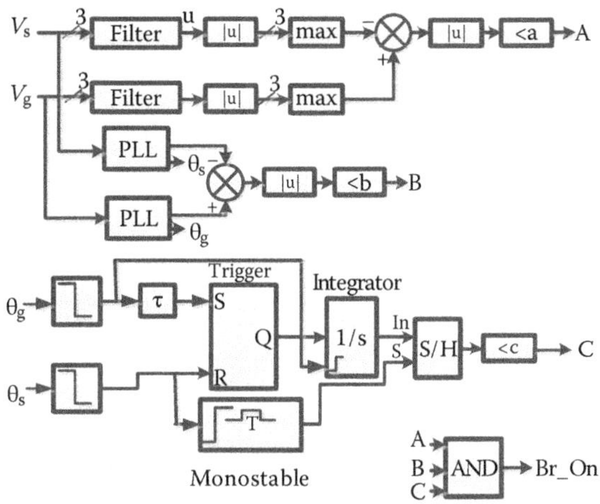

FIGURE 6.29 Block diagram of the subsystem **Start**.

the signal, whose amplitude is equal to 1. When the waveform drops at the PLL output that is connected to the grid, the flip-flop is reset, and the output of the integrator rewrites into the block for sample and storage **S/H**. Therefore, the output of this block is proportional to the phase difference of the considered voltages; when this difference is small, the permission to turn on **Br** is given. The same method of synchronization was used in the model **Photo_m71**. The process of the stator connecting to the grid is fixed in Figure 6.30. It is seen that **Br** closes at $t \approx 3$ s without noticeable current inrush; afterward, DFIG continues to accelerate and increases the power delivered into the grid.

The model **Wind_DFIG_2** differs from the previous model that it uses, for VSI-Ge control, the stator voltage-oriented control system, whose block diagram is depicted in Figure 6.31 [3]. The grid voltage space vector amplitude is U_{ds}; at this, $U_{qs} = 0$. The block PLL measures the angle of vector θ_s. Moreover, rotor angle θ_r is supposed to be known (measured), perhaps, with the error θ_e. By means of the Park transformation with the transformation angle $\theta_\Delta = \theta_s - \theta_r$, the rotor current components i_{dr} and i_{qr} are calculated, which are the feedback quantities for the proper controllers.

To neglect the stator winding resistance [3], then $\Psi_{ds} = 0$, and from Equation 4.14, it follows that

$$i_{dr} = -\frac{L_s}{L_m} i_{ds}. \tag{6.20}$$

The torque in pu, taking into account Equations 4.23 and 4.10, is

$$T_e = -\Psi_{qs} i_{ds} = -\frac{L_m}{L_s} U_{ds} i_{dr}. \tag{6.21}$$

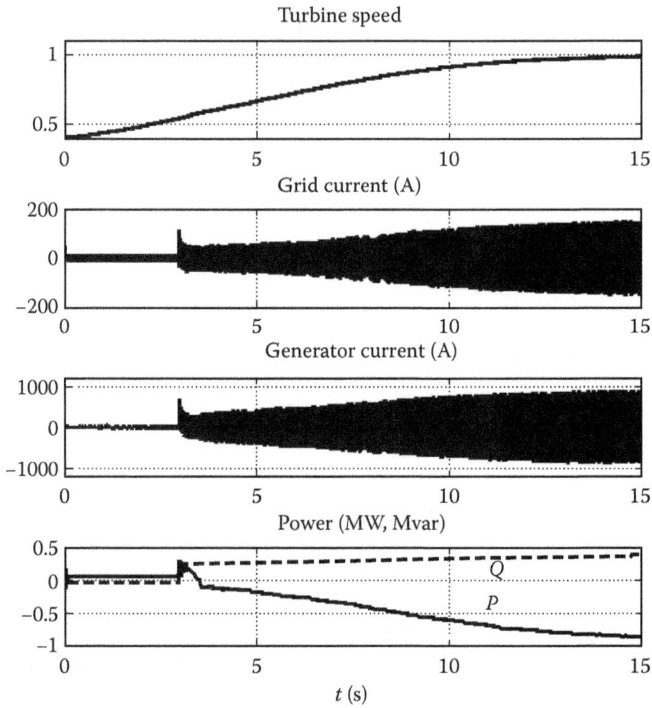

FIGURE 6.30 Process of the stator connecting to the grid in the model **Wind_DFIG_1a**.

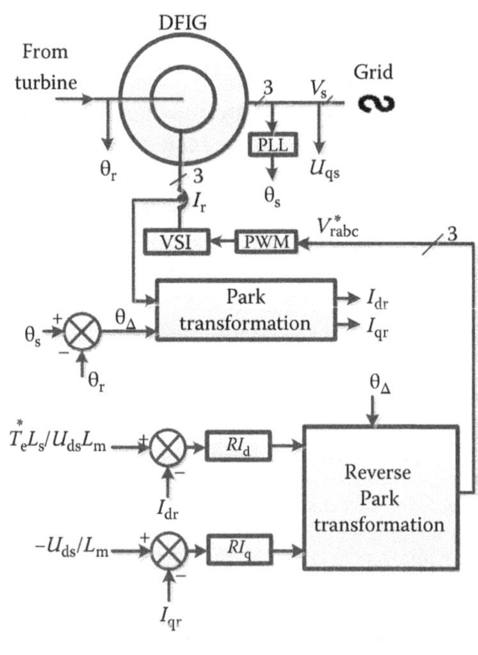

FIGURE 6.31 Block diagram of the stator voltage-oriented control system.

Then, under the given torque T_e^*, the reference for current i_{dr} is

$$i_{dr}^* = -\frac{L_s}{L_m}\frac{T_e^*}{U_{ds}}. \tag{6.22}$$

From the expression for the reactive power

$$Q = U_{qs} i_{ds} - U_{ds} i_{qs}, \tag{6.23}$$

it follows that

$$i_{qs} = -\frac{Q}{U_{ds}}; \tag{6.24}$$

then, regarding Equation 4.10, the first relationship in Equation 4.14 may be written in pu as

$$-U_{ds} = -\frac{L_s Q}{U_{ds}} + L_m i_{qr}, \tag{6.25}$$

so that the wanted value of the current i_{qr} may be written as

$$i_{qr}^* = -\frac{U_{ds}}{L_m} + \frac{L_s}{L_m}\frac{Q}{U_{ds}}. \tag{6.26}$$

(In Figure 6.31, $Q = 0$ is accepted.) The outputs of the current controllers, with the help of the reverse Park transformation, transform into the three-phase signal that controls the VSI-Ge PWM. Instead of PLL, the estimation of the stator flux linkage space vector components may be done with the help of Equation 4.151 for θ_s computation.

Before the stator connection to the grid, the synchronization of the induced stator voltage with the grid voltage is carried out. As the stator voltage, the voltage drop across the resistors connected to the stator is used. Their resistances are selected as 100Ω. When one tries to increase these resistances or exclude them from the model at all, the program gives the report about the error or (and) the integration of the equations stops. The existence of these resistors distorts the process of the synchronization a little.

At simulation start, the stator is disconnected from the grid, the VSI-Gr fabricates across the capacitor in the DC link the given voltage (400 V), the VSI-Ge fabricates the given rotor current i_{qr}^*, and the reference $i_{dr}^* = 0$; the value i_{qr}^* decreases if, during the synchronization, the DFIG will get much more than the grid voltage. WG is speeding up under the influence of the wind force. The conditions for the completion of the synchronization process are the same, as in the previous model (Figure 6.29). After synchronization completion, the reference i_{dr}^* is given.

During the synchronization, a compensation of the error in the mounting of the rotor position sensor θ_r is carried out. It is supposed that this error θ_e is not more than

30°, which is not difficult to realize in practice. An indicator of the correct sensor adjustment is $U_{qg} = 0$; the latter is the q-component of the DFIG voltage space vector V_g that is aligned with the grid space vector V_s, so that the compensating signal θ_{cr} is formed at the output of the PI controller, whose input is U_{qg}.

The error θ_e is assigned in the block **Teta_Err** in the subsystem **Control_VSI-Ge/Delta_Angle**; the circuits for the error *dteta* calculation are in the subsystem **Start**. In the results of simulation given further, it is taken that $H = 3$ s that leads to the essential increase in the simulation time, so that the steady states are not always reached. The reference torque is produced as in the previous model. The processes in the model under $\theta_e = 0$ are shown in Figure 6.32. The reactive power is about zero. In comparison with the previous model, the considered one fabricates much less the higher harmonics in the DFIG and grid currents (THD < 1%). The power distribution among the grid, the stator, and the rotor is displayed in Figure 6.33.

The values of the reactive power under $\theta_e = 0°$, 30°, and −30° when the circuits for correction is active are shown in Figure 6.34. In all cases, the reactive power is not more than 2% of the active one. It may be seen by the scope **Supply** that the connection of the stator to the grid is fulfilled without the perceptible inrushes of stator and grid currents. It can be noted that in the boundary values θ_e, the system is disabled without correction.

The demonstration models of WG with DFIG are included in SimPowerSystems: **Wind Farm DFIG Detailed Model (power_wind_dfig_det)** and **Wind Farm**

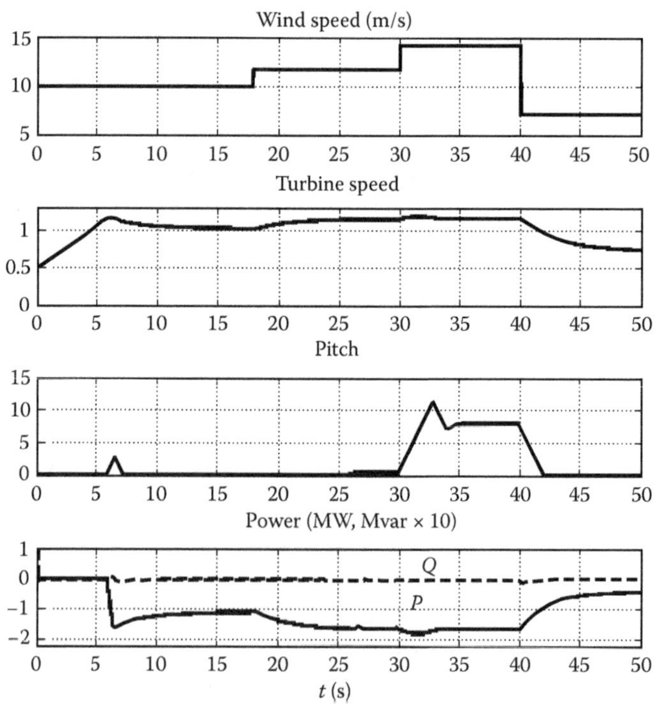

FIGURE 6.32 Processes in the model **Wind_DFIG_2** under $\theta_e = 0$.

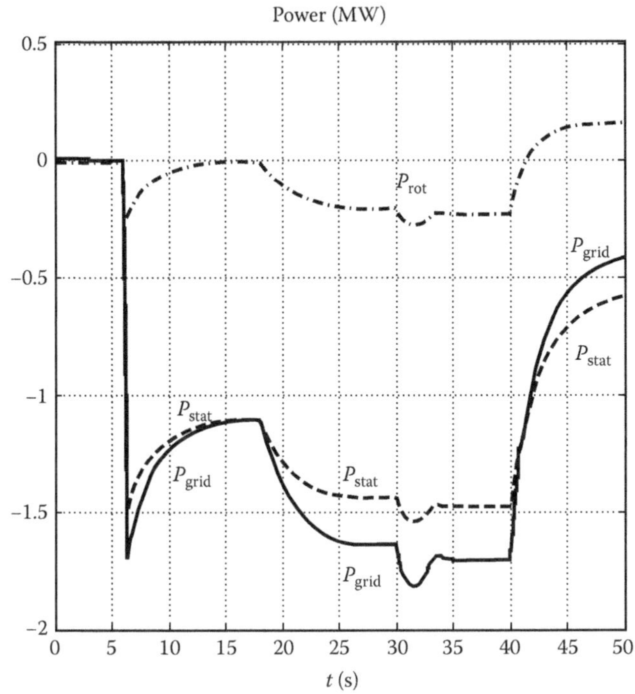

FIGURE 6.33 Power distribution in the model **Wind_DFIG_2**.

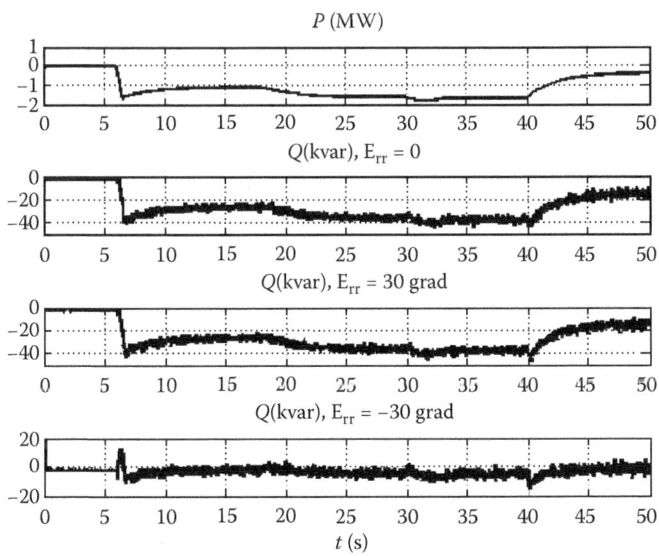

FIGURE 6.34 Reactive power in the model **Wind_DFIG_2** with the sensor position correction.

(DFIG Phasor Model) (**power_wind_dfig**). These models are of certain interest; the more so that first of them is based on the realized WG set of General Electric. However, the specific wind turbine model is utilized in this model. The speed WG reference ω^* is defined in the following way: if the power in pu $P < 0.75$, then

$$\omega^* = 0.51 + 1.42P - 0.67P^2 \tag{6.27}$$

if $P > 0.75$, $\omega^* = 1.2$.

The second model uses the standard wind turbine model but operates only in the Phasor mode.

The developed model **Wind_DFIG_3** unites the properties of both models; at that, the main principles of the control system structure of the model **Wind Farm DFIG Detailed Model** are preserved. Since in SimPowerSystems the description of the control system is lacking, its block diagrams and short descriptions are given in the following; at this, some details of limited interest are omitted.

In principle, the VSI-Ge control system is close to that given in Figure 6.31. At first, the stator flux linkage module is estimated by the following:

$$\Psi_{d(q)s} = \frac{U_{d(q)s} - R_s i_{d(q)s}}{\omega_s} \tag{6.28}$$

$$\Psi_{mod} = \sqrt{\Psi_{ds}^2 + \Psi_{qs}^2}, \tag{6.29}$$

that is, unlike the previous model, the stator winding resistance and the possible deviation of the stator voltage frequency from the standard one are taken into account. The computed value in pu is used instead of U_{ds} in Equation 6.22 for i_{dr}^* calculation.

The reference i_{qr}^* is defined with the reactive power controller (Figure 6.35) with the inner stator voltage controller. As the feedback quantity, the stator voltage positive sequence amplitude is used. Both controllers are of the integral type. We added the opportunity, with the help of the tumbler, to expel of the power controller and to have only the voltage controller.

The current reference value i_{qr}^* is limited as a function of the value i_{dr}^*, and in cases when the module of the stator current space vector exceeds the nominal value. The differences between references and measured values of the currents i_{dr} and i_{qr} come to the current controllers $\mathbf{PI_d}$ and $\mathbf{PI_q}$; at this, the error signal ε is limited smoothly with the help of the function 2/π arctan ε. The decoupling quantities, which improve the system dynamics characteristics, are added to the controller outputs.

Using Equations 4.11 and 4.15, the decoupling quantities may be written as

$$Comp\_q = R_r i_{qr} + \Delta\omega(L_r i_{dr} + L_m i_{ds}) \tag{6.30}$$

$$Comp\_d = R_r i_{dr} - \Delta\omega(L_r i_{qr} + L_m i_{qs}), \tag{6.31}$$

where $\Delta\omega = \omega_s - \omega_r$.

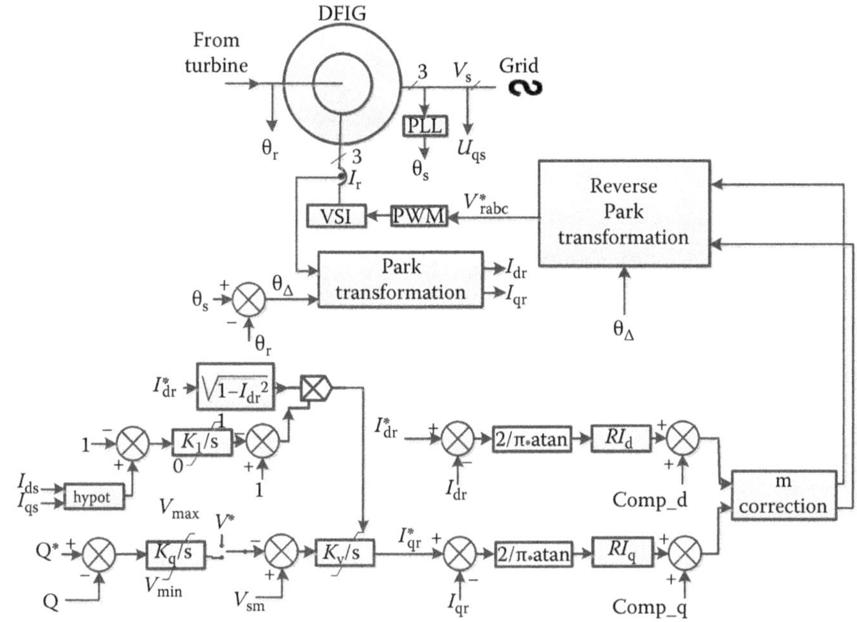

FIGURE 6.35 Block diagram of VSI-Ge control system in the model **Wind_DFIG_3**.

When *Comp_q* and *Comp_d* are computed, not the actual but the reference current values are utilized; for the stator current references, the relationship in Equation 4.10 are used under $d\Psi_{qs}/dt = 0$ and $d\Psi_{ds}/dt = 0$; Equation 4.14 is substituted in them instead of Ψ_{ds}, and Ψ_{qs}. From the first equality received in this way, the expression for i_{qs} under $U_{qs} = 0$ is found, which, afterward, is substituted in the second equality; in this way, the expressions for the stator reference currents using the rotor reference currents are found. The obtained expressions are rather cumbersome and are not given here; they may be understood from the computational scheme shown in the diagram of the subsystem **Control/Rotor-side controls**.

The resulting outputs are corrected in dependence on the actual voltage in the DC link V_{dc} in such a way that the rotor voltage was equal to its nominal value under the modulation factor $m = 1$ to have the rotor nominal voltage. For this, the output quantities are multiplied by $V_{nom-r}/(0.612V_{dc})$ (Equation 3.2), where V_{nom-r} is the rotor nominal voltage. At this, the relation between the output signals is maintained, $m \leq 1$. By means of the reverse Park transformation, the output quantities are converted in the three-phase signal that comes to the PWM block that fabricates the gate pulses for the VSI.

The VSI-Gr control system (Figure 6.36) contains the controller of the voltage across the capacitor in the DC link, whose output defines the reference I^*_{dgc} for the current component at the VSI output that is aligned with the stator voltage; the reference for I^*_{qgc} is zero. The decoupling quantities at the controller outputs are

$$Comp\_qg = -R_L I^*_{qgc} - \omega_s L_L I^*_{dgc} \tag{6.32}$$

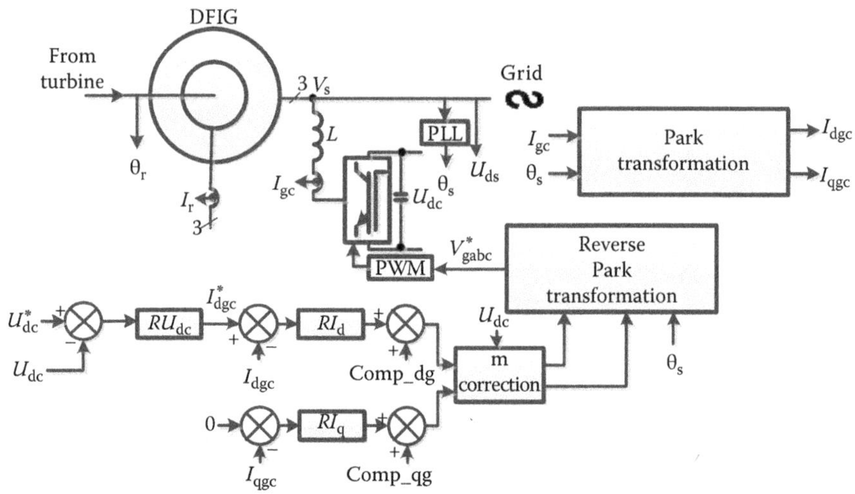

FIGURE 6.36 Block diagram of VSI-Gr control system in the model **Wind_DFIG_3**.

$$Comp\_dg = U_{ds} - R_L I^*_{dgc} + \omega_s L_L I^*_{qgc}, \tag{6.33}$$

where R_L and L_L are the resistance and the inductance, respectively, of the reactor **L** at the VSI-Gr output.

In total, this models a wind farm consisting of 6 WGs connected in parallel as one WG with power of 6 × 1.5 MW. The elastic coupling is taken into account. The multipage dialog box from the model **power_wind_dfig_det** is retained, but the page *Turbine Data* may be ignored. In order to specify the turbine parameters, the option *Look Under Mask* for the subsystem **DFIG Wind Turbine** has to be selected, and the dialog box of the standard block **Wind Turbine** has to be opened.

Three ways to define the WG power are intended (the subsystem **Speed & Pitch Control**). In the first option, the speed controller is employed, whose output defines the torque T_e^*. The speed reference is proportional to the wind speed, so that with the given wind speed, the WG speed would be keeping with the point of the maximum power. The switch **Tumbler1** in the subsystem **Rotor-side Control** is in the top position.

When this tumbler is in the low position, the current component i^*_{dr} is given, whose value is defined by the output of the active power controller in the subsystem **Speed & Pitch Control/Power_Control**. As a feedback, the value of the generated power is used. The switch **Tumbler** in the subsystem **Speed & Pitch Control/Power_Control** selects the setpoint device of the power as a function of the rotating speed: with the simpler characteristic depicted in Figure 6.37a, with the tumbler in the low position, or with more intricate characteristic depicted in Figure 6.37b, in the top tumbler position. The power in pu of the nominal DFIG power that exceeds the turbine nominal power by 10% is given.

The processes in the model with the power characteristic shown in Figure 6.37b are displayed in Figure 6.38. When the other options are used, the wanted WG

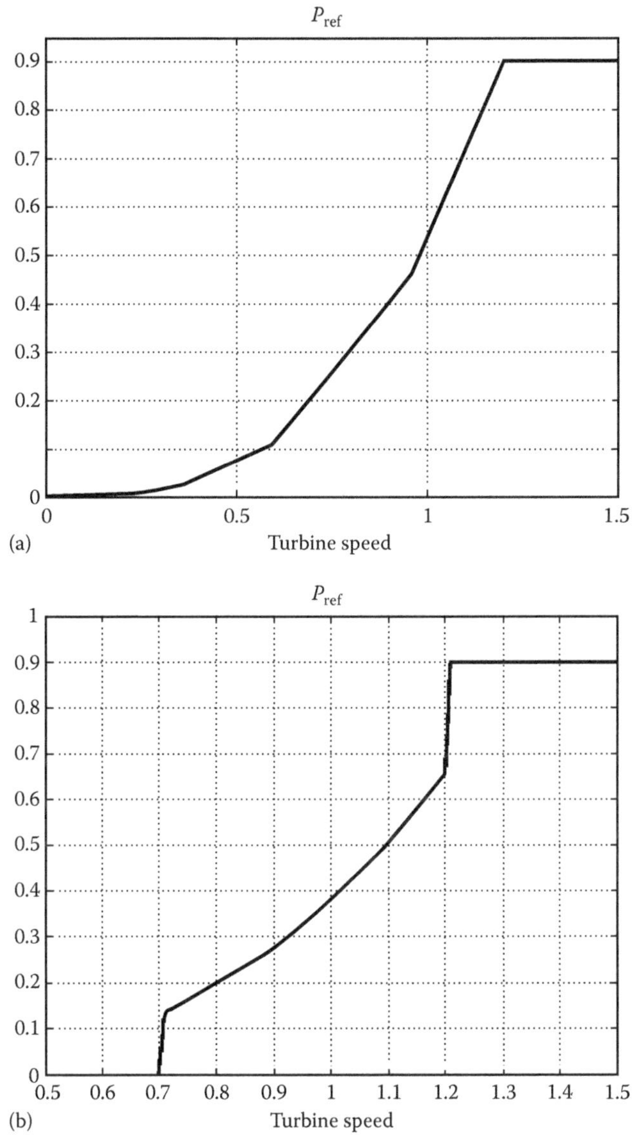

FIGURE 6.37 Reference WG power characteristics: (a) the simple characteristic and (b) the intricate characteristic.

characteristics are obtained, in principle, but some nuances are observed, owing to different parameter adjustments.

Furthermore, WG operation under essential sag of the 120 kV voltage is considered. The voltage amplitude variation by −100% at t = 4–4.15 s is selected in the source dialog box. The other parameters are as follows: the wind speed is 12 m/s, the initial WG speed is 1.2, and the initial DFIG slip is −0.2. The voltages and the

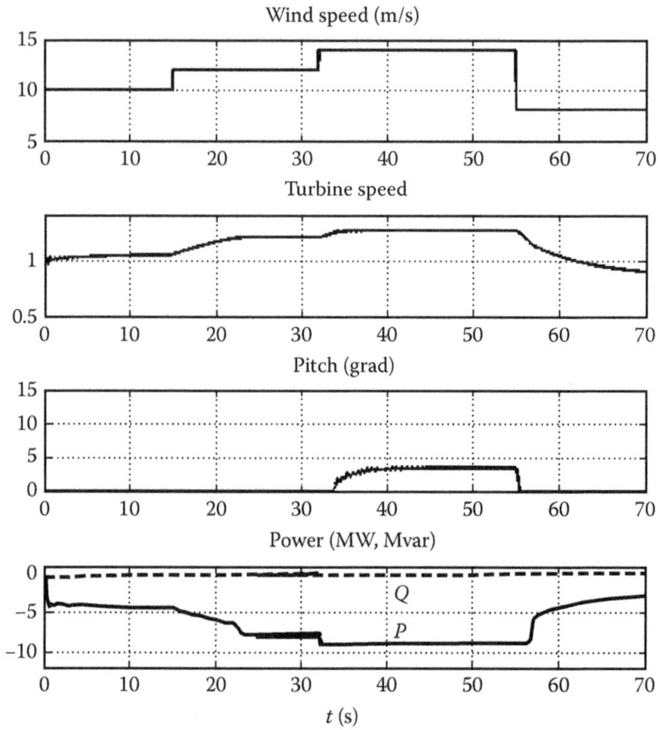

FIGURE 6.38 Processes in the model **Wind_DFIG_3** with characteristic in Figure 6.37b.

currents in this case are shown in Figure 6.39. It is seen that after the recovery of the grid voltage, WG continues to operate.

In the next model **Wind_DFIG_4**, the brushless doubly fed IG described in Chapter 4 is employed. The IG power is 50 kW, and the natural speed is 375 r/min. The WG produces this power at the wind speed of 12 m/s and under the rotating speed of 1.2 × 375 = 450 r/min. The IG and the power circuit parameters and the VSI-Gr control system **Control_Inv** are also the same as in the model **IM_double_2**. The control system block diagram of the converter connected to the IG (subsystem **Control**) is given in Figure 6.40. It differs from that depicted in Figure 4.10, mainly, by the addition of the reactive power controller. The reference for the speed controller is proportional to the wind speed. It is supposed that WG speeds up to the speed close to the natural with the wind force; afterward, the signal *On* is formed and the system goes over to the synchronous speed mode. The processes in the model are shown in Figure 6.41. The active power is close enough to its optimal values (taking into account the losses in the system), and the reactive power is close to zero. Before start, the rotor was set in the random position (45°).

In the model **Wind_DFIG_5**, the DFIG with power of 160 kVA under the voltage of 400 V supplies the isolated load. The initial excitation is provided with the battery having the maximum voltage of 590 V; the nominal DC voltage is 700 V. When WG

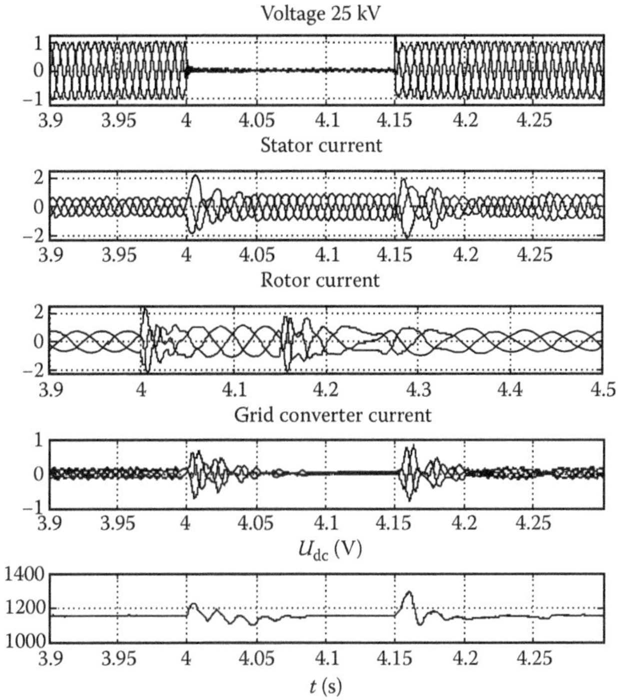

FIGURE 6.39 Processes in the model **Wind_DFIG_3** during grid voltage disappearance.

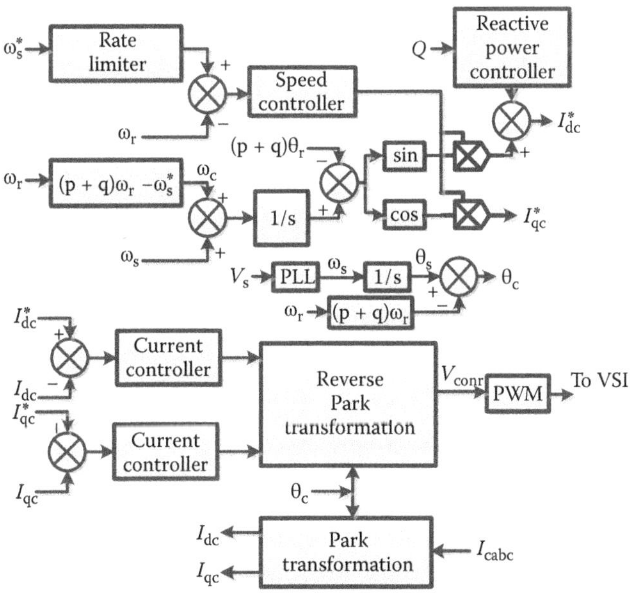

FIGURE 6.40 Control system block diagram of the converter connected to the IG in the model **Wind_DFIG_4**.

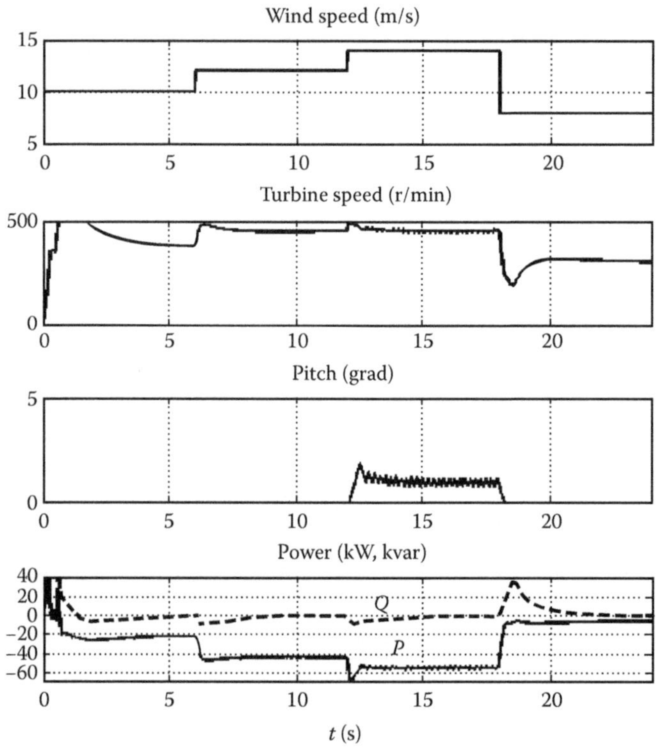

FIGURE 6.41 Processes in the model **Wind_DFIG_4**.

is in operation, the battery is recharged and the corresponding circuits are not shown, because they were considered in Chapter 5.

In the isolated installed WG, it is necessary to have equality of produced and consumed power. The former is controlled by the change of the blade position; the latter is changed with the employment of the dummy (ballast) load and with the switching over of the parts of the main load. In the considered model, the full power of the main load is taken as 108 kW (≈ 0.7 × 160 kW); it is possible to turn off its main part equal to 71 kW when the rotating WG speed decreases with wind speed reduction. The maximum power of the ballast load is 40 kW. Since at the base wind speed of 10 m/s, the rotating speed in pu is 1, and the power at that is 0.7, under WG speed decreasing to 0.9, the load 71 kW is turned off; it is turned on again, when, with the wind speed rise, the rotating speed reaches the value of 1.1 with the fully used ballast load (signal B1 = 1), which proves the availability of the WG sufficient power.

The model of the ballast load is taken from the demonstration model **Wind-Turbine Asynchronous Generator in Isolated Network**. Each phase has eight resistors, whose resistances are distributed by binary progression, so that, with the proper control, it can realize 255 steps of the resistance. The diagram of the ballast load control is depicted in Figure 6.42.

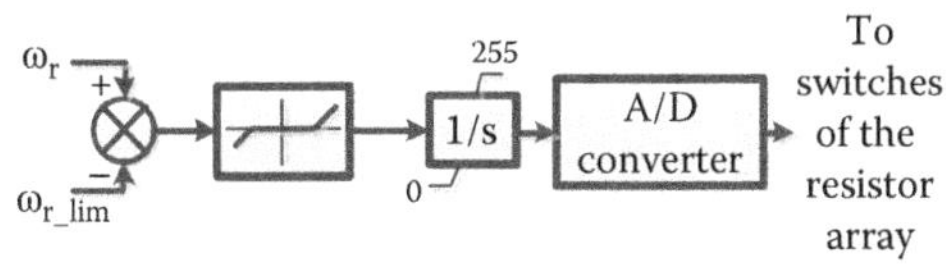

FIGURE 6.42 Diagram of ballast load control.

The amplitude and the frequency of the DFIG stator voltage are defined with the processes in the rotor. The rotor current components i_{dr} and i_{qr} are computed with the help of the Park transformation with the angle $\theta_\Delta = \theta_s - \theta_r$; they are the feedback quantities for the controllers. The angle θ_s is an integral of the given frequency $\omega_s = 2\pi f_s$; the block **Virtual PLL** in the subsystem **Control_VSI-Ge** is used for θ_s assignment. The stator flux linkage-oriented control method is utilized. According to Equation 4.14, in order to have $\Psi_{qs} = 0$, the reference for i_{qr} must be

$$i^*_{qr} = -\frac{L_s}{L_m} i_{qs}. \tag{6.34}$$

The reference for i^*_{dr} is a sum of the two terms; one of them defines the flux linkage of the no-load DFIG and is equal to $\Psi_{ds}/L_m = V_s/L_m$ in pu; another is the output of the stator voltage controller. The control system block diagram is depicted in Figure 6.43.

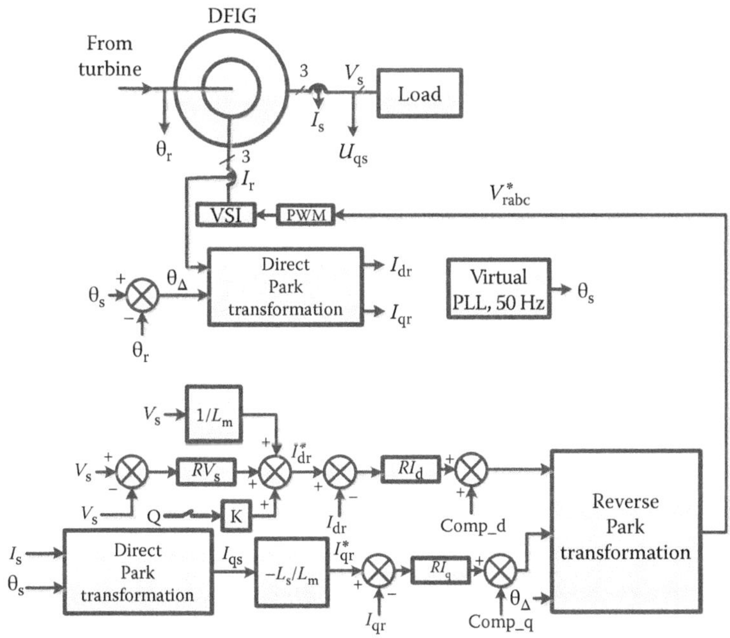

FIGURE 6.43 Block diagram of the control system **VSI-Ge** in the model **Wind_DFIG_5**.

The VSI-Gr control system differs from the one considered earlier for the determination of the angle of the Park transformation θ_s; the virtual PLL mentioned earlier is employed; moreover, because of $\Psi_{qs} = 0$, $V_s = U_{sq}$ and the output of the DC voltage controller defines the VSI-Gr current component i_q, not i_d, as in the previous models.

The processes under wind speed change are shown in Figure 6.44. The load voltage is kept equal to the reference; the WG rotating speed is limited. The variation of the powers in different system units is shown in Figure 6.45.

The active load has been supposed above. The load voltage is shown in Figure 6.46a, when the main load 71 kW has the inductive component of 30 kvar. The brief but visible voltage deviations during the transients may be seen, which are made much less with the utilization of the addition action on the rotor current component i_{dr} that is proportional to the reactive power Q; the tumbler **Manual Switch** in the subsystem Control_VSI-Ge is in the top position (Figure 6.46b).

When the anemometer is available, the switching over of the load parts can be carried out as a function of the wind speed. Such system is realized in the model **Wind_DFIG_5a**. The wind speeds, at which the load parts are turned on/turned

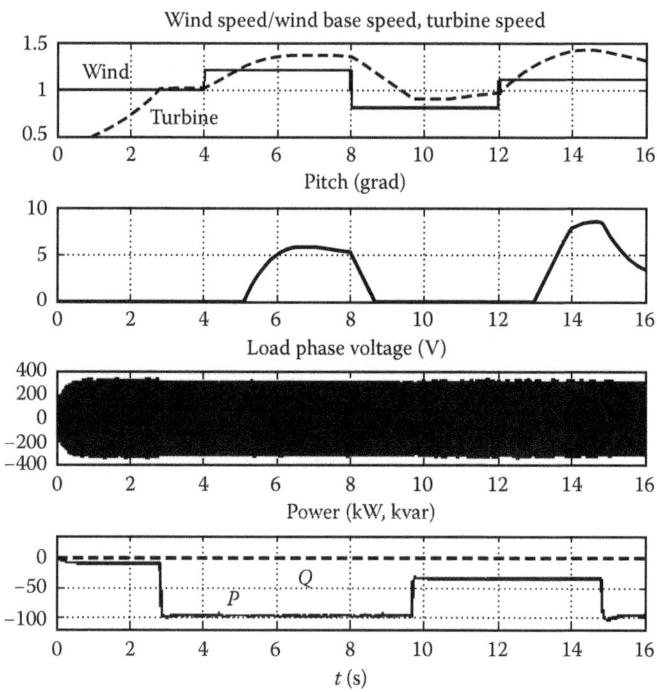

FIGURE 6.44 Processes in the model **Wind_DFIG_5**.

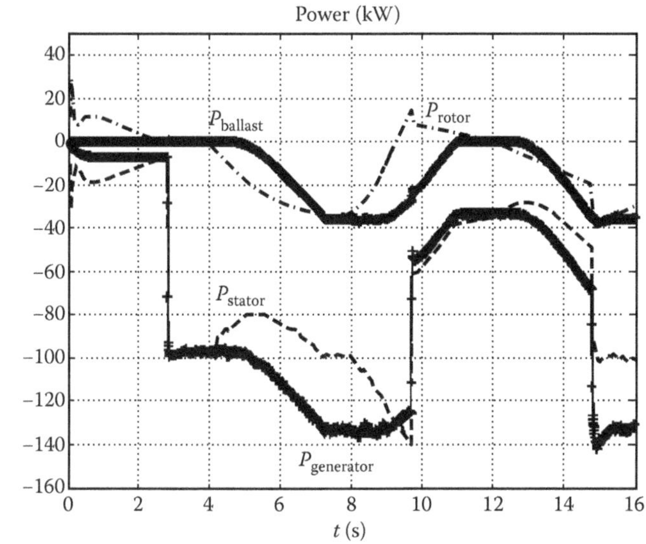

FIGURE 6.45 Power distribution in the model **Wind_DFIG_5**.

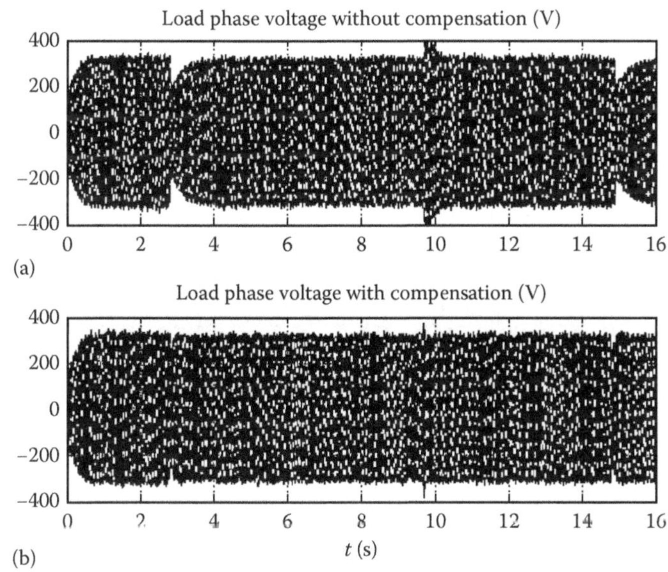

FIGURE 6.46 Load voltage in the model **Wind_DFIG_5** with the active–inductive load: (a) without additional action and (b) with additional action from Q.

off, are accepted as $V_i/V_0 = 7/6$, 9/8, and 11/10 m/s. Since the power of the given part, at the wind speed V_0, must match the WG power, the following powers of the parts are taken: 15, 31, and 51 kW, with the permanent load of 8 kW. The processes in the model are shown in Figure 6.47.

The model **Wind_DFIG_6** is intended for the investigation of the WG operation with unbalanced load. When WG operates in isolation, the load is often single phase. In these cases, the three-phase generators with the accessible stator neutral are used. Because the model of such a generator is absent in the SimPowerSystems, the virtual transformer with the accessible neutral and with the transfer ratio 1:1 is set at the DFIG stator output; the parameters of the transformer are close to the ideal ones, and its primary current is considered as the stator current.

Although one tries to get near the balanced condition by load distribution in the phases, it does not come out fully; the asymmetry causes the stator current negative sequence that leads to a number of unwanted occurrences, including generator torque pulsations, which cause increased wear of the WG mechanical parts and the appearance of acoustic noises. For the neutralization of that phenomenon, the additional action on the VSI-Gr or VSI-Ge can be utilized.

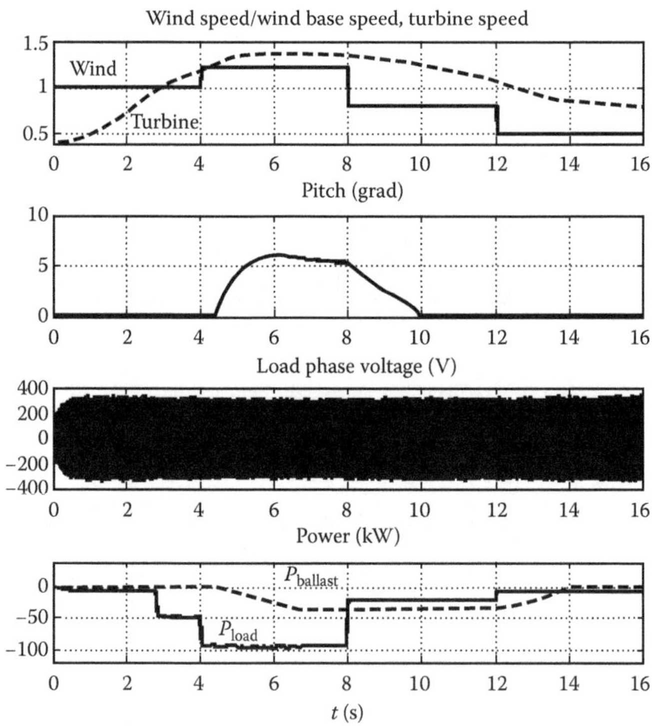

FIGURE 6.47 Processes in the model **Wind_DFIG_5a**.

The first option is used in the considered model. The idea is to form the current at the output of VSI-Gr that contains the negative sequence, which is equal to that in the load. At this, the negative sequence in the stator current will be absent. The control system diagram is depicted in Figure 6.48. The reference for the current controller I_g at the VSI-Ge output consists of two terms; the first one is defined by the output of the DC voltage controller, and the other compensates the negative sequence in the stator current I_s. The fabrication of this sequence is carried out with the Park transformation, whose axis rotates in the direction that is opposite to the usual. The output quantities have a harmonic component with frequency of 100 Hz that is filtered off with the low-pass filters. The filter outputs come to the PI controllers $\mathbf{PI}_{dn}$ and $\mathbf{PI}_{qn}$ having zero references. The controller outputs are converted into a three-phase signal with the help of the reverse Park transformation, whose axis rotates in the direction opposite to the usual as well.

The following process is simulated: during start, it is the balanced mode with each phase load of 35 kW; at $t = 1.2$ s, the power in phase B increases by 7 kW and decreases in phase C by 7 kW; at $t = 2.2$ s, the compensating circuit is turned on. The changes of the negative sequences in currents of the load, the stator, and the VSI-Gr are displayed in Figure 6.49, together with the DFIG torque and the load voltage. The torque fluctuations decrease essentially, although they are bigger than with the balanced load; it may be explained with the nonideality of the system elements and the errors that are inherent to the three-phase current controller. The load voltage is kept constant.

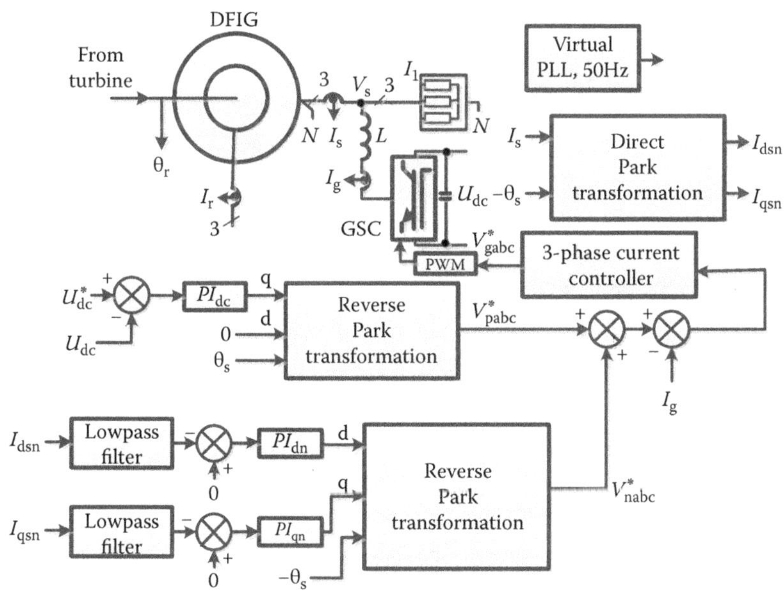

FIGURE 6.48 Diagram of the control system for the load unbalance to compensate.

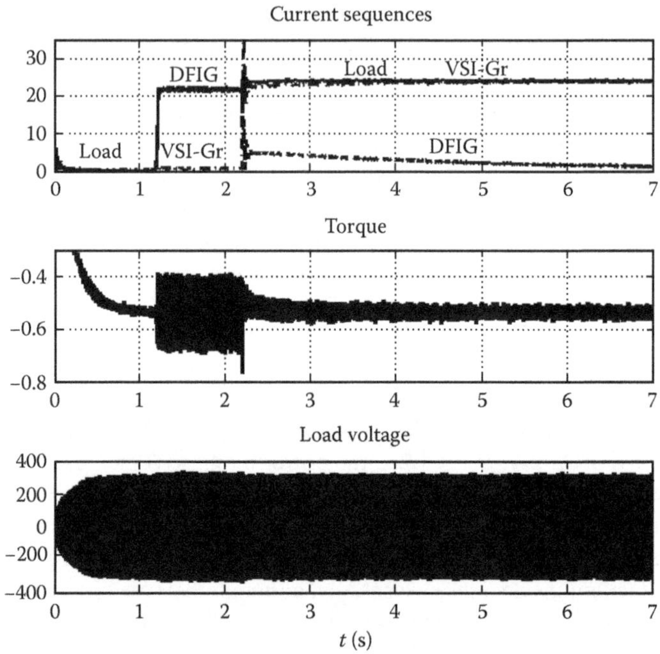

FIGURE 6.49 Processes in the model **Wind_DFIG_6**.

6.4 WG WITH SYNCHRONOUS GENERATORS

The SGs with the excitation winding, unlike the IG, do not need the initial excitation; the possibility to change the rotor flux gives an opportunity to control the reactive power. The generators are connected to the network directly or via the controlled converters. The first possibility is employed in the model **Wind_SG_1**. The turbine, at a wind speed of 12 m/s, has a nominal power of 2 MW; at that wind speed, the optimal rotating speed in pu is 1, the SG voltage is 750 V, and the power is 2.2 MVA. The SG supplies the local load with power of 1 MW via the 35 kV/750 V step-up transformer; the power surplus is sent to the network having a short-circuit power of 40 MVA via the line that is 50 km in length and the 230 kV/35 kV group transformer. At low wind speed, both sources supply the load. SG is excited from the standard exciter ST2A. Two operation modes are intended: when the tumbler **Switch** is in the low position, the SG constant voltage is assigned; when the tumbler is in the top position, the SG reactive power is controlled.

Before SG connection to the network, the amplitude, the frequency, and the phase of the SG voltage must be approximately equal to the parameters of the network voltage that is provided with the subsystem **Start**. Its diagram is about the same, as depicted in Figure 6.29. The WG start is shown in Figure 6.50. The turbine is speeding up under the influence of the wind to the rotating speed close to the nominal one. Under start, angle β is taken to be equal to 20°. The breaker is closed at t = 6.7 s. Some overshoot can be seen in the SG current, owing to the wide tolerances accepted

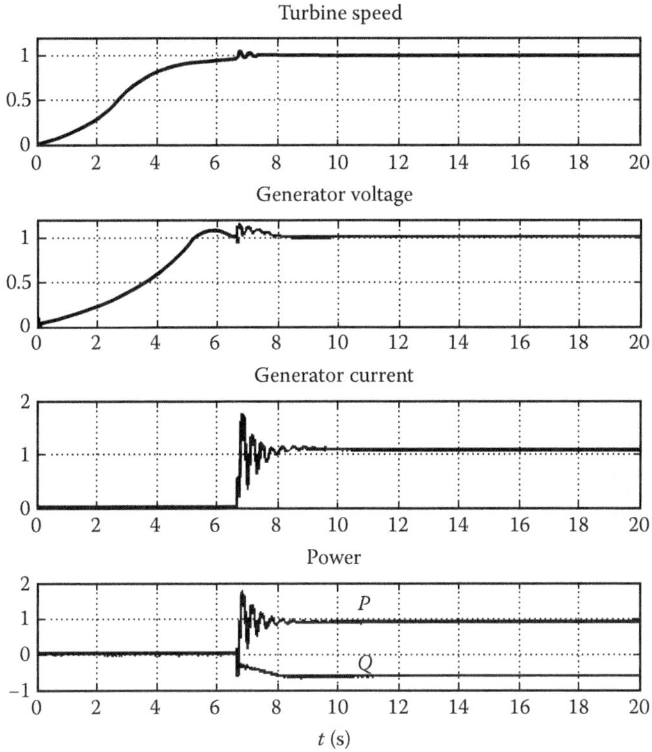

FIGURE 6.50 Processes in the model **Wind_SG_1**.

in the subsystem **Start**: by the frequency of 2 Hz and by the voltage amplitude of 10%. After the closing of the breaker, the WG rated duty is set.

The model **Wind_SG_1a** contains the same system, but for the speeding up of the simulation, the start circuits are excluded, and the simulation sampling time is increased to 50 μs. The power changes under wind speed V_w alterations are shown in Figure 6.51, with voltage control. It is seen that at $V_w = 10$ m/s and $V_w = 6$ m/s, the WG power is not sufficient for the load supply; the deficient power is taken from the network; at the bigger wind speed, there is a power surplus that is delivered to the network. The same process is shown in Figure 6.52, but under the assignment of the zero reactive power of SG. It is seen that it is really equal to zero, but the network reactive power essentially increases. The system response to the sag of the 230 kV network voltage by 50% with the duration of 120 ms is fixed in Figure 6.53. After the network voltage recovery, WG returns to the normal operation.

But more often, the SG in WG is used together with the converters, according to the diagram depicted in Figure 6.3c. The converter connected to the grid is usually VSI (VSI-Gr), whereas the converter connected to the SG can be either the VSI or the DC/DC boost converter with the diode rectifier, whose diagram is depicted in Figure 6.54. The latter is simpler, but the rectifier input current contains higher harmonics, causing the SG additional heating.

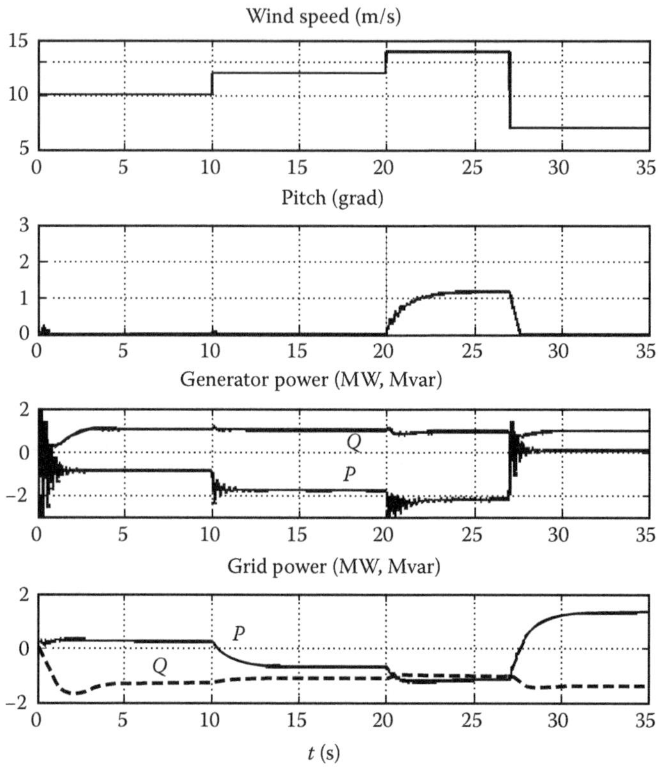

FIGURE 6.51 Power changes under wind speed V_w alteration for the model **Wind_SG_1a**.

The model **power_wind_type_4_det** is included in SimPowerSystems that models the system that consists of the SG, the generator converter according to the diagram in Figure 6.54, and the VSI-Gr. The equivalent SG with the power of 10 MW models five WGs having the power of 2 MW operating in parallel. The turbine model, the power circuits, and the investigated operating conditions are the same as for the model **power_wind_dfig_det** considered earlier. The reference for the reactor current I is defined with the SG speed controller output; the reference for the speed controller is

$$\omega^* = 0.425 + 1.183P - 0.5551P^2, \tag{6.35}$$

at $P < 0.75$; if $P > 0.75$, $\omega^* = 1$, which corresponds to the optimal tracking characteristic. Here P is the actual power at the VSI-Gr output. This model is not considered here in detail.

The six-phase SG described in Chapter 4 is utilized in the model **Wind_SG_2**. The SG power is 4 MVA; the voltage of each three-phase winding is 2400 V. The program SM_MODEL_6ph4 is executed before simulation start, in which the SG parameters are specified and the model parameters are computed. The power circuits

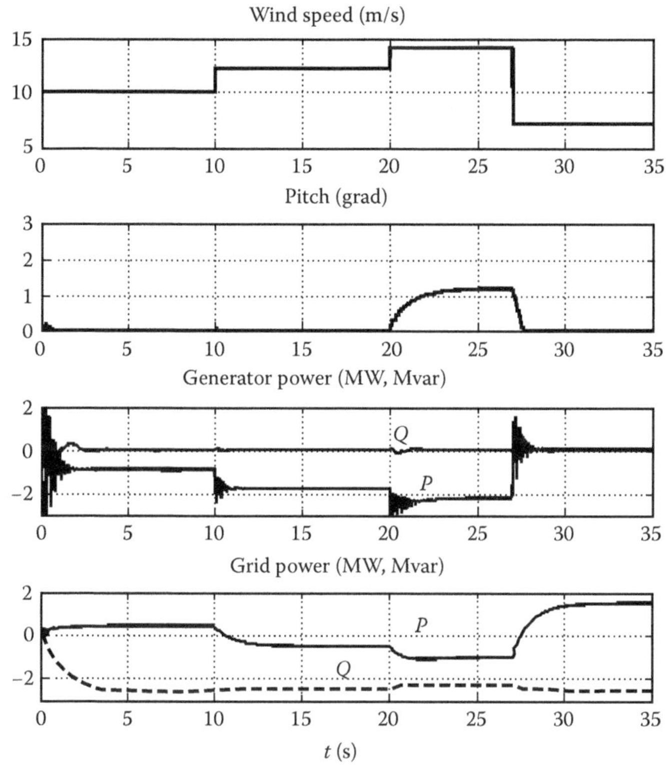

FIGURE 6.52 Power changes in the model **Wind_SG_1a** under SG zero reactive power.

of each three-phase winding are arranged as in the model **power_wind_type_4_det** mentioned earlier: the diode rectifier, the DC/DC boost converter, and the VSI-Gr. The VSI-Gr outputs are connected parallel to the low-voltage winding of the WG step-up transformer with the voltage of 35 kV/4 kV. The high-voltage winding via the line that is 20 km in length is connected to the group transformer having the power of 100 MVA and the voltage of 120 kV/35 kV and, afterward, to the network with the power of 2 GVA (Figure 6.55).

The turbine has the nominal power of 3.6 MW at the base wind speed of 12 m/s. The excitation control system produces the stator flux linkage nominal value at a rotating speed that is not more than the nominal value, and the flux linkage lowering under the speed excess. For flux linkage calculation, the same circuits as in the model **Wind_IG_7** are used, utilizing Equation 4.151. The exciter is modeled to be simplified, as a first-order low-pass filter. The switches K of the DC/DC converters are controlled, in order to maintain the average value of the current in the reactor L, which for each converter is

$$I_1^* = \frac{0.5T_e\omega_r}{U_{dc}}, \tag{6.36}$$

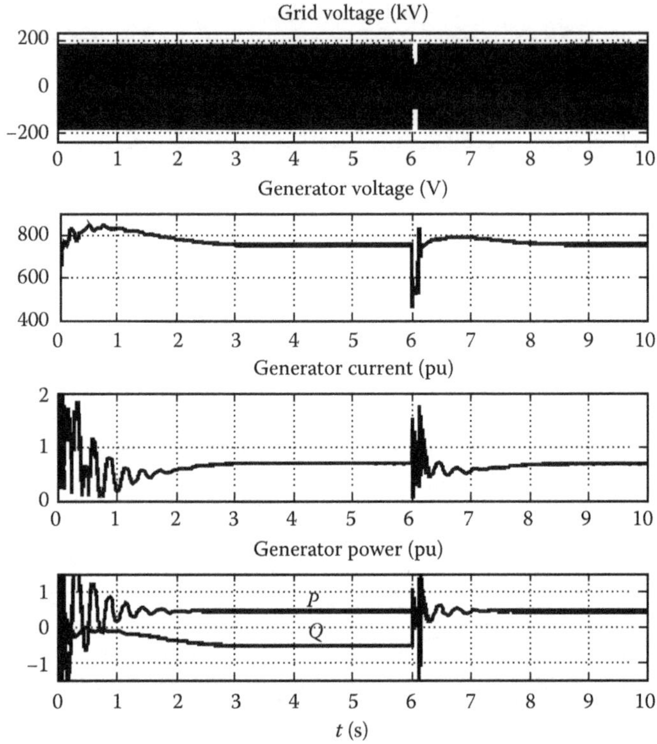

FIGURE 6.53 Response of the model **Wind_SG_1a** to the network voltage sag.

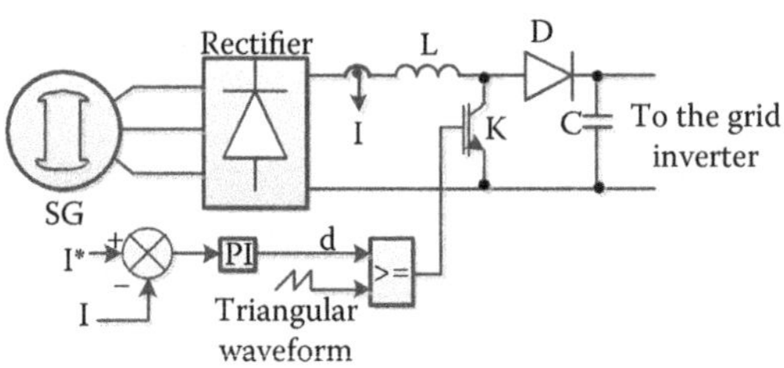

FIGURE 6.54 SG and DC/DC boost converter.

where T_e is defined by Equation 6.17, and U_{dc} is the average value of the rectifier voltage. Therefore, at every value of the rotating speed, the SG load corresponds to the maximum possible power at this speed; if there is a discrepancy between this speed and the optimal speed at the given wind speed, the rotating speed will change until it reaches its optimal value, as it has been already described.

The VSI-Gr control system diagram is depicted in Figure 6.56. Three controllers of the phase currents obtain the references from the block of the reverse Park transformation, whose angle of transformation is defined with the help of the PLL

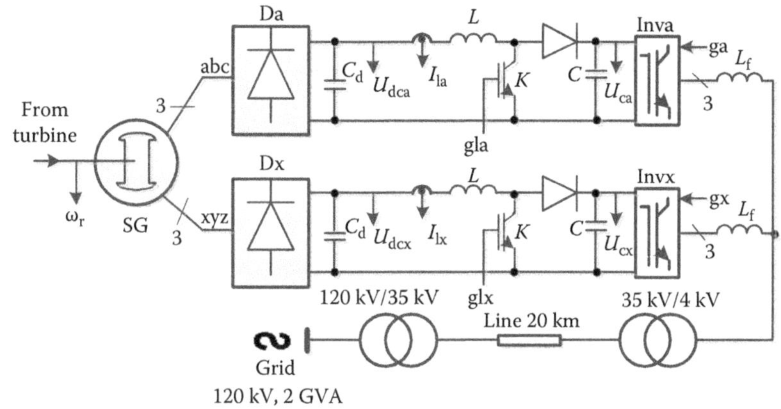

FIGURE 6.55 Arrangement of the power circuits in the model **Wind_SG_2**.

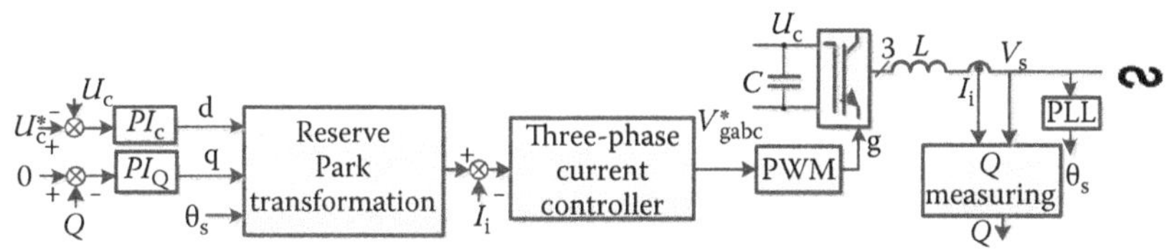

FIGURE 6.56 VSI-Gr control system diagram of the model **Wind_SG_2**.

connected to the network voltage. The I_d component is defined by the DC voltage controller, and the I_q component by the controller of the network reactive power. In order to decrease the content of the higher harmonics in the network current, the triangular waveforms of both PWM units are shifted mutually by one-half of their period. The DC voltage is taken to be equal to 6 kV.

The operation at the wind speed of 12 m/s is simulated, with the subsequent increase to 14 m/s and the following decrease to 7 m/s. The change of the number of the variables is displayed in Figure 6.57. The optimal values of the power are reached practically. So, for instance, at the wind speed of 12 m/s, the turbine torque referring to the SG is 0.9 (1 pu of the turbine nominal torque), and the powers of the SG and the network are 0.88 × 4 and 3.4 MW, respectively, whereas at the wind speed of 7 m/s, they must be by $(7/12)^3$ times lower, that is, they must be equal to 0.7 and 0.67 MW; we have in fact 0.68/0.67 MW, respectively. At the wind speed of 14 m/s, the turbine rotating speed is limited effectively. The scope **Capacitance_Voltage** proves that the voltages across the capacitors in the DC links are maintained to be equal to 6 kV with good accuracy; the scope **Grid** shows that the content of the higher harmonics in the network current is not much: in the nominal condition, THD ≈ 2%.

The elasticity of the kinematic train is additionally taken into account in the model **Wind_SG_2a**, as in the earlier considered model **Wind_DFIG_3**. The program `SM_MODEL_6ph4a.m` is executed before simulation start, in which, additionally

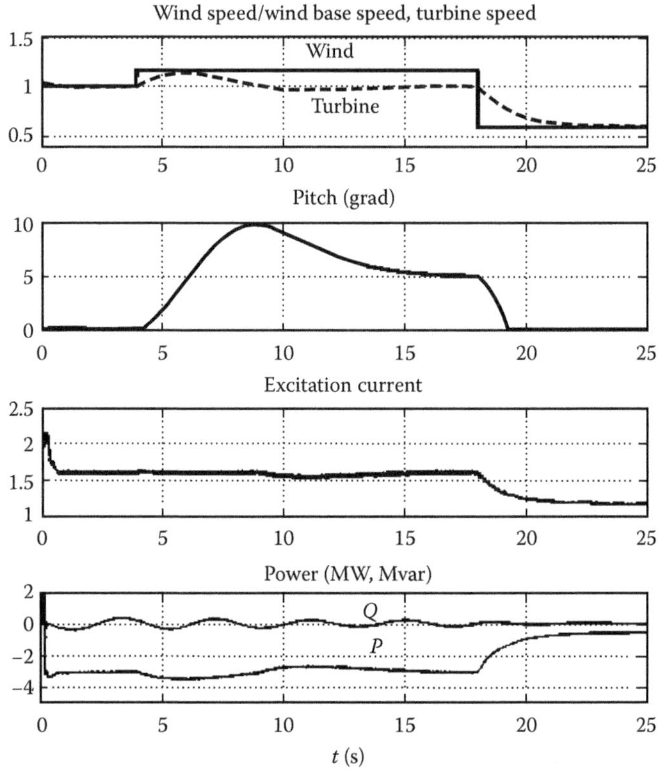

FIGURE 6.57 Processes in the model **Wind_SG_2**.

the coefficients of stiffness K (torque in pu/rad) and of damping D_m (torque in pu/speed change in pu) are specified, and also separately the moments of inertia of SG J and the turbine J_1 (the value $J + J_1$ is equal to the value J in the program `SM_MODEL_6ph4.m`).

The promising WG set with a power of 9 MW is modeled in the model **Wind_SG_3**. The 12-phase SG is utilized, whose model was considered in Chapter 4.

With a total power of 10 MVA, each winding has the nominal voltage of 3200 V; the SG parameters are specified in the program `SM_MODEL_12ph2.m`. Two VSIs are connected back-to-back to the circuit of the winding, as shown in Figure 6.58. The outputs of the VSI-Gr are connected in parallel and are connected to the 35 kV/3200 V step-up transformer. The transmission line is the same, as in the previous model.

The VSI-Ge controls the SG rotating speed. The output of the speed controller assigns the current components I_q, and the reference is zero for components I_d. The orientation is carried out by the rotor position. The speed reference is proportional to the wind speed: when the latter is 12 m/s, the former is 1 pu. The VSI-Gr controls the voltages across the capacitors in the DC link and the reactive power, as in the previous model (Figure 6.56).

The simulation of the considered system is carried out very slowly, owing to the many converters used, so a number of simplifications are made, in order to have a

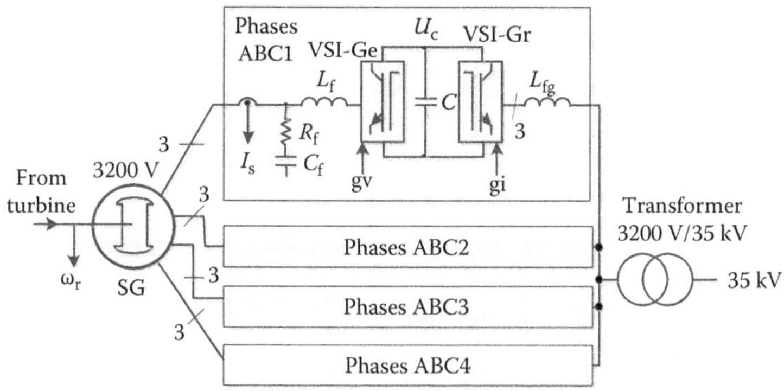

FIGURE 6.58 WG structure in the model **Wind_SG_3**.

reasonable simulation duration. The constant of inertia is essentially decreased in the model. Moreover, it is accepted that the information about the stator flux linkage is derived from the SG model; for this, the additional elements are put into the subsystem **SM_12/SM_Electr/Stator_Rotor**. The simulation of the converters is executed with the option *Average–model based VSC* that does not give the possibility to

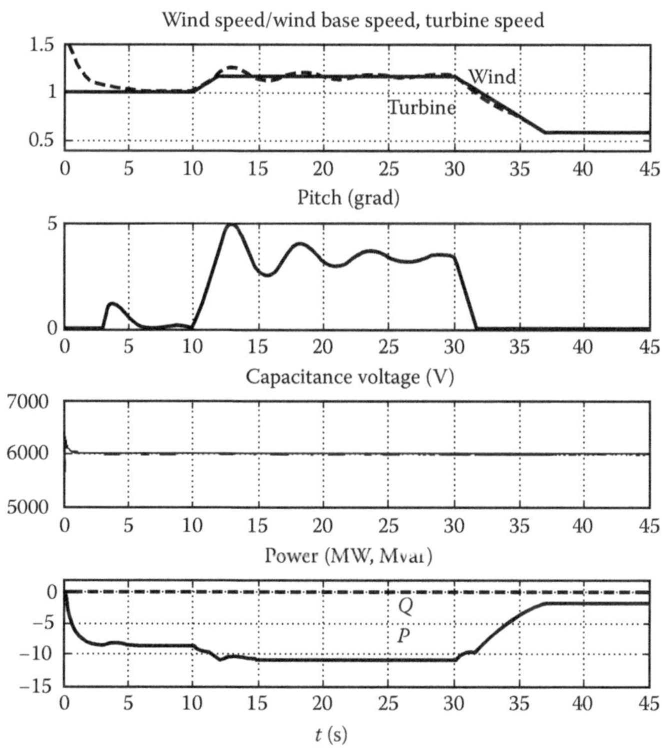

FIGURE 6.59 Processes in the model **Wind_SG_3**.

estimate the content of the higher harmonics. Some results of simulation are shown in Figure 6.59. When the wind speed is less than the base wind speed, the power is close to optimal (bearing in mind the losses). At the wind speed of 14 m/s, the power and the SG speed exceed the nominal a little, because in the block that controls the blade position **beta_contr**, the controller integral component is excluded. If it is

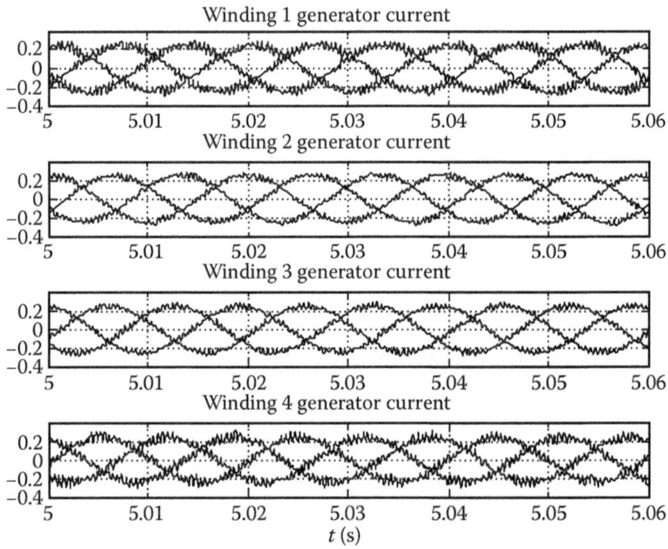

FIGURE 6.60 SG winding currents in the model **Wind_SG_3a**.

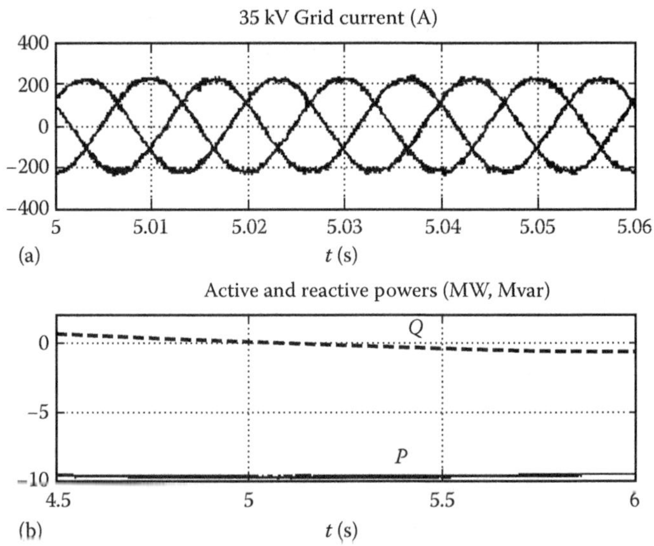

FIGURE 6.61 (a) Network current and (b) powers in the model **Wind_SG_3a**.

inadmissible, the adjustment or (and) the structure of the controller must be changed. The voltages across the capacitors are kept equal to 6 kV.

The model **Wind_SG_3a** is intended for the investigation of the steady states. The turbine is not modeled: proportional to the selected wind speed, the SG rotating speed (with the proportional factor 1/12) and the optimal torque are assigned. The IGBT transistors are utilized in the converters; four blocks for the estimation of the flux linkage components are put into the subsystem **Flux_Control** that are taken from the model **Wind_SG_2**. The triangular waveforms of the VSI-Gr PWM are shifted relative to each other by one-quarter of their period, which is equal to 1 ms. The simulation runs slowly. The SG winding currents at the nominal conditions are fixed in Figure 6.60. The network current at the voltage level 35 kV and the powers are displayed in Figure 6.61. For the network current, THD < 5%.

6.5 WG WITH PMSG

The PMSGs are considered today to be the most promising generators for WG. Unlike IG, they do not need the initial excitation, and unlike DFIG and SG with the winding on the rotor, they do not have the contact rings on the rotor. Their disadvantages are the higher cost (that reduces gradually) and the possibility to be demagnetized with large inrush currents.

The principles of PMSG control were considered in Chapter 4. The surface magnet generator (SPMSG) is utilized in the model **Wind_PMSM_1**, in which $L_d = L_q$; therefore, the generator current component I_q is defined with the wanted torque, and the component I_d must be equal to zero. The following PMSG parameters are taken under simulation: the transparent power S = 2.2 MVA under the nominal voltage V_n = 690 V and the frequency of 9.75 Hz and Z_p = 26. Then the nominal rotating speed is $\omega_n = 2 \times \pi \times 9.75/26 = 2.355$ rad/s, the nominal current is $I_n = \sqrt{2} \times 2.2 \times 10^6/690/\sqrt{3} = 2606$ A (amplitude), and the nominal torque is $T_{en} = 2.2 \times 10^6/2.355 = 934.2$ kNm. With the flux of the permanent magnets Ψ_r = 9.18 Wb, the nominal torque is obtained under

$$I_{qn} = \frac{934,200}{1.5 \times 9.18 \times 26} = 2609 \text{ A.}$$

The turbine has a power of 2 MW at the base wind speed of 12 m/s, having the nominal rotating speed. VSIs connected back-to-back utilize IGBT and PWM with the frequency of 2160 Hz. The VSI-Ge control system contains the rotating speed controller, and the reference speed is defined as

$$\omega_r^* = K_m \sqrt[3]{P_g}, \tag{6.37}$$

where P_g is the measured PMSG power, and K_m is the coefficient that depends on the accepted units and on the turbine parameters (in the case under consideration, in pu, K_m = 1). Therefore, the value ω_r^* is optimal if P_g is equal to the optimal value at the given wind speed. If the latter rises, P_g also increases, and the rotating speed increases until it reaches the value that corresponds to the maximum at the new wind speed.

The VSI-Gr is connected to the power system via the 35 kV/690 V step-up transformer. The model of the power system is the same, as in the previous model. As for the VSI-Gr control system, it contains the controller of the voltage across the capacitor in the DC link and the reactive power controller—as in the previous model.

The following process is simulated: the operation under the nominal conditions (wind speed is 12 m/s), the increase in the wind speed to 15 m/s at $t = 7$ s, and its abatement to 7 m/s at $t = 20$ s. The transients of the PMSG rotating speed, the capacitor voltage, and the active and reactive network powers are displayed in Figure 6.62. At a wind speed of 15 m/s, the power and the rotating speed are limited by the change of the blade angle position. It may be seen by the scope **Grid Power** that THD is not more than 2% in the network current.

The WGs considered earlier demand, for their function, the transducers for the rotor position/speed measurement and often the wind speed meter. These mechanical devices have little reliability: sensor failures contribute to more than 14% of failures in WG [6]; therefore, the development of WG without the meters mentioned earlier is very topical.

The turbine power P_m depends on the wind speed V_w and the rotating speed ω (Equation 6.1); therefore, by knowing the power and the rotating speed, the wind

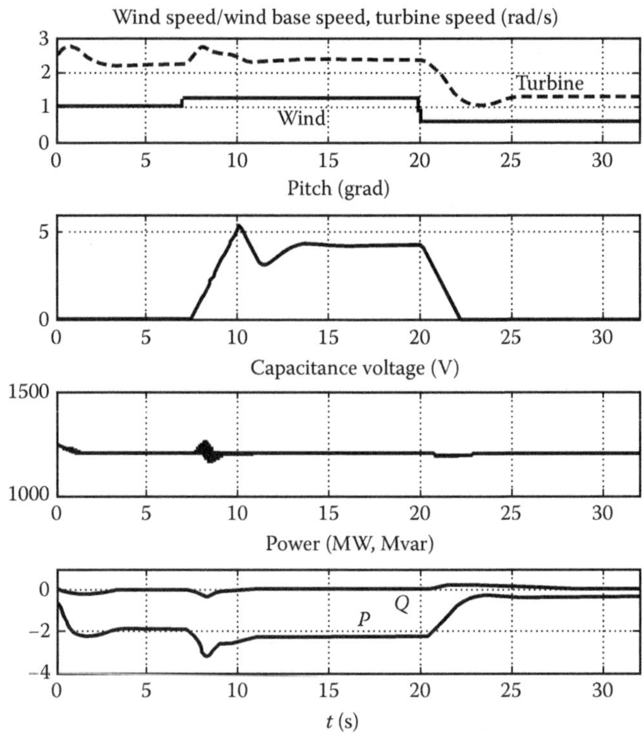

FIGURE 6.62 Processes in the model **Wind_PMSM_1**.

speed may be estimated. It is assumed that the turbine characteristics obtained by either calculation or experiment are available. The quantity P_m is not measured, but the generator electric power P_{el} can be computed, and afterward the estimation P_{me} of P_m may be calculated as

$$P_{me} = P_{el} + P_{din} + P_{loss}, \tag{6.38}$$

where P_{din} is the dynamic power during the changing of the rotating speed, and P_{loss} is the power of the losses.

The second quantity is computed as

$$P_{din} = J\omega \frac{d\omega}{dt}; \tag{6.39}$$

the third quantity is a sum of the losses in the generator winding, of the core losses, of the ventilation losses, and of the friction losses.

The sum of the copper, the friction (and other constant losses), and the ventilation losses may be computed as

$$P_{loss} = 3/2\left(I_d^2 + I_q^2\right)R_s + (D + F\omega_r)\omega_r, \tag{6.40}$$

where I_d and I_q are the components of the stator current space vector directed along the axes d and q, and R_s is the stator winding active resistance.

Then the wind speed can be found by using the circuits with the feedback, depicted in Figure 6.63.

The different types of the observers can be used for the estimation of the PMSG position and the rotating speed; the observers that use the sliding mode seem promising. Earlier, such observers were utilized in the PMSM [7,8]; their employment for WG is considered in Ref. [6].

For SPMSG, the following equations in the stationary reference frame are valid:

$$L\frac{d\mathbf{i}_s}{dt} = \mathbf{A}_{11}\mathbf{i}_s + \mathbf{A}_{12}\Psi_r + \mathbf{u} \tag{6.41}$$

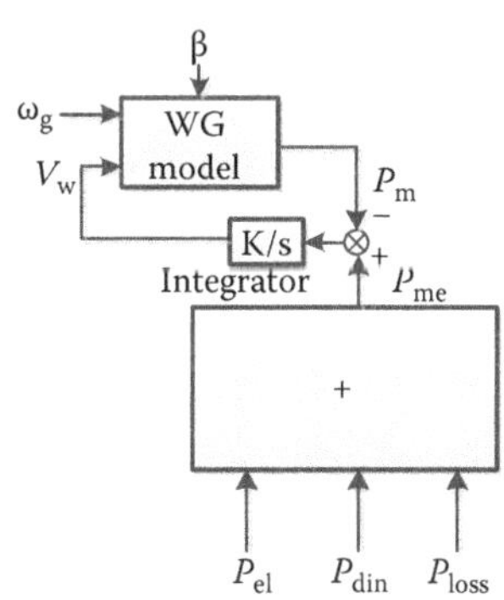

FIGURE 6.63 Block diagram for wind speed estimation.

$$\frac{d\Psi_r}{dt} = \mathbf{A}_{22}\Psi_r \tag{6.42}$$

$$\mathbf{A}_{11} = a\mathbf{I},\ a = R_s$$

$$\mathbf{A}_{12} = c\mathbf{J},\ c = -\omega_r$$

$$\mathbf{A}_{22} = d\mathbf{J},\ d = \omega_r,$$

where

$$\mathbf{J} = \begin{vmatrix} 0 - 1 \\ 1 \ldots 0 \end{vmatrix},$$

R_s is the stator winding resistance, L is the stator winding inductance, and ω_r is the rotor rotating speed (electrical).

The vector Ψ_r has the components $-F_r \sin\theta$ and $F_r \cos\theta$, F_r is the rotor flux, and θ is the rotor position. Only the current $\mathbf{i}_s$ is measured.

For Ψ_r estimation, the sliding mode observer is employed. The following current estimation $\hat{\mathbf{i}}_s$ is used:

$$L\frac{d\hat{\mathbf{i}}_s}{dt} = \mathbf{A}_{11}\mathbf{i}_s + \mathbf{u} - \mathbf{z} \tag{6.43}$$

$$\mathbf{z} = K_1\mathbf{sign}(\mathbf{e}_i), \mathbf{e}_i = \hat{\mathbf{i}}_s - \mathbf{i}_s$$

The equation for the error $\mathbf{e}_i$ is

$$L\frac{d\mathbf{e}_i}{dt} = \mathbf{A}_{12}\hat{\Psi}_r - \mathbf{z}. \tag{6.44}$$

If $K_1 > \max|W|$, where W is the first term in Equation 6.44, the sliding mode for $\mathbf{e}_i$ is realized, and it is possible to take $\mathbf{e}_i = d\mathbf{e}_i/dt = 0$ [9], so that

$$\mathbf{z} = \omega_r[\Psi_{r\beta}, -\Psi_{r\alpha}]. \tag{6.45}$$

After averaging the obtained values of the vector $\mathbf{z}$ components, we find $e_\alpha = \omega_r F_r \cos\theta$ and $e_\beta = \omega_r F_r \sin\theta$, so that

$$\theta = \tan^{-1}\frac{e_\beta}{e_\alpha}. \tag{6.46}$$

The rotating speed may be found as

$$\omega_r(k) = \frac{\theta(k) - \theta(k-1)}{T_c}, \tag{6.47}$$

where T_c is the sampling period.

The considered system is modeled in **Wind_PMSG2_sensorless**. WG has the power of 250 kW at the base wind speed of 11 m/s; the nominal rotating speed is 22 rad/s. SPMSG has the power of 280 kVA and has 18 pole pairs, F_r = 1.052 Wb, so that the nominal voltage is 22 × 18 × 1.052 = 416 V, and the frequency is 22 × 18/(2π) = 63 Hz.

The WG and the grid connected in parallel supply the load of 200 kW. At the big wind speed, the power surplus is sent to the grid; at the low wind speed, the load power shortage is made up with the grid. The latter is modeled to be simplified, as the source of 400 V with the inner impedance. The high-pass filter with a tuning frequency of 350 Hz is set in parallel to the load. Two back-to-back VSIs are set at the PMSG output, as in the previous model, having the same control systems; but at the inputs of the VSI-Ge control system, instead of the quantities of PMSG position and speed, their estimations are given.

For WG operation optimization, it contains the speed controller; the reference for this controller is proportional to the wind speed (at V_w = 11 m/s, the reference is 22 rad/s).

The estimation of the wind speed is carried out in the subsystem **Vwest** that is made according to the diagram in Figure 6.63. The relationships for the estimation of the rotor position and speed given earlier are realized in the subsystem **Slide_Contr**. The estimations of the stator current components I_α and I_β are formed at the integrator outputs. In order to decrease oscillations, owing to the discreteness of computation (jogging), instead of the function *Sign*, the block **Saturation** with limits ±1 is employed; the error of 1A leads to saturation. The parameter K_1 is accepted to be equal to 100.

The estimations of the EMF components e_α and e_β are formed at the outputs of the first-order filters with the time constant T_f = 1 ms. The circuits for the estimation of the rotor position θ_e (*Tetae*) by Equation 6.46 and of the rotating speed by Equation 6.47 with T_c = 1 ms are placed in this subsystem as well. The calculated speed value is filtered with the time constant of 10 ms.

Because, for receiving initial estimations, some steady-state conditions have to be reached in the system, the start of operation and the turning on of the different blocks are moved apart in time.

The results of simulation, the wind speed equal to 11 m/s increases to 14 m/s at t = 2.5 s and falls to 7 m/s at t = 7.5 s, are shown in Figure 6.64. It is seen that the wanted regimes are obtained. For example, with the measured power of 236 kW at the wind speed V_w = 11 m/s, at V_w = 7 m/s, the power has to be 236 × $(7/11)^3$ = 61 kW; the scope **Power** shows 63.5 kW. The currents of the WG, the grid, and the load in the different instants (at different wind speeds) are displayed in Figure 6.65. At V_w = 14 m/s, the part of the WG power is sent to the grid (the voltage and the grid current are in antiphase), but at V_w = 7 m/s, the WG and the grid supply the load together (the voltage and the grid current are in-phase).

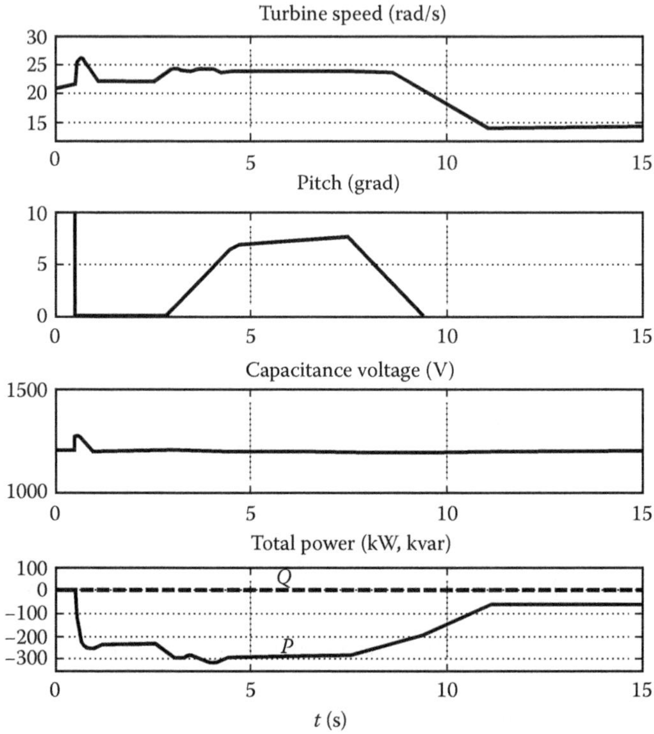

FIGURE 6.64 Processes in the model **Wind_PMSG2_sensorless.**

The model **Wind_PMSG3** is a copy of the model **Wind_PMSG1**, with the difference that, instead of SPMSG, the interior magnet generator (IPMSG) is utilized ($L_d \neq L_q$). The power, the voltage, and the rotating speed are preserved, and the number of the pole pairs is higher, so that the rotor flux is reduced. In order to have the minimum IPMSG current with the same torque, the current component I_d must be controlled so that the relationship in Equation 4.161 will be fulfilled. The IPMSG parameters necessary for that are specified in the option *Model Propeties/Callbacks/InitFcn*. The blocks that compute the reference I_d^* are added in the subsystem **VSI-Ge_Reg**: the block **Gain1** recalculates the actual reference for i_{sq} into the relative unit I_q^* as it is indicated in Section 4.3.1; the block **Fcn1** solves the quadratic equation 4.161 for I_d^*; the block **Gain2** carries out the calculations inverse to the block **Gain1**. It may be seen after simulation that at V_w = 12 m/s, the rms of the SG current is 1686 A. To repeat the simulation with $I_d^* = 0$, this value turns out to be 1771 A, by 5% more.

The six-phase IPMSG considered in Chapter 4 is utilized in the model **Wind_PMSG4**. The IPMSG power is 3 MVA, and the voltage of each winding is 4 kV. The IPMSG has six pole pairs; the output voltage nominal frequency is 40 Hz, so that the nominal rotating speed is

$$\omega_{rnom} = (40/50) \times 100 \times \pi/6 = 41.9 \text{ rad/s} = 400 \text{ r/min}.$$

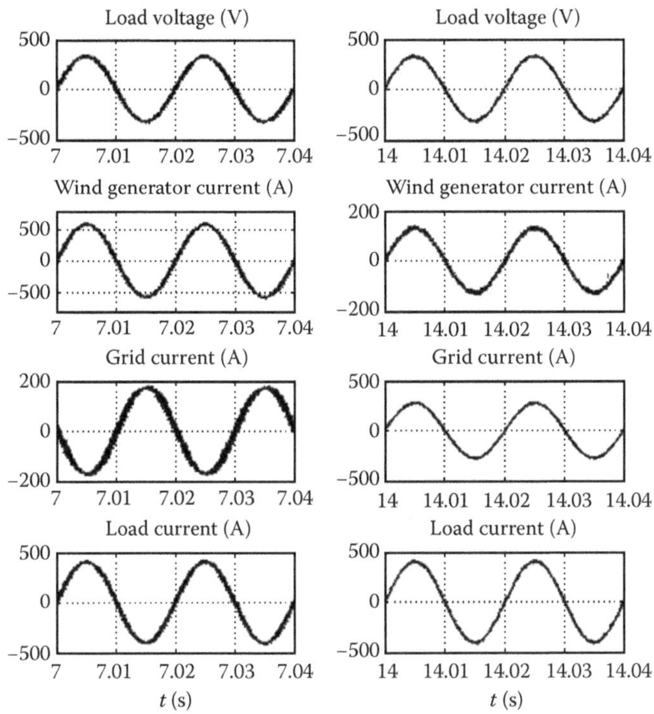

FIGURE 6.65 Currents in the model **Wind_PMSG2_sensorless.**

The turbine model is the same, as in the previous model, and its power is 2.5 MW. Two back-to-back VSIs are set in the circuit of each winding, as has already been made in the model **Wind_IG_6**; the outputs of both VSI-Gr are connected in parallel and are connected to the 35 kV/4 kV step-up transformer; the triangular waveforms for PWM are mutually shifted by 1/2 of their period. The network model is the same, as in the previous models. The program `Contr_PMSM_T2.m` has to be executed before simulation start.

The optimization of the WG operation is reached with employment of relationship 6.17 in the subsystem **Control**; the losses in the stator windings are taken into account. After the calculation of the wanted torque, this value is converted into the relative value *m*; afterward the reference currents I_d^* and I_q^* are computed, as shown in Section 4.3.2 (Figures 4.32 and 4.33). The VSI-Gr has the same control system as in the previous model. The reactive power controller is placed in the subsystem **Control**.

It should be said about the ballast resistors that are set in the PMSG model parallel to the current sensors. Under their absence, the simulation runs very slowly or stops altogether. However, they cause the additional losses of power that can distort the simulation results. In the considered model, because of the high generator voltage, their resistance is taken as 1 kΩ; however, with the simulation sampling period of 10 μs taken initially, the simulation runs very slowly; with sampling period reduced to 5 μs, the simulation runs much faster (by three times!), so that the resistances and the sampling period have to be defined by the simulation experiment.

Some simulation results are displayed in Figure 6.66. The rotating speed changes proportional to the wind speed; the values of the power and of the speed are limited at high wind speed (remember that for simulation speed-up, the turbine constant of inertia is reduced, and the rate of the wind speed change is increased about the real values). The currents in the IPMSG windings and in the grid at the voltage level of 35 kV in the nominal conditions are shown in Figure 6.67.

A relatively high level of the IPMSG voltage makes the employment of the three-level VSI reasonable, instead of a two-level VSI. Such inverters are used in the model **Wind_PMSG4a**. The processes in this model do not differ from those in the previous model; the currents are shown in Figure 6.68. The grid currents contain fewer harmonics.

Wind_PMSG5 models the wind park that consists of two clusters; they contain N and N_1 of the identical WGs, respectively; every WG is made as in the model **Wind_PMSG1**. Each cluster is modeled as a generalized WG with corresponding parameters. The clusters are connected in parallel to the 120 kV/35 kV step-up transformer. The clusters are located so that their wind speeds can be different. The processes in the model with $N = 10$ and $N_1 = 8$ under wind speed changing are depicted in Figure

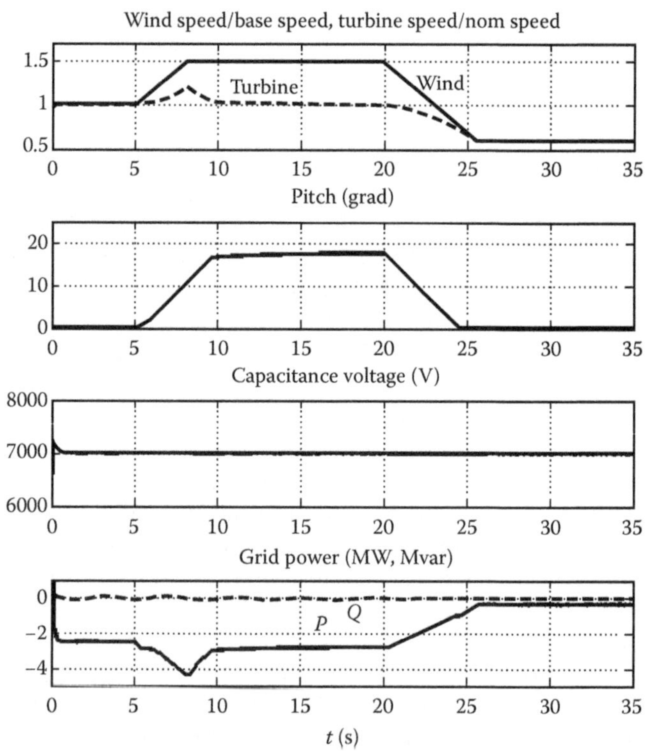

FIGURE 6.66 Processes in the model **Wind_PMSG4.**

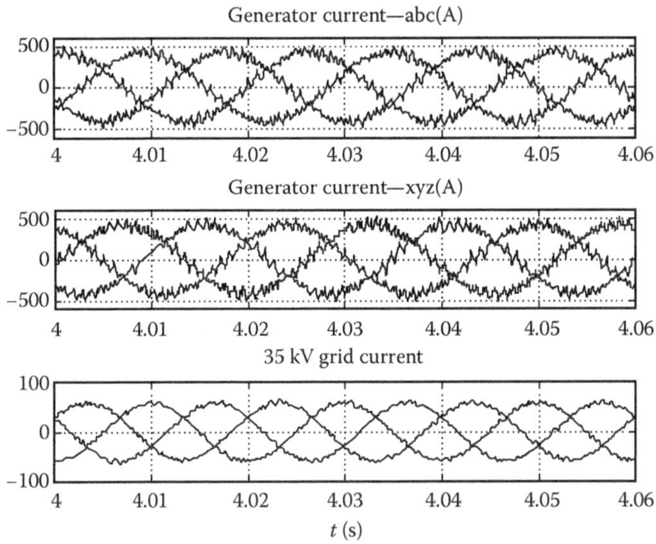

FIGURE 6.67 Generator and grid currents in the model **Wind_PMSG4**.

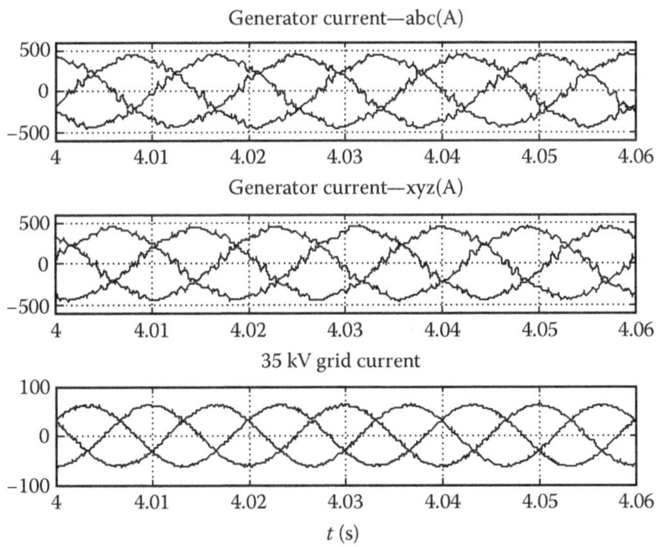

FIGURE 6.68 Generator and grid currents in the model **Wind_PMSG4a**.

6.69. The processes when the voltage of the source 120 kV at $t = 4$ s during 100 ms drops by 25% are displayed in Figure 6.70.

It has been already said about the tendency to place the wind parks in the sea—offshore. For the majority of the operating offshore wind parks, the connection with the power system is fulfilled with the AC transmission line by underwater cables.

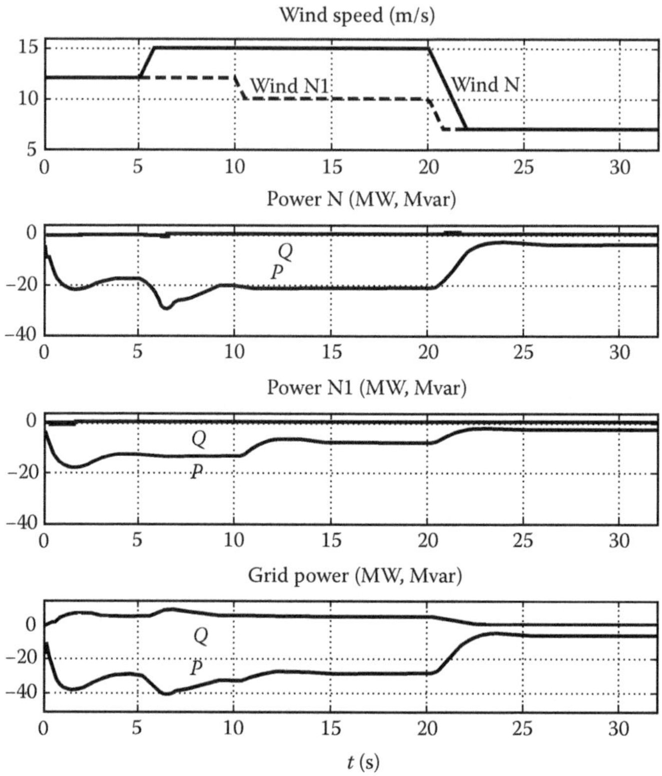

FIGURE 6.69 Processes in the model **Wind_PMSG5**.

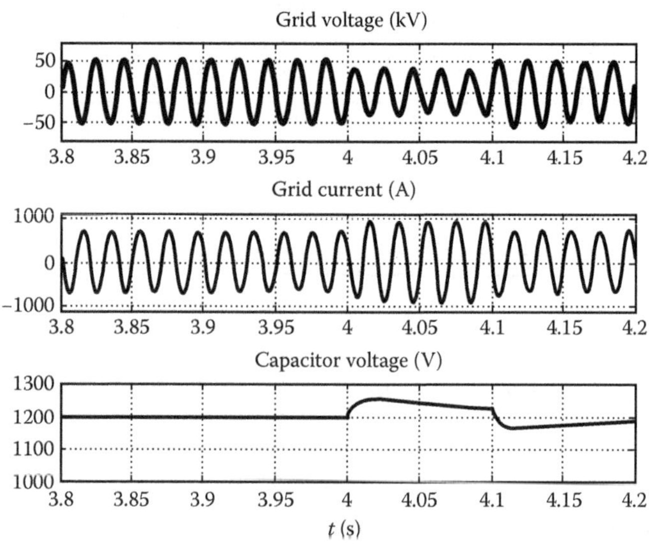

FIGURE 6.70 Processes in the model **Wind_PMSG5** under the grid voltage sag.

These cables have a distributed capacitance that essentially exceeds the parameter of the air transmission lines; this fact causes big leakage currents and limits the possible distance of the wind park from the coast. The compensators of the reactive power (CRP) are usually set at the receiving terminal. The model **Wind_PMSG6** is the modification of the previous model for the offshore placement. The clusters are connected with cables 5 km in length to the offshore step-up transformer, which, via three underwater cables that are 50 km in length, is connected to the onshore network. In the point of connection with the network (onshore), the CRP is set, which is the inductive part of SVC model (Chapter 3).

The processes in the system are shown in Figure 6.71. CRP is turned on at $t = 10$ s. Before this time, the large reactive power is sent to the network that comes to naught after CRP is turned on. At this, the installed reactive power at the voltage level of 16 kV is (the phase CRP inductance is 60 mH)

$$Q_{in} = \left(\frac{16}{\sqrt{3}}\right)^2 \frac{3}{2\pi \times 50 \times 60 \times 10^{-3}} = 13.6 \text{ Mvar}.$$

Bearing in mind the mentioned feature of the AC power transfer, the transfer of the electric power from the offshore wind park to the onshore with the employment of HVDC seems to be very promising [10]. Unlike the onshore HVDC, HVDC employment for energy transfer from offshore to onshore has a number of special features. In the first case, the source and the recipient of the sufficient large power are supposed; the operator controls the flow of the transmitted power with the change

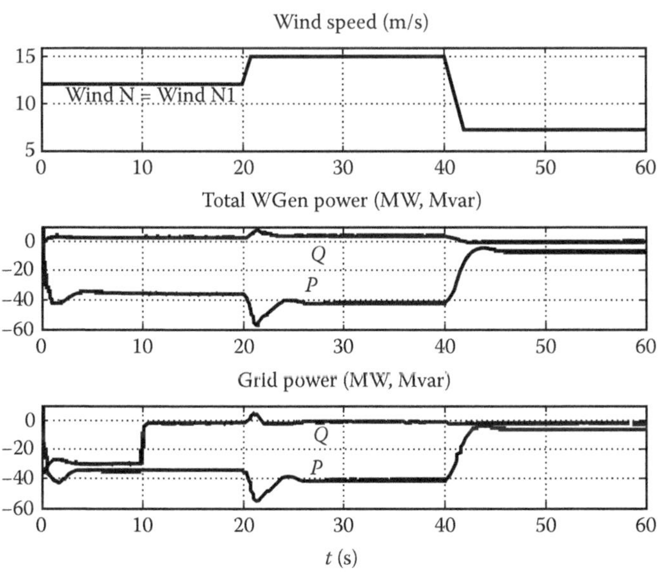

FIGURE 6.71 Processes in the model **Wind_PMSG6.**

of the current quantity in the direct current circuit (see model **HVDC**). In the second case, it is impossible to increase the flow of the transmitted energy at the given wind speed, and the control system must transfer the maximum power. Moreover, the possibility to change the rotating speed of the individual WG for maximum efficiency must be provided.

There are a lot of different proposals about the HVDC transmission structure for offshore systems [11], which, however, do not reach the level of realization. Furthermore, the models of some possible variants are suggested. Only the systems with VSI are considered, although a number of proposals with current source inverters and with thyristors exist.

The design with the series connection of the VSI-Ge is depicted in Figure 6.72. The VSI-Gr and the matching transformer are installed onshore. The VSI-Ge changes the WG rotating speed in the optimal way; the VSI-Gr maintains the DC voltage equal to the assigned value. This configuration contains minimum elements, but a number of the shortcomings are inherent to it; a main shortcoming is the high standards of the WG insulation. When one WG fails, the capacitor this WG is bypassed, which leads to the rise of the voltages across the remaining capacitors. Because the capacitor total voltage is maintained, the voltage deviation of the individual capacitors from the average value equal to $U_m = U_c/n$ can take place.

In the following two models, the considered system is simulated with the different VSI-Gr. Since the number of individual WGs is rather large, the simulation runs very slowly. Therefore, in the developed models, the number of WGs is taken to be essentially smaller in comparison to the real sets that lead to the decrease of the voltage level in the system.

Four WGs are connected in series in the model **Wind_PMSG_7**; each WG has a power of 3 MVA under the voltage of 4 kV; the turbine nominal power is 2.5 MW, and the nominal rotating speed is 60.3 rad/s that under $Z_o = 6$ gives the nominal frequency of 57.6 Hz.

The VSI-Ge control system is taken from the previous model, but the circuits that influence I_q^* are added (see further); the DC voltage is taken as 7 kV. The VSI-Gr control system is taken from the previous model as well; the assigned transmission line voltage is $4 \times 7 = 28$ kV. The VSI-Gr is the two-level one and is connected to the power

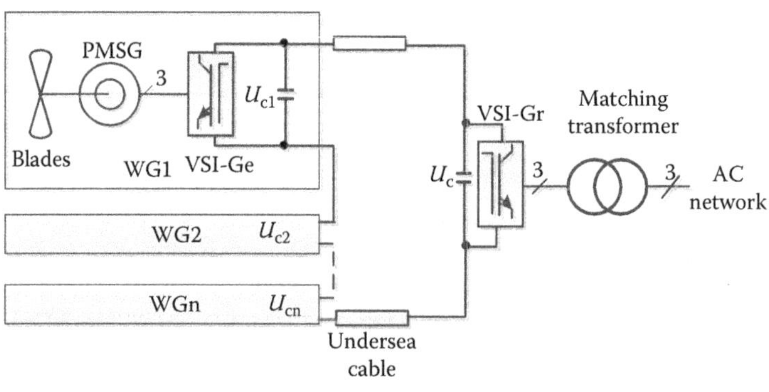

FIGURE 6.72 Offshore wind park with DC transmission with series connection DC links.

system via the 16 kV/120 kV step-up transformer. The block **Three-phase harmonic filter** tuned for suppression of 5th and 7th harmonics is set at the VSI-Gr output.

We have already discussed the influence of the resistors that are set at the PMSG outputs. In the case under consideration, they are taken as 300Ω that with a nominal voltage of 4 kV cause notable losses of active power. At the same time, in an attempt to increase their resistances, the simulation slows down sharply or stops altogether. In order to compensate for the error that is caused by the losses in these resistors, the power of losses is measured; the torque applied to the PMSG is increased by the proper value.

Since all WGs are located not far one from the others, the wind speeds for all WGs are approximately the same; nevertheless, some distinctions can exist caused by gusts or with the WG mutual influence.

The process is simulated when, at common wind speed of 12 m/s, the wind speed for WG4 drops by 1 m/s at a time interval of 4–8 s. At this, the voltage across its capacitor decreases and across others increases, since the sum remains constant. Since it is not possible to increase the WG power, the system is employed in the model, at which, when the voltage across the WG capacitor exceeds the assigned level, the WG rotating speed increases a little that leads to the reduction of the fabricated power and to the limitation of the capacitor voltage; this limits the voltage reduction across the capacitor of WG with the small wind speed too. Unfortunately, the total power of WG park decreases at that. The process is shown in Figure 6.73. The voltages across the capacitors WG1–WG3 increase by 360 V, and the voltage across the capacitor WG4 reduces by 1190 V; the power delivered to the network decreases by 0.7 MW. To repeat the simulation without the described action on WG rotating speeds, it may be seen that the voltages across the capacitors WG1–WG3 increase by 430 V, and the voltage across the capacitor WG4 reduces by 1350 V, but the power delivered to the network decreases only by 0.5 MW.

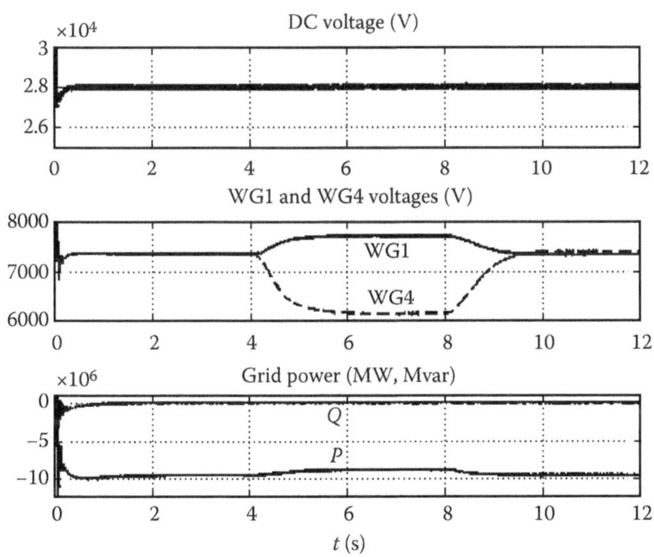

FIGURE 6.73 Processes in the model **Wind_PMSG_7**.

The three-level VSI-Ge instead of the two-level one is utilized in the model **Wind_PMSG_7a**; the processes in it do not differ practically from one shown earlier.

Because of the shortcomings of the arrangement with the series WG connection mentioned earlier, the configurations with their parallel connection are developed, that is, with the underwater DC power system. Such a system must have the step-up transformers for transformation of the relatively low WG voltage into the transmitted high voltage and for the WG insulation. These transformers can operate at a low frequency of 20–60 Hz or at a middle frequency of about 1 kHz. The former are more bulky but need smaller elements of the power electronics.

The diagram with the parallel connection of the high-voltage VSIs is depicted in Figure 6.74. The step-up transformers are set between the PMSG with the nominal voltage of 4 kV and the VSI-Ge; the step-up transformer is an integral part of the WG unit (Figure 6.3). WGs can be controlled individually to obtain the maximum power. Since at the wind speed equal to one-third of the designed one, the WG power is 4% of nominal; it may be supposed that the minimal rotating speed is not smaller than 30% of the nominal. Because the PMSM voltage reduces proportionally to the rotating speed decrease, and in line with this, its current also decreases (the power decreases as the cube of the value of the rotating speed), such range for transformer frequency change is permitted. For the transformer utilized in the previous model, the nominal frequency is 57.6 Hz, so that in the given WG set, the transformer minimal frequency is 20 Hz.

In the model **Wind_PMSG_8**, four WGs with the power of 10 MW are connected in parallel. Each WG can be a promising WG of indicated power or imitate a cluster of WGs with less power [5]. The WG has an SPMSG with the power of 12 MVA under the voltage of 4 kV and the frequency of 57.6 Hz; it is equipped with a three-level VSI. The VSI-Ge control system is the same as in the previous model.

The three-level VSI is utilized at the receiving end as well, whose control system is also adopted from the previous model. The filters of the 3rd, 5th, and 7th harmonics are set at the input of the low side of the matching transformer.

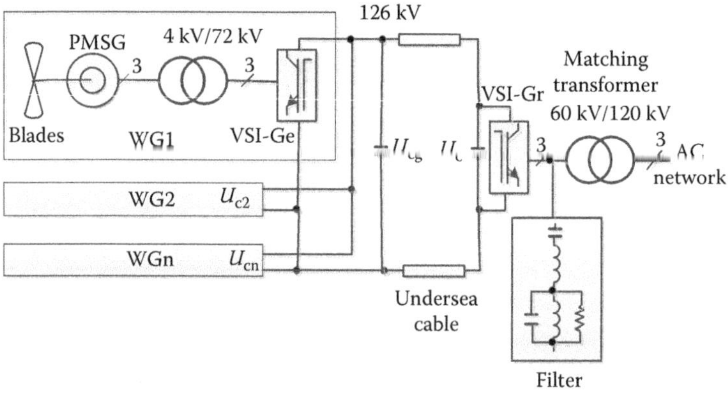

FIGURE 6.74 Configuration of the offshore park with low-frequency transformers and HVDC transmission.

The process is simulated, when, at the initial common wind speed of 12 m/s, the wind speed of the WG4 drops by 1 m/s in the time interval (2–4 s), and at $t = 7$ s, the common wind speed increases to 15 m/s. The start process is not simulated. Since the turbine constant of the inertia is decreased and the rate of wind speed change is increased relative to the real values, the deviations of the control variables during the transients seem bigger as in reality. The process is fixed in Figure 6.75.

The different possibilities to build an HVDC offshore transmission of electrical energy with medium frequency transformers are considered in a number of researches [10]. The system, whose structure is shown in Figure 6.76, is used in the model **Wind_PMSG_9**. The DC/DC boost converter is made, as shown in Figure 6.54. The reference for the inductor current is defined as

$$I^* = \frac{K_m \omega_r^2 - D - F\omega_r}{U_{dcx}} \omega_r P_{nom}, \tag{6.48}$$

where ω_r is the turbine rotating speed in pu, P_{nom} is the turbine nominal power, and U_{dcx} is the voltage of the diode rectifier. At this, as it has been already said, MPPT is fulfilled. With the help of the VSI **Inv**, the voltage U_{cx} at the output of the DC/DC

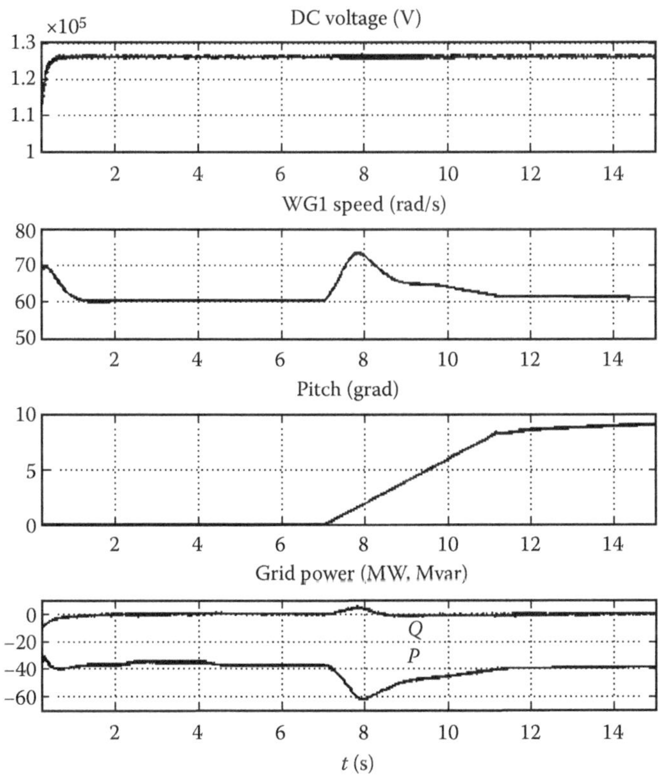

FIGURE 6.75 Processes in the model **Wind_PMSG_8**.

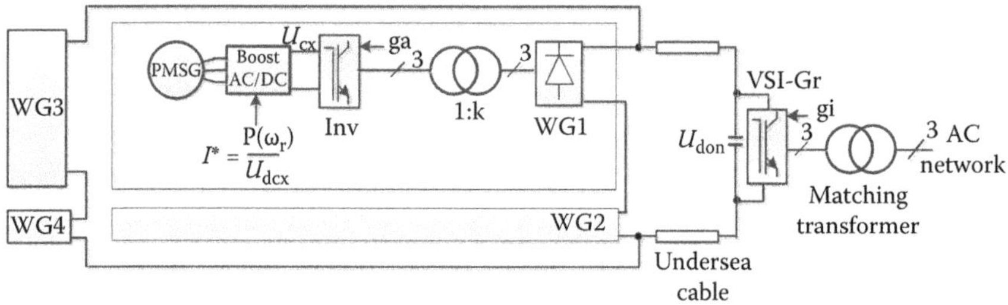

FIGURE 6.76 Configuration of the offshore park with medium-frequency transformers and HVDC transmission.

converter ($U_{cx} > U_{dcx}$) converts into the three-phase voltage of the rectangular waveform with a frequency of 1 kHz that comes to the step-up transformer ($k > 1$) and is afterward rectified with the diode rectifier. Several identical systems can be connected in series, in order to step up the transferred voltage, and in parallel, in order to increase the transferred power. Onshore, the received DC voltage is converted, with the VSI-Gr, into the three-phase voltage that, via the matching transformer, is delivered in the power system.

In the model, the turbine power is taken to be equal to 5 MW, and the PMSG power is 6 MVA; the rest of the parameters are the same as in the previous model, $k = 3$. Two identical WGs are connected in series and are connected to the underwater transmission line. The three-level VSI is utilized onshore; its control system is the same, as in the previous models. The assigned DC line voltage is taken to be equal to $21 \times 3 \times 2 = 126$ kV, that is, the voltage at the output of each converter AC/DC in the nominal condition has to be 21 kV.

The process is simulated, when, at an initial common wind speed of 12 m/s, the wind speed of the WG2 drops by 1 m/s in the time interval (2–4 s), and at $t = 7$ s, the common wind speed increases to 15 m/s. The simulation runs slowly, and the start process is not simulated; the turbine constant of the inertia is decreased. The process is shown in Figure 6.77. The essential losses of the power are seen in the system. A more detailed simulation shows that the main losses occur in the high-frequency DC/DC converters. Although the simulation proves the operability of such a system, the development of the devices that can switch over the high voltages and currents with high frequency is up-to-date without the difficult technical problem; moreover, there are problems with the load distribution under series and parallel connections of the individual WG.

It is possible to build the offshore park with the underwater AC bus [10,11] (Figure 6.78). Each WG has the DC/DC boost converter and the VSI, as the previous model, but the converter functions change: the first converter maintains the voltage across the capacitor at the VSI input (20 kV in the given case); the VSI carries out the optimal changing WG speed with the control of current component I_d at the output of VSI according to

$$I_d^* = \sqrt{2}\,\frac{K_m\omega_r^2 \quad D \quad F\omega_r}{\sqrt{3}V_1}\,\omega_r P_{nom}, \tag{6.49}$$

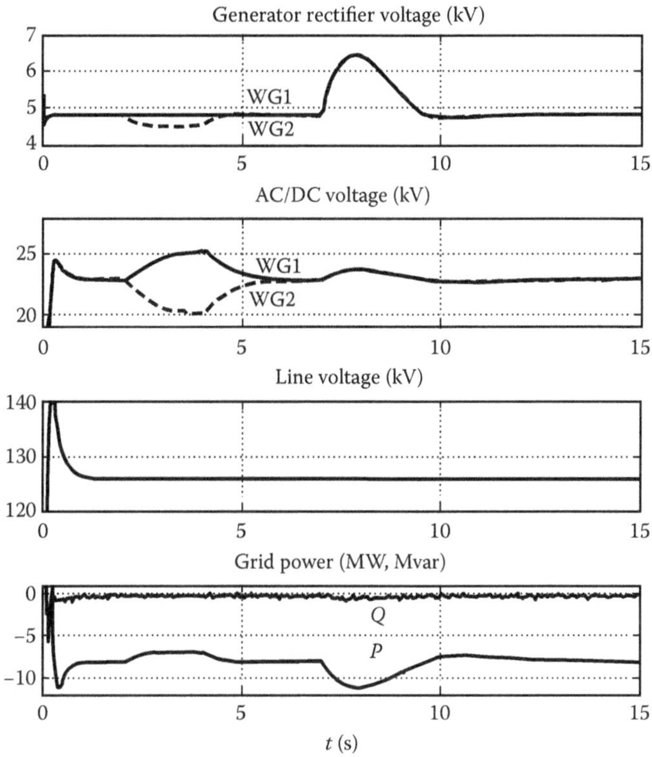

FIGURE 6.77 Processes in the model **Wind_PMSG_9**.

where V_l in the case is 12 kV (because the current at the 35 kV/12 kV transformer low sides is measured).

All WGs are connected to the underwater bus via the 35 kV/12 kV step-up transformers. The synchronizing voltage for the bus is produced with the converter **Rect** that is connected to the underwater bus via the 105 kV/35 kV transformer and can operate in two modes. When the switch **Sw** is in position *a*, the square gate pulses of the assigned frequency (50 Hz in the case) come to the converter control input; it is the so-called six-step operation. The DC voltage in the transmission line has to be equal to

$$U_d = \frac{\pi V_{leff}}{\sqrt{6}} = \frac{\pi \times 105}{\sqrt{6}} = 135 \text{ kV}.$$

When the switch **Sw** is in position *b*, with the aid of PWM, the voltage at the **Rect** input is controlled: the three-phase voltage with the frequency of 50 Hz, coming to PWM, is modulated by the amplitude with the output of the voltage controller.

VSI-Gr maintains the assigned line voltage with the system that was repeatedly employed in the previous models.

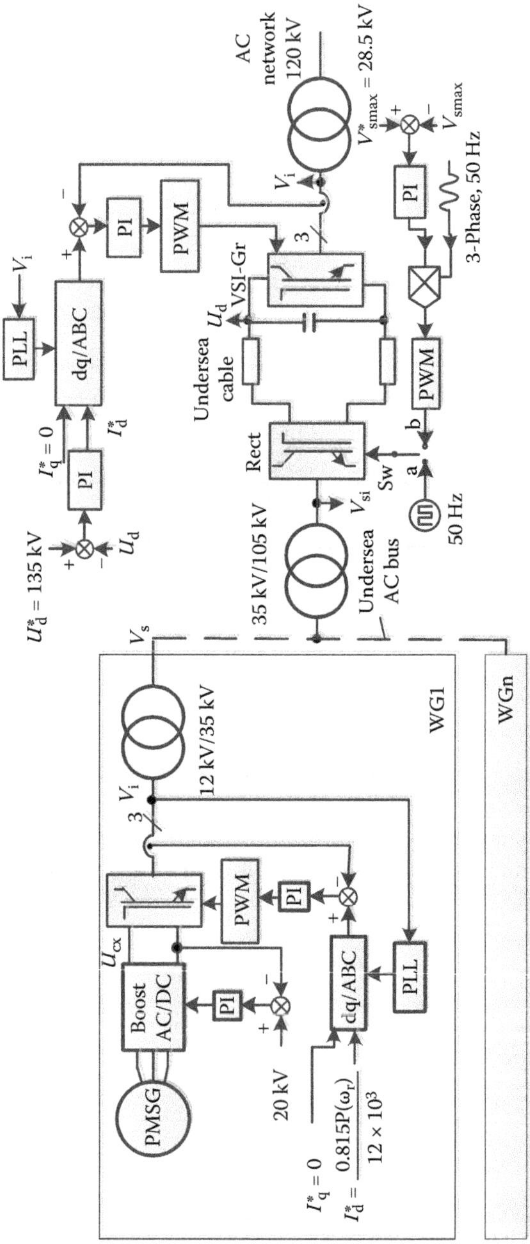

FIGURE 6.78 Configuration of the offshore park with the medium-frequency transformers, underwater AC bus and HVDC transmission.

The described system is simulated in the models **Wind_PMSG_10** (with two-level VSI) and **Wind_PMSG_10a** (with three-level VSI). Three WGs each with a power of 5 MW are connected in parallel. The process is simulated, when, at the initial common wind speed of 12 m/s, the wind speed of WG3 drops by 1 m/s in the interval (5–7 s); at $t = 7$ s, the common wind speed increases to 15 m/s, and at $t = 12$ s, it drops to 7 m/s. The start process is not simulated; it is supposed that the capacitors of the line have been already charged with the help of the VSI-Gr. Some results for the model **Wind_PMSG_10a**, with the **Sw** in the position *a*, are shown in Figure 6.79.

Although most WGs send the produced electric energy to the power system, there are many units operating without the grid, permanently or from time to time. Because the demanded wind speed does not always exist, such units operate together with other sources of electric energy, for instance, with batteries or diesel generators, which begin to operate when the wind speed drops lower than the necessary level. When WG works in isolation, the load control is of essential importance, as has been already said in the beginning of this chapter. At this, the WG must fabricate the wanted voltage, frequency, and power by itself.

In the model **Wind_PMSG_11**, the DC/DC boost converter is set at PMSG output (Figure 6.80) that controls the voltage U_{ci} across the capacitor C_i, as in the previous model. At the VSI PWM input, the three-phase quantity with the frequency of 50 Hz that is modulated by amplitude with the output of the output voltage controller comes.

The maximum power that the WG can pick up depends on the present wind speed V_w. Therefore, the load must be divided into parts; the parts have to be turned on or turned off depending on the V_w. Of these parts, the unswitched (or permanent)

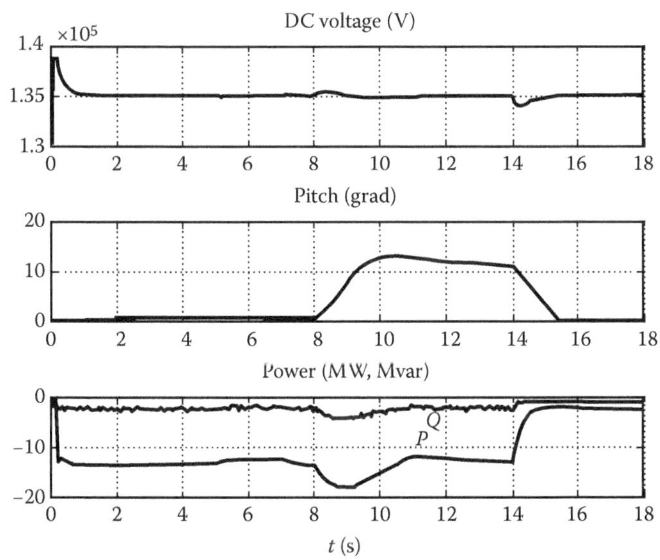

FIGURE 6.79 Processes in the model **Wind_PMSG_10a**.

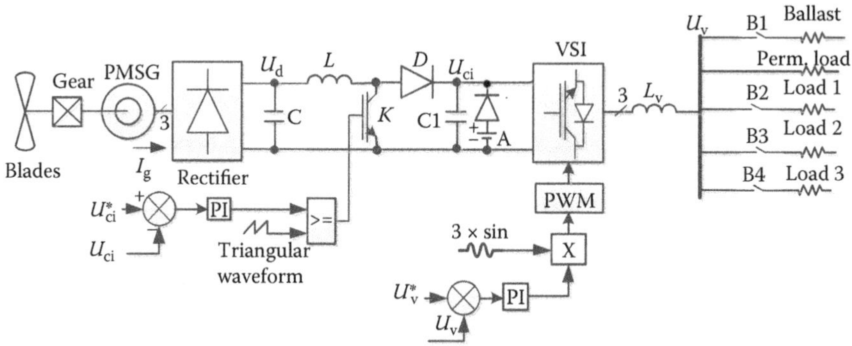

FIGURE 6.80 Diagram of the control system of the model **Wind_PMSG_11**.

part can be distinguished, whose power must be kept up in any case (the systems of emergency lighting and communication, the life support system). The battery **A** is intended to continue the supply. The battery voltage is less than the assigned voltage U^*_{ci}, and under U_{ci} decrease, the battery supplies to the VSI. The battery capacity must be sufficient in order to supply the permanent load during the time that is necessary to turn on the emergency source of the electric power supply. The battery can recharge from the VSI DC voltage during the time when the wind speed is in the operating range (the circuits for recharge are not shown). The battery also helps to keep up the output voltage under the abrupt fall of the wind speed—until the time of the switching off of the proper parts of the load.

WG usually operates on the right, descending branch of its characteristic (Figure 6.7). The rotating speed corresponds to the load power (plus losses) at the present wind speed. With the load increase, the rotating speed decreases and the WG power increases, until the new equilibrium point will be reached. Such a process takes place while the WG power is less than the maximum value in the present wind speed. Hence, in the system under consideration, there is no need to control the WG, in order to reach the maximum power, because this condition will be automatically obtained when required. Either the pitch control or the ballast control is used to limit the WG rotating speed, when the wind speed increases essentially.

Since the anemometers are not always set, it is desirable to have the WG without them. In order to estimate the wind speed, the scheme depicted in Figure 6.63 can be employed; after the wind speed estimation, the maximum available load can be defined. Since the number of available load steps is limited, it is possible to dispense with the wind speed estimation, namely, to calculate several dependences of the power from the rotating speed at certain wind speeds and to compare the current value of the power with the calculation results, determining, in this way, the power range that can be obtained at the existing wind speed.

In the model **Wind_PMSG_11**, WG has the nominal power of 60 kW and the nominal rotating speed of 22 rad/s. The circuits for the control of the voltage across the capacitor C_i are placed in the subsystem **Uc_contr**; the assigned voltage value is 500 V; the operation of the DC/DC boost converter was considered above. The circuits for current protection of the switch K (which is usually IGBT) are provided:

when the current through K exceeds 350 A, it is cut off; the consequent gate pulse generation is possibly 0.1 s later after the current drops to 250 A.

The modulating signal of the inverter has a frequency of 50 Hz. The 3rd harmonic is added. The amplitude of the modulating signal is defined by the output of the load voltage controller (subsystem **Control**). The voltage reference corresponds to the rms phase-to-phase voltage of 220 V. When the WG is turned off, the battery **Ak** with a rated voltage of 400 V supplies the permanent load. The LC filter is set at the inverter output.

The load (the subsystem **Load**) consists of the active–inductive parts with the following powers: the permanent load, 10 kW; and the switching on and switching off loads 13, 19, and 14 kW. The following options of load part switching are considered: with direct wind speed measurement; with estimation of the wind speed; and with discrete estimation of the available load power. The choice is carried out with the switches **Mode_Switch** and **Switch2** in the subsystem **Load**. The ballast (dummy) load with power of 25.5 kW is used in the model that is analogous to the one used in the model **Wind_DFIG_5** (with different power).

Let us carry out the calculations of the wind speeds, at which the supply of the switched load parts is ensured. To repeat the simulation many times with different load powers, at the base wind speed, until the load power is found, at which the steady state is obtained with the WG rotating speed of 22 rad/s, one may make sure that it takes place under $P_{ln} = P_e = 52.6$ kW; at this, the turbine power is 60 kW. Therefore, the WG efficiency is $\eta = 52.6/60 = 0.877$.

Furthermore, we find under $P_e = 56$ kW (10 + 13 + 19 + 14) that the demanded turbine power is 56/0.877 = 63.9 kW = 1.064 pu. Since in the point of the maximum power it must be $(V_w/10)^3 = 1.064$, the wind speed, at which the operation with such a load is possible, is $V_{w3} = 10.2$ m/s. With the switching off of the last part $P_e = 56 - 14 = 42$ kW, and the demanded turbine power is 42/0.877 = 47.9 kW = 0.8 pu; from the condition $(V_w/10)^3 = 0.8$, the wind speed is found as, at which the operation with such a load is possible, $V_{w2} = 9.3$ m/s. With the switching off of the second part $P_e = 42 - 19 = 23$ kW, the demanded turbine power is 23/0.877 = 26.6 kW = 0.437 pu; from the condition $(V_w/10)^3 = 0.437$, the wind speed is found as, at which the operation with such a load is possible, $V_{w1} = 7.6$ m/s. Under the wind speed, at which the condition $(V_w/10)^3 = 10/0.877/60 = 0.19$, that is, $V_w = 5.75$ m/s, the WG stops.

The circuits for the wind speed estimation V_{we} are placed in the subsystem **Load**. They are the simplified version of the scheme used in the model **Wind_PMSG2_sensorless**. When the system uses the wind speed V_w or its estimation V_{we}, the maximum WG power at this wind speed is computed in pu as $P_m = (V_w/10)^3$, and afterward, this power is compared with the powers of the load parts, taking the efficiency into account. The parts are switched off when the WG power calculated this way exceeds the part power a little; they are switched on when the calculated power values are 5–10% more than the switching-off values.

The changes of a number of system variables are displayed in Figures 6.81 and 6.82. The wind speed was estimated, and the load voltage was kept constant (**Switch1** in the subsystem **Control** is in the low position; in the subsystem **Load**, the tumbler **Mode_Switch** is in the top position, and **Switch2** is in the low position). The wind speed is decreasing, beginning from the instant $t = 2$ s, from the initial value of 11 m/s

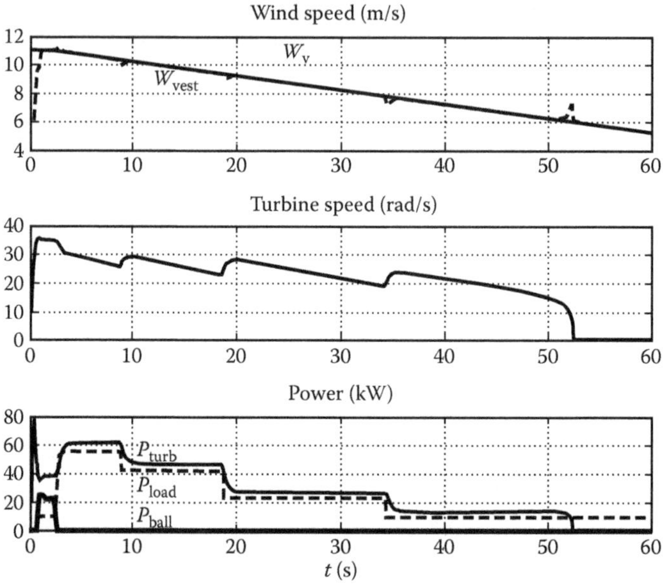

FIGURE 6.81 Speeds and powers in the model **Wind_PMSG_11**.

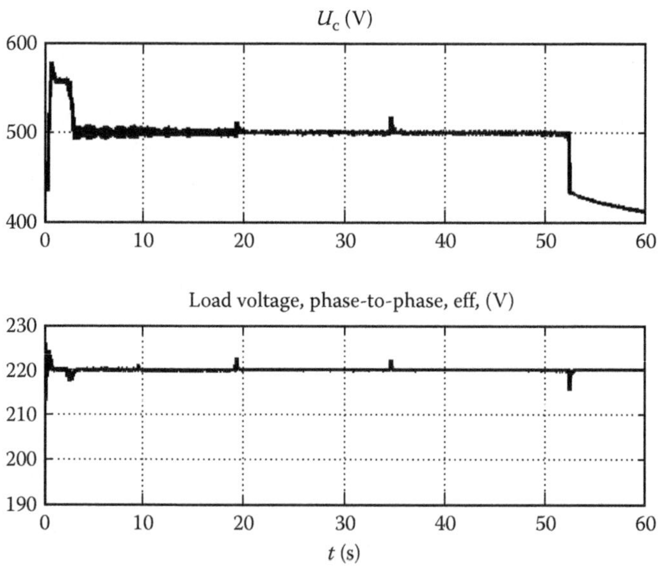

FIGURE 6.82 Voltages in the model **Wind_PMSG_11**.

with the deceleration of 0.1 m/s^2. It may be seen that after the initial fitting, the quantities V_w and V_{we} are practically equal until the WG stops, when it does not matter. The load parts are turned successively off at the instants 8.8, 18.6, and 34.3 s. At t = 52 s, the wind speed is not sufficient to supply the permanent load. The WG stops, and the battery powers this load. The load voltage is maintained with good precision.

When the method of estimation of the permissible power is utilized, the signals for the selection of the load parts are fabricated in the subsystem **Load_contr**. In the subsystem **Compute**, with the employment of the same formula that is used in the turbine model, the powers *P1*, *P2*, and *P3* at wind speeds 8.8, 9.8, and 10.7 m/s, respectively, under the present rotating speed are calculated. These values are compared with the present value of the PMSG power *P*. If $P < P1$, only the permanent load 10 kW remains; if $P1 < P < P2$, the load 13 kW is turned on in addition; if $P2 < P < P3$, the load 19 kW is added; if $P3 < P$, the last load 14 kW is turned on. It may be found after simulation that the turning-off times are 9, 17.8, and 34.2 s, that is, about the same as with the wind speed value usage.

In order to increase the gap between the wind speeds, when a certain load part is turned on and turned off, it is possible, with V_w dropping, to decrease the load voltage in the permissible limits. The effect of these steps depends on the load power on the applied voltage. It is made in the subsystem **V-comp**. When a certain load part is turned on, the voltage across the load is set to be equal to 220 V; with wind speed dropping, the voltage decreases by 10%. To run the simulation with a known wind speed (the tumblers **Mode_Switch** and **Switch2** in the subsystem **Load** are in the top positions, the tumbler **Switch1** in the subsystem **Control** is in the top, the tumbler **Switch2** is in the low positions, and the blocks **Relay** in the subsystem **Step Control** must be replaced with the blocks that are displayed on the subsystem scheme above the inscription *U_load* = *var*), the turning off times will be 14.3, 23.4, and 37.2 s, that is, the turning off of load parts occurs noticeably later, as with the constant load voltage. The change of the load voltage is shown in Figure 6.83.

When the method of estimation of the permissible power is utilized, the tumbler **Switch2** in the subsystem **Control** is set in the top position, the blocks **Relay** and **Constant** are changed for the indicated ones on the subsystem scheme in the subsystem **Compute**. The processes that are obtained after the simulation execution are about the same as in the previous case.

The WG of relatively small power (60 kW) is utilized in the considered model; the power control in conformity with the load power is carried out with the ballast resistors. However, the isolated operating WG with large power is also in demand, and in this case, the ballast is very bulky and costly. Without the ballast, the only means to maintain this conformity is controlling the blade position β. But it is necessary to keep in mind that this process is slower than the employment of the ballast load.

In the model **Wind_PMSG_11a**, the WG with power of 400 kW is utilized; the base wind speed is 11 m/s. The powers of the load parts are taken as 90, 140, and 70 kW; the permanent load is 20 kW; the wind speeds, at which the loads are switched, are calculated as shown earlier. In the control system, the steps are taken in order for the load parts to be switched on consecutively in time with delay, so that the blade position has time to change. The resistor that is put in parallel to the capacitor at the VSI input is turned on during transients or under failure for the short time,

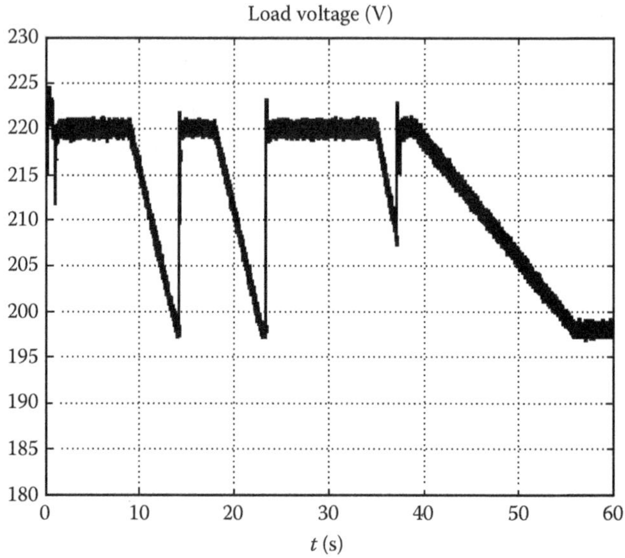

FIGURE 6.83 Load voltage during a decrease of the wind speed for the variant with voltage change as a function of the wind speed.

so that its installed power is relatively small. With the assigned capacitor voltage of 500 V, the resistor circuit is closed at the voltage of 650 V and is disclosed when the voltage is getting lower as 550 V. Only the variant with direct wind speed measurement is considered.

In the control system of angle β (subsystem **beta_contr** in the model **Turbine**), the reference for the WG rotating speed, under which this angle begins to increase, is defined as the minimum of two quantities: 1.25 of the nominal rotating speed and 1.25 of the optimal rotating speed under the given wind speed. β_{max} = 20° is accepted, and the maximum rate of changing is 4°/s.

The following process is fixed in Figures 6.84 and 6.85 (tumbler **Switch** is in the top position): the start at a wind speed of 11 m/s, its reduction to 5 m/s at t = 7 s, its rise to 7.9 m/s at t = 25 s, and its subsequent increase to 12 m/s at t = 35 s. The rate of the wind speed change is 1 m/s/s. The processes under the gradual rise of the wind speed from 4.5 m/s with the rate of 0.4 m/s/s are shown in Figures 6.86 and 6.87. It is seen that the WG unit operates properly.

This WG unit can be made with the inner current controller (model **Wind_PMSG_11b**). The processes in this model do not practically distinguish from the one given earlier.

Since the wind speed can be low for a long time, the battery capacity may not be sufficient for powering the permanent load; in these cases, other permanent sources of electric energy must be employed. The diesel generator sets with SG are usually utilized. They can be connected permanently or as required. In the first case, it is necessary to have in mind that when the diesel rotates at the nominal speed, its power must not be less some value, for example, 20%–30% of nominal. The advantage of this

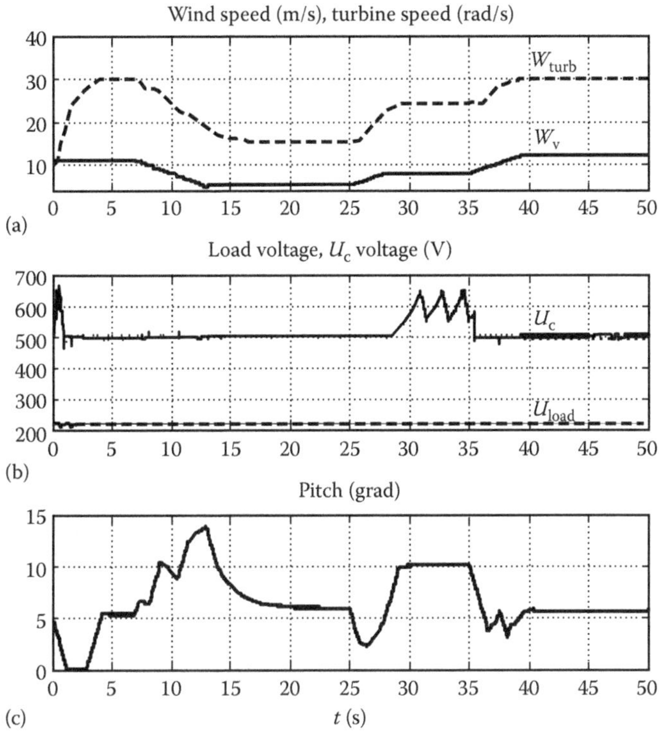

FIGURE 6.84 (a) Speeds, (b) voltages, and (c) pitch in the model **Wind_PMSG_11a** at different wind speeds.

scheme is that the system without interruption delivers the demanded load power, and the SG maintains the wanted voltage and frequency at the load. In the second case, the seamless connection of the SG to the load with the operating WG or of the WG to the load, when it is supplied from the SG, has to be carried out.

The first case is modeled in **Wind_PMSG_12**. The diagram of the set is depicted in Figure 6.88. The turbine power is 30 kW at the base wind speed of 11 m/s. PMSG has the following parameters: P_n = 33 kVA, ω_{mnom} = 22 rad/s, Z_p = 18, and Ψ_r = 0.526 Wb. The block **Synchronous Machine pu Fundamental** (42.5 kVA; 230 V; 50 Hz; 1500 r/min) is utilized as a model SG. The maximum load power is 35 kW, so that both SG individually and WG with the help of 20% SG power can supply the load.

The subsystem **Wind_Syst** models the WG with its electrical part that is connected to the load via the inductor of 4 mH and is made as shown in Figure 6.88. The VSI-Ge changes the PMSG speed proportionally with the wind speed; at this, ω_{mmax} = 24 rad/s. The VSI connected to the load (VSI-L) maintains the voltage across the inverter capacitor equal to 500 V with the same control system that was used in the previous models. The subsystem **Ballast** is the same, as in the model **Wind_PMSG_11**. The ballast power begins to increase, when the diesel generator power P_{diz} reduces to 0.2 of its nominal power P_{diznom}, and it continues to increase until P_{diz} reaches the value of 0.3 P_{diznom}. When P_{diz} reaches 0.4 P_{diznom}, the ballast power reduces to zero.

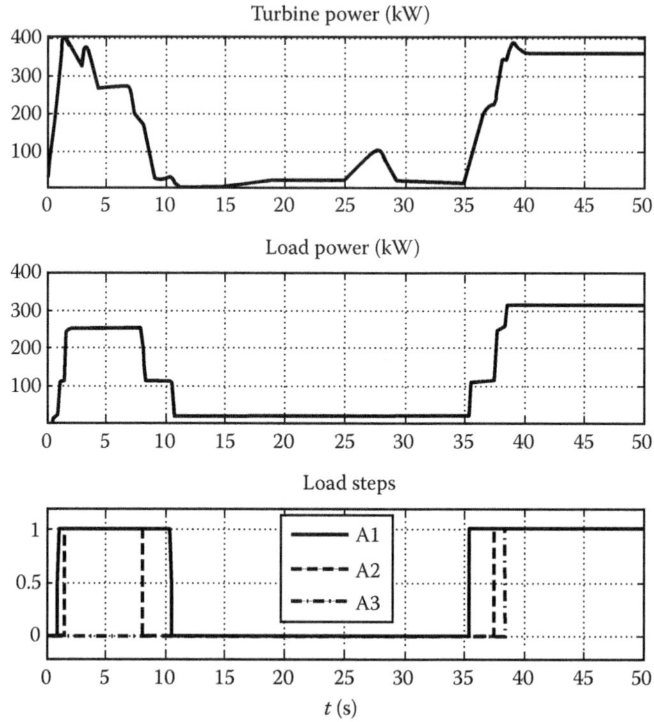

FIGURE 6.85 Powers, load part switching in the model **Wind_PMSG_11a** at different wind speeds

The SG is driven with diesel, whose model is the subsystem **Diesel Engine Governor**, which is practically the same as that used in the model **Diesel_IG**, with the difference that not the torque but the mechanical power is the model output. The block AC5 described in Chapter 4 is utilized for the SG voltage control.

A number of scopes give an opportunity to observe the system operation. The scope **PMSM** fixes the current, the speed, and the torque of the WG, and the scope **Scope_Uc** shows the voltage across the capacitor of the VSI-L. The five-axes scope **Currents** shows the load voltage on the first axis, and on the others—the currents in PCC in the following order: WG, SG, load, and ballast. The scope **SM** records the input power of SG, its excitation voltage, the amplitude of the output voltage, and the SG speed. The scope **Power** shows the powers of the previously mentioned four circuits in the same order (WG—yellow, SG—magenta [red purple], load—cyan, ballast—red). The power is calculated as a product of the instantaneous voltage and current values of one phase averaged for 20 ms and multiplied by 3.

The following process is simulated: System starts with zero wind speed ($V_w = 0$; the mechanical system braked); WG starts at $t = 2$ s with $V_w = 5$ m/s; at $t = 5$ s, V_w increases to 13 m/s; and V_w decreases to 5 m/s at $t = 9$ s. The load, being equal to 30 kW, in the time interval of 8–11 s increases by 5 kW.

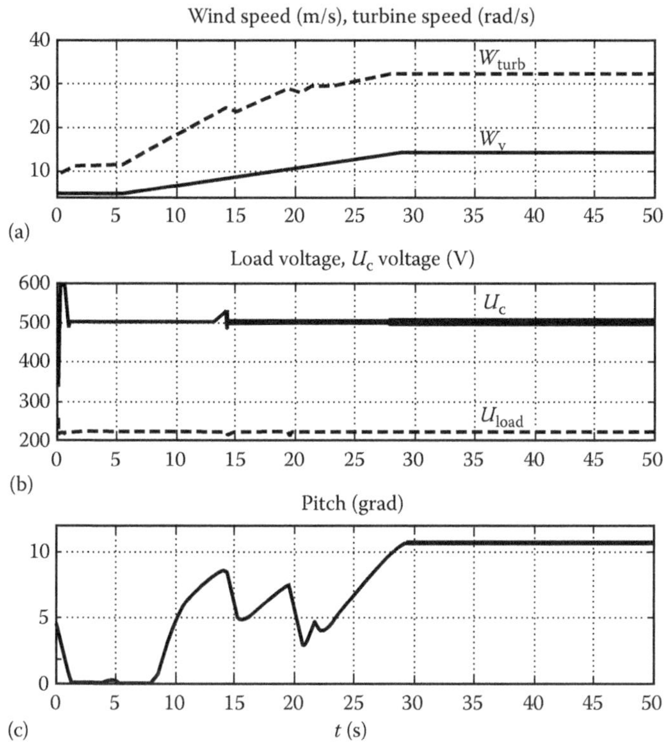

FIGURE 6.86 (a) Speeds, (b) voltages, and (c) pitch in the model **Wind_PMSG_11a** under the gradual rise of the wind speed.

Some results of simulation are shown in Figure 6.89. It follows from studying these figures and the scope data that the load voltage during all conditions remain, equal to 232 V, and the frequency is limited by the permissible boundaries (the short time deviation in dynamics does not exceed 1%–2%). The deviation of the capacitance voltage from the value of 500 V does not exceed 1%. With the rising wind speed and WG rotating speed, the SG and diesel powers decrease but do not fall lower than 11 kW.

In the model **Wind_PMSG_13**, the load supply is carried out from both the grid and the WG; at this, both sources can operate together or separately. The WG model with power of 400 kW is the same as in the model **Wind_PMSG_11b**; from the latter, the load model is taken as well, with a difference that by the signal *All*, all load parts are turned on. The relatively weak grid with the short-circuit power of 10 MVA is connected to the breaker **Br2**.

The control system of the DC/DC boost converter connected to the WG keeps the voltage across the capacitor equal to 500 V. The VSI-L control system contains the generator of the three-phase voltage (Gen-3) with the possibility to change the amplitude, the frequency, and the phase. The Gen-3 output is the reference for the controller of the three-phase current in the inductor at the VSI-L output. When the grid is

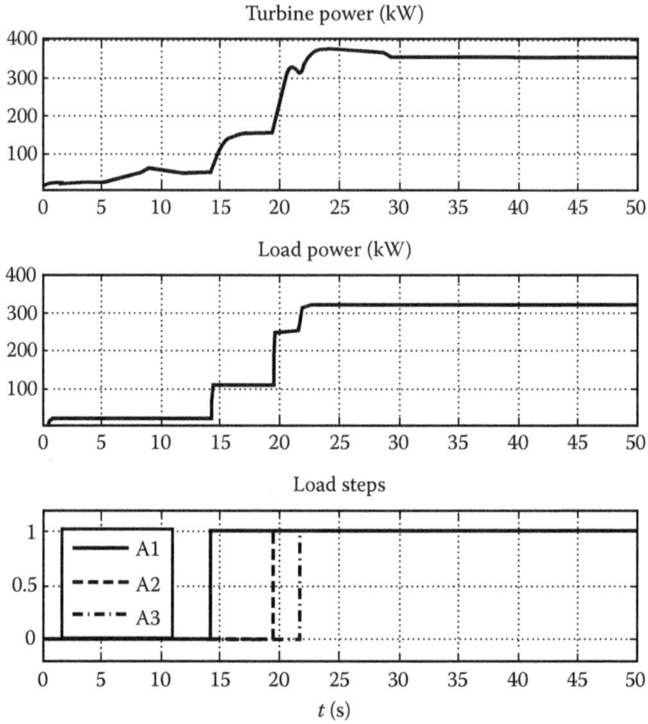

FIGURE 6.87 Powers, load part switching on in the model **Wind_PMSG_11a** under the gradual rise of the wind speed.

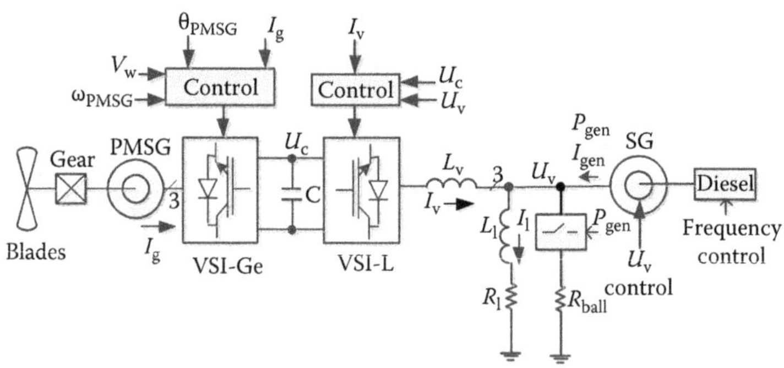

FIGURE 6.88 Diagram of the model **Wind_PMSG_12**; diesel-SG and WG operate in parallel.

turned off, the amplitude of the current reference is defined by the output of the load voltage controller, as in the model **Wind_PMSG_11b**; when the grid is turned on, the current reference is defined according to the relationship in Equation 6.49 (for simplicity, it is taken as $D = F = 0$) for harvesting maximum wind energy.

The diagram of the current reference formation is depicted in Figure 6.90. The main element is the integrator with reset **Integr** that integrates the quantity f_s, which

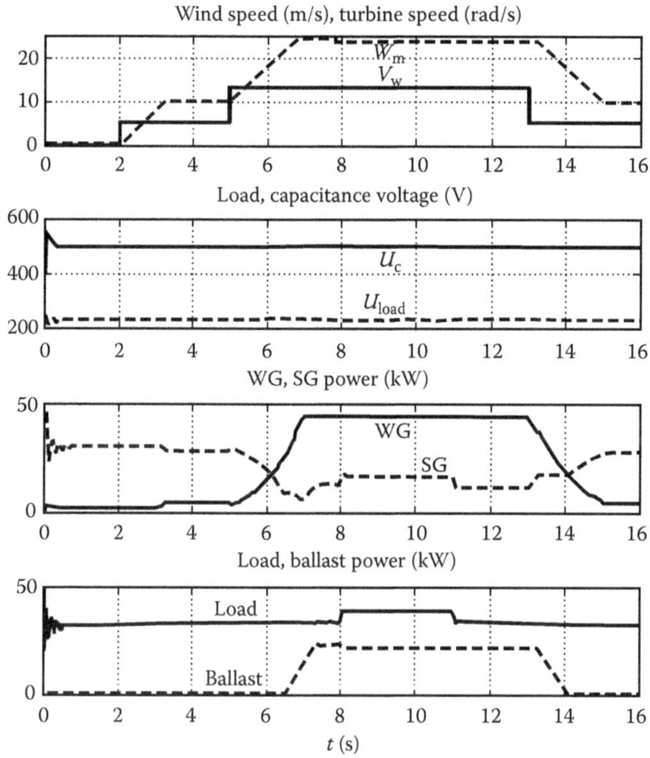

FIGURE 6.89 Processes in the model **Wind_PMSG_12**.

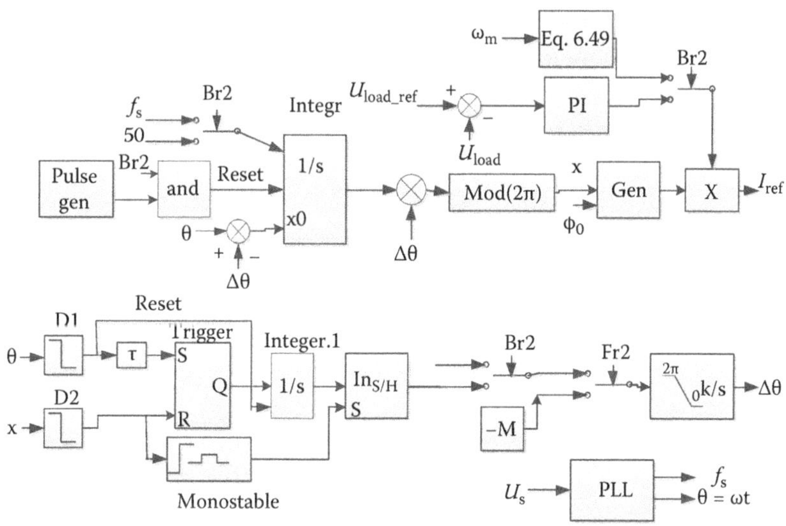

FIGURE 6.90 Control system diagram of the model **Wind_PMSG_13**.

is proportional to the frequency. When the grid is disconnected, the frequency is 50 Hz; when the grid is turned on, this value is defined by the output of PLL that is connected to the grid voltage. The source phase outputs are computed as $U_i = \sin x \cos(\varphi_0 + 2\pi/3i) + \cos x \sin(\varphi_0 + 2\pi/3i)$, where $i = 0, 1, 2$, and the quantity $x = 2\pi \int f_s \, dt$.

During the isolated WG operation, the generator of the three-phase signal **Gen** works independently; when the grid is turned on, its work is synchronized with the grid voltage. It is obtained with the equality of the saw-tooth waveforms [0 2π] at the outputs of the integrator and of the block PLL θ. When **Br2** is closed, quantity θ (with the correction Δθ, see further) comes to the integrator initial condition input x_0, and the pulses of rather high frequency (it is taken as 10 Hz) come to the input *Reset*, so that every 0.1 s, the quantity *x* is becoming equal to quantity θ (if the difference existed).

Quantity Δθ is intended for the initial matching of the phases of *x* and θ, when the grid is turned on with the operating WG. When the grid is ready to be turned on, but before closing the **Br2**, the signal *Fr2* is produced and the process of synchronization of the VSI-L voltage with the grid voltage begins. At this, the instants are fixed, when the saw-tooth waveforms of θ and *x* change by steps from 2π to 0. The circuits that are shown in the low area of Figure 6.90 are used. The **Trigger** is set during the

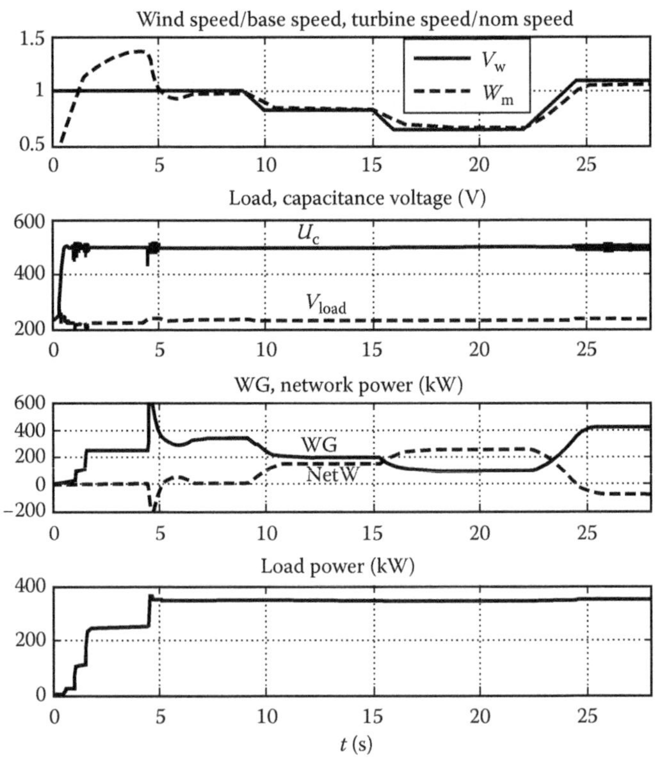

FIGURE 6.91 Processes in the model **Wind_PMSG_13.**

time interval between these two instants. Correspondingly, the output of the integrator **Integr1** in the moment, when x changes from 2π to 0, is proportional to this time interval. In this instant, the integrator output is stored in the block of sample and storage **S/H**, so the output of this block is proportional to the difference of the angle positions of the grid and the VSI voltage space vectors. This difference is integrated with the integrator **Integr2** that has the output limitations of 0 and 2π and is added to quantity x, changing in this way its phase until phase equality is reached. After phase equality is obtained (controlled by time), the breaker **Br2** is closed, and the output of **Integr2** is kept. This integrator is set to zero for the preparation for the following synchronization cycle by giving to its input the negative quantity M, after turning off **Br2** and **Fr2**. We note that the analogous principle and circuits were utilized in the model **Wind_DFIG_1a** (Figure 6.29).

The following process is simulated: isolated WG operation with $V_w = 11$ m/s, fabrication of the signal **Fr2** at $t = 4.1$ s, closing of **Br2** at $t = 4.5$ s, wind speed reduction to $V_w = 9$ m/s at $t = 9$ s, subsequent reduction of V_w to 7 m/s at $t = 15$ s, and V_w increase to 12 m/s at $t = 22$ s. Some results of the simulation are shown in Figure 6.91. It is seen how the power that is consumed from the grid for maintenance of the constant load power 343 kW changes when the wind speed changes. At the wind speed of 12 m/s, the WG power surplus appears, and the part of this power is delivered to the grid.

It is seen from Figure 6.92a that the grid connection takes place seamlessly. But to repeat the simulation without circuits for synchronization described earlier, the

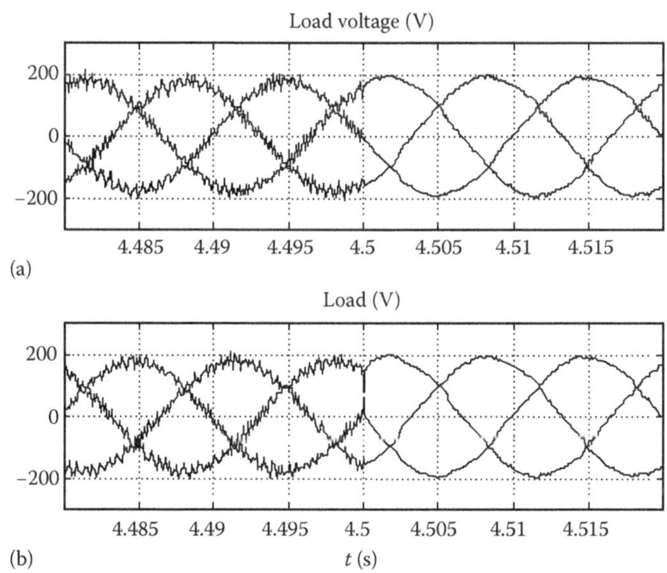

FIGURE 6.92 Load voltage during grid connection: (a) with developed synchronization circuits and (b) without synchronization circuits.

load voltage will change as depicted in Figure 6.92b, which proves the utility of the employed method.

REFERENCES

1. Global Wind Energy Council (GWEC). *Global Wind Report 2014*. GWEC, Brussels, 2015, March.
2. World Wind Energy Association (WWEA). *Half-Year Report*. WWEA, Bonn, 2014.
3. Wu, B., Lang, Y., Zargari, N., and Kouro, S. *Power Conversion and Control of Wind Energy Systems*. John Wiley & Sons, Inc., Hoboken, NJ, 2011.
4. Ackermann, T. (ed.). *Wind Power in Power Systems*. John Wiley & Sons, Ltd, West Sussex, 2005.
5. Slootweg. J. G., and Kling, W. L. Aggregated modelling of wind parks in power system dynamics simulations. *2003 IEEE Bologna PowerTech Conference*, Bologna, Italy, 2003, vol. 3, 1–6.
6. Qiao, W., Yang, X., and Gong, X. Wind speed and rotor position sensorless control for direct-drive PMG wind turbines. *IEEE Transactions on Industry Applications*, 48(1), 2012, January/February, 3–11.
7. Perelmuter, V. M. *Direct Control of the AC Motor Torque and Current* (in Russian). Osnova, Kharkov, 2004.
8. Chi, S., and Xu, L. Position sensorless control of PMSM based on a novel sliding mode observer over wide speed range. *Proceedings of the Fifth International Power Electronics and Motion Control Conference*, 2006, August, vol. 3, 1–7.
9. Utkin, V. Sliding mode control design principles and applications to electrical drives. *IEEE Transactions on Industrial Electronics*, 40(1), 1993, February, 23–36.
10. Molinas, M. Offshore wind farm research: Quo vadis? *Wind Seminar 2011, Norwegian University of Science and Technology (NTNU), Trondheim*, 2011, 1–47.
11. Liserre, M., Cárdenas, R., Molinas, M., and Rodríguez, J. Overview of multi-MW wind turbines and wind parks. *IEEE Transactions on Industrial Electronics*, 58(4), 2011, April, 1081–1095.

List of the Appended Models

Battery_Example: *Simple model with the Battery*
Battery_Example1: *Battery and Grid with Voltage Control*
Battery_Example2: *Battery with Galvanic Separation with Middle-Frequency Transformer*
Cable-3Ph.m
Contr_PMSM_DTC1.m
Contr_PMSM_T1.m
Contr_PMSM_T2
Diesel_IG: *Diesel and IG, 1st Version*
Diesel_IG1: *Diesel and IG, 2nd Version*
Double_Im.m
Example 3b
Example 3c
Example_Phasor
Example2
Example3
Example3a
Excitation_examples: *Excitation Models*
Filter_Ex_1: *Employment of the Second-Order Filter*
Filter-Ex: *Employment of the Filters and the Measuring Blocks*
Four_legs1: *4-Legs Two-Level Inverter*
Four_legs2: *4-Legs Inverter with Voltage Control*
Four_legs3: *4-Legs Inverter, Source Voltage Variation*
Fuel_PEMFC: *Simple Model of the Fuel Cell PEMFC*
Fuel_PEMFC2: *Fuel Element with AC Output*
Fuel_PEMFCB: *Fuel Cell PEMFC and Battery*
Fuel_PEMFCB1: *Fuel Cell and Battery with reversible DC/DC Converter*
Fuel_SOFC: *Simple model of the Fuel Cell SOFC*
Graf_Photo.m
HNPC_5L: *5-Level HNPC Inverter*
HVDC: *DC Transmission Line*
HydroM1: *Small Hydraulic Turbine and Synchronous Generator supplying an Isolated Load, with Ballast*
HydroM2: *Small Hydraulic Turbine and Induction Generator supplying an Isolated Load, with Ballast*
HydroM3: *Small Hydraulic Turbine and Induction Generator supplying Network*
HydroM4: *Small Hydraulic Turbine and IG, with Inverter and Rectifier*
HydroM5: *Small Hydraulic Turbine and IG with Wounded Rotor, DTC*
HydroM6: *Small Hydraulic Turbine and IG with Wounded Rotor, DTC, simplified Model*

HydroM7: *Small Hydraulic Turbine and IG with wounded Rotor supplying Isolated Load*
HydroP1: *Hydrogenerator and isolated Load*
HydroP2: *Two Units Hydraulic Turbine-Generator supplying the Common Load*
HydroP3: *Two Units Hydraulic Turbine-Generator and the Power Grid*
IG_6Ph: *Diesel and 6-Phase IG*
IG_switchpole: *WG with IG with Pole Number Changing*
IM_double_1: *Brushless Doubly-Fed IG, Simple Version*
IM_double_2: *Brushless Doubly-Fed IG*
IM_MODEL_0.m
IM_MODEL_0.m
IM_MODEL_01.m
IM_MODEL_1.m
IM_MODEL_1a.m
IM_MODEL_6Ph.m
IM_MODEL_6ph_1200.m
IMatr_Conv1: *Indirect MC, the Special PWM, Switches Commutate without Current*
IMatr_Conv2: *Indirect MC and IG*
M2CAC: *Modular Multi-Level Inverter, DC-AC, with PWM2*
M2CACR: *Modular Multi-Level Inverter as Rectifier*
M2CDCR: *Modular Multi-Level Inverter, DC-LR, without PWM*
M2CDCR1: *Modular Multi-Level Inverter, DC-RL, with PWM*
M2CHVDC: *HVDC Transmission Line with Modular Converters*
Matr_Conv1: *Matrix Converter with IGBT*
Matr_Conv2: *Matrix Converter with IGBT and Speed Control*
Модель SG_12ph1: *12-Phase SG and Network*
MODEL_SM_12ph.m
MODEL_SM_12ph1.m
MODEL_SM_12ph2.m
MTURBO1: *Microturbine, simple Model*
MTURBO2: *Microturbine and Grid Inverter*
MTURBO3: *Generator of Microturbine with constant Speed, Power Electronics and Grid*
MTURBO4: *Microturbine with Speed Control by Generator Converter (idealized Model)*
MTURBO5: *Microturbine with Speed Control (sensorless Model)*
myfun.m
myfun1.m
myfun2.m
OLTC_trans: *Transmission Line with the On-Load Tap-Changing Transformer*
Ph3-Cable: *Cable Transmission Line and SVR*
Photo_m1: *Photocell Model*
Photo_m2: *Photoarray under partly Shadowing*
Photo_m3: *Photocell, Dependence Power from Voltage*

Photo_m4I: *MPPT with Current Measurement*
Photo_m4S: *MPPT with Perturb & Observe Approach*
Photo_m4V: *MPPT with Voltage Measurement*
Photo_m4V1: *MPPT with Circuit Interuption*
Photo_m5: *MPPT with Incremental Conductance Method*
Photo_m6: *Photovoltaic and Battery with Ballast control*
Photo_m6_1: *Photovoltaic and Battery without Ballast control*
Photo_m6_2: *Photovoltaic and Battery with AC Output*
Photo_m7: *PV and single-phase grid working parallel*
Photo_m71: *PV and single-phase grid working parallel and separately*
Photo_m8: *PV and three-phase grid working parallel*
Photo_m81: *PV and three-phase grid working parallel, PWM current control*
Photo_m9: *Power PV set and Network*
PMSMFLY: *PMSM with Flywheel*
Sat_trans: *Transformer with Saturation and Hysteresis*
Sel_tab_3.m
Sel_tab_5_1.m
Sel_tab_5_2
Selection_invest: *To_SHEM_5Modules*
SG_12ph: *12-Phase SG and resistive Load*
SG_12ph1_det: *12-Phase SG with 3-level Inverters and Network*
Simple_SG: K_d *Calculation of the Simplified SG*
SixphPMSM: *Six_Phase PMSM Generator*
Sixphsynchmachine: *Six_Phase Synchronous Generator*
SixphSynMachine_det: *Six_Phase SG with Id, Iq Controllers*
SM_12phase_drive: *Power Drive with 12-Phase SM*
SM_MODEL_6ph2.m
SM_MODEL_6ph4
SM_MODEL_6ph4a.m
SRG: *Switch Reluctance Generator*
SRG1: *Switch Reluctance Generator with mechanical Coupling*
SRGFLY: *SOFC and Switch Reluctance Generator (SRG)*
STATCOM_3_SHEM: *STATCOM with Cascaded H-bridge Multilevel Inverter with SHEM, 3 Modules*
STATCOM_3_SHEM_DC: *STATCOM with three-stage cascaded H-bridge multilevel inverter with SHEM, DC Sources*
STATCOM_5_SHEM: *STATCOM with Cascaded H-bridge Multilevel Inverter with SHEM, 5 Modules*
STATCOM_Hinv1: *STATCOM with Multilevel Cascaded Inverter*
STATCOM_modinv: *STATCOM with the Modular Inverter*
STATCOM_modinv1: *STATCOM with n = 10, without Transformer*
SVC1: *Reactive Power Stabilization by SVC with Switched Capacitors*
Three_level1: *Three-Level Voltage-Source Inverter*
Three_level2: *Forming of the AC Voltage with Low THD*
Three_Ph_Contr_Source: *Three-Phase Programmable Voltage Source*
Two_level1: *Two-level Voltage-Source Inverter*

Two_level2: *Two-level Voltage-Source Inverter, Averaging Structure*
Var_load: *Dynamic Load*
Wind_DFIG_1: *WG with wounded rotor IG and power network operating parallel, DTC*
Wind_DFIG_1a: *Start of WG with Wounded Rotor IG and DTC*
Wind_DFIG_2: *WG with Wounded Rotor IG, Vector Control*
Wind_DFIG_3: *Wind Farm with DFIG*
Wind_DFIG_4: *WG with Brashless Doubly-fed Generator*
Wind_DFIG_5: *WG with Wounded Rotor IG and Isolated Load*
Wind_DFIG_5a: *WG with Wounded Rotor IG and Isolated Load, with Wind Speed Measurement*
Wind_DFIG_6: *WG with Wounded Rotor IG and isolated unbalanced Load, Standard IG Model*
Wind_IG_1: *Wind Turbine and Induction Generator with Constant Speed supplying a Power Grid*
Wind_IG_2: *Wind Turbine and Induction Generator with Switching Number of Poles*
Wind_IG_2a: *Wind Turbine and Induction Generator with Switching Number of Poles (transients)*
Wind_IG_3: *Wind Park with Induction Generators*
Wind_IG_4: *Off-Shore Wind Park with Induction Generators*
Wind_IG_5: *Wind Turbine and Induction Generator with mechanical Coupling supplying a Power Grid*
Wind_IG_6: *Wind Turbine with 6-Phase IG*
Wind_IG_7: *Wind Turbine with IG and Three-Level Inverters*
Wind_PMSG_1: *WG with PMSG and Power Grid*
Wind_PMSG_2_sensorless: *WG with PMSG, Sensorless Control with Sliding Mode Estimation*
Wind_PMSG_3: *WG with Salient Pole PMSG and Power Grid*
Wind_PMSG_4: *WG with 6-Phase PMSG*
Wind_PMSG_4a: *WG with 6-Phase PMSG and 3_Level Inverters*
Wind_PMSG_5: *Wind Park with PMSG*
Wind_PMSG_6: *Off-Shore Wind Park with PMSG, with SVC, AC Connection*
Wind_PMSG_7: *Off-Shore Park with series DC Connected WG*
Wind_PMSG_7a: *Off-Shore Park with series DC Connected WG and Three-Level Inverters*
Wind_PMSG_8: *Off-Shore Park with parallel DC Connected WG*
Wind_PMSG_9: *Off-Shore Park with DC Connected WG and Boost Transformer 1 kHz*
Wind_PMSG_10: *Off-Shore Park with HVDC and AC undersea Grid*
Wind_PMSG_10a: *Off-Shore Park with HVDC and AC undersea Grid, with Three-Level Inverters*
Wind_PMSG_11: *WG with PMSG and Isolated Load*
Wind_PMSG_11a: *WG with PMSG 400 kW and Isolated Load*
Wind_PMSG_11b: *WG with PMSG 400 kW and Isolated Load, with Inner Current Control*

Wind_PMSG_12: *WG with PMSG and Diesel-Generator working parallel*
Wind_PMSG_13: *WG with PMSG and Network working together and separately*
Wind_SG_1: *Wind-Synchronous Generator with Constant Speed*
Wind_SG_1a: *Wind-Synchronous Generator with Constant Speed, Simplified Model*
Wind_SG_2: *Wind-6-Phase Synchronous Generator*
Wind_SG_2a: *Wind-6-Phase Synchronous Generator with Drive Train*
Wind_SG_3: *Wind-12-Phase Synchronous Generator*
Wind_SG_3a: *Wind-12-Phase Synchronous Generator, Simplified Model*
Z_inv: *Simple Model with Z-Inverter*
Z_invV: *Z-Source Inverter with Voltage Control 1*
Z_invV1: *Z-Source Inverter with Voltage Control 2*
Zig_zag_trans: *Thyristor Rectifier with 4 Phase-Shifting Transformers*
Zig_zag_transN: *Thyristor Rectifier with 4 Phase-Shifting Transformers with the new blocks*

Index

Page numbers followed by f and t indicate figures and tables, respectively.

D

E

F

G

H

K

L

M

T

U

V

Z